Springer Series in Computational Mathematics

7

Dietrich Braess

Nonlinear Approximation Theory

With 38 Figures

Springer-Verlag
Berlin Heidelberg New York
London Paris Tokyo

Professor Dr. Dietrich Braess
Fakultät für Mathematik, Ruhr-Universität Bochum
Postfach 10 21 48, D-4630 Bochum 1

Mathematics Subject Classification (1980):
41-02, 41A15, 41A20, 41A21, 41A25, 41A30, 41A50, 41A52, 41A55, 41A65, 65D10, 65D15, 65D32

ISBN 3-540-13625-8 Springer-Verlag Berlin Heidelberg New York
ISBN 0-387-13625-8 Springer-Verlag New York Berlin Heidelberg

Library of Congress Cataloging-in-Publication Data
Braess, Dietrich, 1938–
Nonlinear approximation theory.
(Springer series in computational mathematics; 7)
Bibliography: p. Includes index.
1. Approximation theory. I. Title. II. Series.
QA221.B67 1986 511'.4 86-10101
ISBN 0-387-13625-8 (U.S.)

Printed in Germany

Typesetting: Asco Trade Typesetting Ltd., Hong Kong
Printing and bookbinding: Graphischer Betrieb Konrad Triltsch, Würzburg
2141/3140-543210

To Anneliese

Preface

The first investigations of nonlinear approximation problems were made by P.L. Chebyshev in the last century, and the entire theory of uniform approximation is strongly connected with his name. By making use of his ideas, the theories of best uniform approximation by rational functions and by polynomials were developed over the years in an almost unified framework. The difference between linear and rational approximation and its implications first became apparent in the 1960's.

At roughly the same time other approaches to nonlinear approximation were also developed. The use of new tools, such as nonlinear functional analysis and topological methods, showed that linearization is not sufficient for a complete treatment of nonlinear families. In particular, the application of global analysis and the consideration of flows on the family of approximating functions introduced ideas which were previously unknown in approximation theory. These were and still are important in many branches of analysis.

On the other hand, methods developed for nonlinear approximation problems can often be successfully applied to problems which belong to or arise from linear approximation. An important example is the solution of moment problems via rational approximation. Best quadrature formulae or the search for best linear spaces often leads to the consideration of spline functions with free nodes. The most famous problem of this kind, namely best interpolation by polynomials, is treated in the appendix of this book.

The monograph grew out of lectures which the author gave on numerous occasions to fourth year students at the Ruhr-Universität Bochum. The prerequisites consist essentially of a good basic knowledge of analysis and functional analysis. A short description of an elementary part of nonlinear approximation suitable for a course for post-graduate students is given in the "Note to Students". The note indicates sections to which a student may restrict himself during a first reading.

It is hoped that this book will prove to be useful for researchers interested in advanced aspects of approximation theory. In this respect the book has been organized so that the discussions of rational functions (Chap. V), exponential sums (Chap. VI and VII), and spline functions with free nodes (Chap. VIII) is in each case independent. We assume only that the reader is accustomed to the basic methods from functional analysis (described in Chap. II, § 1) and the central ideas of critical point theory (Chap. III). The latter parts of Chap. VII and Chap.

VIII require topological arguments and techniques from global analysis which are not elementary. These are presented in Chap. IV.

We have deliberately abandoned a unified theory for the special families described above, since it would conceal more facts than it would make transparent. Usually, there are three ingredients in the proofs: 1) a local argument from differential calculus, 2) a global argument which often involves a topological argument, 3) a compactness argument. Frequently, the compactness part is the most strongly involved. In the families under consideration only weak compactness conditions are fulfilled. The methods for overcoming this difficulty depend on the particular nature of the special case. The exercises indicate which results for a particular family carry over to the others. Moreover, the exercises contain some curiosities and interesting examples or counterexamples which could not be given in the text.

In attributing proper names to the theorems, we have generally followed the common usage in the western literature. The bibliographical notes should not be considered as being complete. We hope for example, that the reader will excuse the many gaps in the Russian literature apparent in our bibliography.

I wish to thank most of all Helmut Werner, to whom I personally owe so much. He created a very stimulating scientific atmosphere at the Institute für Numerische Mathematik of the University of Münster, and this not only because of his deep knowledge of nonlinear analysis. I express my sincere gratitude to Ward Cheney for inviting me to participate in discussions and seminars at the University of Texas at Austin. These were important for the development of critical point theory as it is presented here. A number of my friends and colleagues were kind enough to read large parts of the manuscript and made improvements and valuable suggestions. For this I am especially indebted to Frank Deutsch, Hubert Jongen, Allan Pinkus, Robert Schaback, Josef Stoer, Richard Varga, Luc Wuytack, and A. Zhensykbaev. I also wish to thank Timothy Norfolk for linguistic improvements. For their expert and diligent typing and retyping of the manuscript, I thank Mrs. I. Voigt and Mrs. M. Schulz. I am grateful to Ludwig Cromme, Immo Diener, Kurt Jetter, Manfred Müller, and Robert Schaback for their help in proofreading. Finally, I would like to thank Springer-Verlag for their friendly cooperation.

Bochum, Spring 1986 Dietrich Braess

Note to Students

Large parts of the text are based on lectures which the author gave to fourth year students at German universities. Though this monograph is directed to researchers interested in approximation theory, it is also our intention to provide a text for students. In this respect, the book was designed so that the student may restrict himself to the basic parts of nonlinear approximation theory. The following hints are intended to explain which sections may be skipped.

The first chapter should be regarded as a review of well-known results from the linear theory with which the reader should be acquainted. For later applications it is convenient to consider this in the framework of convex approximation. The concept of strong uniqueness, which is described in the framework of Chebyshev approximation will turn out to be essential. On the other hand, the characterization of nearest points in Haar cones in § 3C and the L_1-theory in § 4 will only be used in some special situations.

The theorem that each boundedly compact Chebyshev set in a smooth Banach space must be convex, is a main result of Chapter II. To understand this one should look at the general tools which are useful for existence proofs (Section 1A), proceed with the discussion on suns and the Kolmogorov Criterion in § 2 until Theorem 2.5, continue in § 3A until Corollary 3.2, and in § 3B until Corollary 3.6.

Methods of local analysis are described in Chapter III. The first section is crucial for the understanding of the rest of the book, but the student may restrict himself to manifolds without boundaries and skip all generalizations which refer to boundaries. Next, the basic facts on varisolvent families from sections 3A and 3B will often be applied in nonlinear Chebyshev approximation. Additional tools and improvements for the differentiable case are provided in 4A–C.

Chapter IV is intended to give some introduction to the use of global methods. In § 1 where some basic concepts are briefly discussed, it is sufficient to understand the definition of critical points of mountain pass type. In lectures the author has only presented either § 2 or § 3. This is enough to give the student some insight into global analysis. For this purpose, the student may choose between the general uniqueness theorem in § 2A–B and a concrete example in § 3.

The basic facts on rational approximation in Chapter V, §§ 1–2, familiarize the reader with the most frequently studied nonlinear family which behaves "almost" like a linear one.

The first contacts with Descartes' rule of signs are encountered in the analysis of exponential sums in Chapter VI, § 1. The theory in § 2 shows that existence proofs in nonlinear problems may require hard analysis and that general existence theorems may support, but cannot replace the individual proofs.

The γ-polynomials in Chapter VII are generalizations of exponential sums. Thus the student may skip § 1 by taking Descartes' rule for exponential sums from Chapter VI. The basic results are developed in §§ 2 and 3.

The investigations in Chapter VIII are only aimed at those who are particularly interested in spline functions or monosplines and Gaussian quadrature formulas. Although the analysis up to Section 3B is not difficult, one always has the added technical complication caused by the interlacing condition.

The appendix on optimal Lagrangian interpolation and the non-elementary sections of Chapters VII and VIII make extensive use of advanced methods of global analysis.

Finally, we mention that the book contains two small complete theories which are suitable for seminars.

A seminar for students who are interested in functional analysis may be based on Chapter II, augmented by Section I.2A for an introduction.

A seminar concerned with the connection between rational approximation and classical constructive function theory could focus on Chapter V. The only prerequisites here are Lemma II.1.4, the lemma of first variation in connection with Exercise III.1.23, and the well known theorem of de la Vallée-Poussin.

Contents

Chapter I. Preliminaries

§ 1. Some Notation, Definitions and Basic Facts

A. Functional Analytic Notation and Terminology

Unless otherwise specified, E will be a real or complex normed linear space. A complete normed linear space is called a Banach space. E' is the dual space, i.e., the Banach space of continuous linear functionals, equipped with the norm $\|l\| = \sup_{\|x\|=1} |l(x)|$.

Let M be a non-empty set in E. $\overline{M}$, $\mathring{M}$ or int M, ∂M, co M, span M denote, resp., the closure, interior, boundary, convex hull, and linear hull of M. The convex hull of two points $x \neq y$ is said to be an interval and denoted as $[x, y]$. If one or both endpoints of the interval are excluded, we use the symobls $[x, y)$, $(x, y]$, or (x, y) respectively.

The set $B_r(x) := \{z \in E, \|z - x\| \leq r\}$ is called a (closed) ball of radius r with center at x. Its interior $\mathring{B}_r(x)$ is called an open ball. If the center is 0, we shall write B_r instead of $B_r(0)$. $S[S', \text{resp.}]$ is the unit sphere in $E[E', \text{resp.}]$.

A sequence shall be denoted by (x_n), $(n = 1, 2, \ldots)$. The set consisting of the points of a sequence is denoted in the same way. A net is denoted by $(x_\alpha)_{\alpha \in A}$, or simply by (x_α), where A is the indexing set. We write $x_n \to x$ or $\lim x_n = x$ if the sequence converges in the strong topology, i.e., if $\|x_n - x\| \to 0$ (as $n \to \infty$). We write w-lim $x_n = x$, if the sequence converges weakly to x, i.e., if $\lim l(x_n) = l(x)$ for each $l \in E'$.

The following classes refer to spaces for which the unit ball has specific properties.

E is called *strictly convex* or *rotund*, if $x, y \in S$, $x \neq y$, implies $\|x + y\| < 2$.

E is called *uniformly convex*, if given $\varepsilon > 0$, there is a $\delta = \delta(\varepsilon) > 0$ such that $x, y \in S$, $\|x + y\| > 2 - \delta$ implies $\|x - y\| < \varepsilon$ (Clarkson 1936).

E has the *CLUR* property (compact locally uniform rotundity property) if $x, y_n \in S$ for all n, $\|x + y_n\| \to 2$ implies that there is a convergent subsequence (y_{n_k}) of (y_n).

Obviously, each uniformly convex space is strictly convex and has the CLUR property. On the other hand each strictly convex finite-dimensional space is uniformly convex.

A functional $l \in S'$ is said to be a *peak functional* to $x \in E$ if $l(x) = \|x\|$.

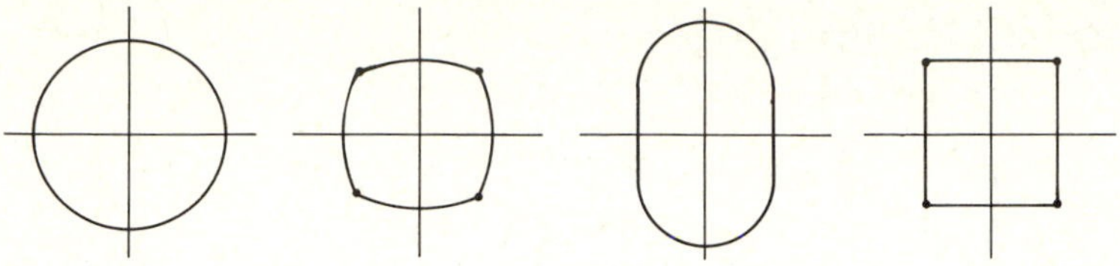

Fig. 1. Unit spheres of strictly convex (a and b) and non strictly convex (c and d), smooth (a and c) and non-smooth spaces (b and d)

E is called *smooth*, if at each point $x \in S$ there exists a unique supporting hyperplane, or equivalently, there exists a unique peak functional to $x \neq 0$. If E is reflexive, then the smoothness of E is equivalent to the strict convexity of E'.

Let M be a linear subsequence of E. The orthogonal complement is $M^\perp = \{l \in E': l(u) = 0 \text{ for } u \in M\} = \{l \in E': M \subset \ker l\}$.

In Fig. 1 unit spheres of strictly convex and non strictly convex, smooth and non-smooth 2-dimensional spaces are illustrated.

An *inner product* (x, y) is a bilinear form from $E \times E$ into $\mathbb{R}$ or $\mathbb{C}$, which is linear in the second argument, satisfies the symmetry relation $(x, y) = \overline{(y, x)}$, and for which $(x, x) > 0$ unless $x = 0$. If the norm is induced by an inner product $\|x\| = \sqrt{(x, x)}$, then E is called an *inner-product space*. A complete inner-product space is said to be a *Hilbert space*.

Let X be a compact set. The set $C(X)$ of (real or complex) functions on X is a Banach space if endowed with the *sup-norm* (or *uniform norm*):

$$\|f\| = \|f\|_\infty = \sup_{x \in X} |f(x)|.$$

In all cases, where the norm on $C(X)$ is not specified, it is assumed to be the sup-norm. Obviously, $C(X)$ is neither smooth nor strictly convex.

Let X be a measure space with positive measure μ and $1 \leq p < \infty$. The space of (equivalence classes of) functions f, for which $|f|^p$ is integrable, is denoted as $L_p(X, \mu)$, when equipped with the L_p-norm:

$$\|f\|_p = \|f\|_{p,\mu} = \left\{\int_X |f|^p \, d\mu\right\}^{1/p}.$$

If X is a subset of $\mathbb{R}^k$ and μ is the Lebesgue-measure, we write $L_p(X)$ instead of $L_p(X, \mu)$. If μ is the counting measure on a countable set, we write l_p. An element of l_p may be regarded as a real or complex sequence $x = (\xi_n)$ and

$$\|x\|_p = \left\{\sum_{n=1}^{\infty} |\xi_n|^p\right\}^{1/p}.$$

The spaces $L_2(X, \mu)$ are Hilbert spaces with the inner product $(f, g) = \int_X \bar{f} g \, d\mu$. More generally, the L_p-spaces, $1 < p < \infty$, are uniformly convex. This follows from Clarkson's inequalities:

$$\|f+g\|_p^p + \|f-g\|_p^p \le 2^{p-1}[\|f\|_p^p + \|g\|_p^p], \qquad p \ge 2$$

$$\|f+g\|_p^{p/p-1} + \|f-g\|_p^{p/p-1} \le 2[\|f\|_p^p + \|g\|_p^p]^{1/p-1}, \qquad 1 < p \le 2$$

Moreover, if $1 < p < \infty$, the dual space of $L_p(X,\mu)$ may be identified with $L_q(X,\mu)$, where $1/p + 1/q = 1$, and $L_p(X,\mu)$ is a smooth space.

On the other hand, $L_1(X,\mu)$ is neither strictly convex nor smooth.

B. The Approximation Problem. Definitions and Basic Facts

The starting point of approximation theory is the concept of best approximation. Let M be a non-empty set in a normed linear space E. Given $f \in E$, the *distance* of f from M is

$$d(f,M) := \inf_{u \in M} \|f-u\|. \tag{1.1}$$

An element $u_0 \in M$ is said to be a *best approximation* or a *nearest point* to f from M, if $\|f-u_0\| = d(f,M)$. The set of all best approximations to f from M is denoted by $P_M f$, or simply by Pf, if this does not cause confusion.

The following definitions were introduced by Efimov and Stechkin.

A set M is called an *existence set* (*uniqueness set*, resp.) if, to each $f \in E$, there is at least one (at most one, resp.) best approximation to f from M. A set M is called a *Chebyshev set* if it is an existence set and a uniqueness set, i.e., if, for each $f \in E$, Pf is a singleton.

Usually, P_M is understood to be a set-valued mapping, $P_M: E \to 2^M$, and is called the *metric projection* of E onto M. If, in particular, M is a Chebyshev set, then the metric projection can also be understood as a mapping from E onto M.

Obviously, each existence set is a closed set. Indeed, if $f \in \overline{M}$ then $d(f,M) = 0$, and thus f is the only best approximation. Hence, $f \in M$ and $M = \overline{M}$.

On the other hand, closedness in itself, is not sufficient for a set M in an infinite dimensional space to be an existence set. Let $\dim E = \infty$, then there is a sequence (e_n) such that $\|e_n\| = 1, n = 1,2,\ldots$ and $\|e_n - e_m\| \ge 1$ whenever $n \ne m$. Obviously, $M := \{u_n : u_n = (1+n^{-1})e_n, n = 1,2,\ldots\}$ is closed. But there is no nearest point to the origin and thus M is not an existence set.

The existence of best approximations can however, be guaranteed by compactness assumptions.

In this context, the approximation problem is considered as an optimization problem. The *metric function* $\|\cdot\|: M \to \mathbb{R}$, given by $u \mapsto \|f-u\|$ is continuous and attains its minimum on a compact set. By definition M is an existence set if and only if for each $f \in E$, the metric function attains its minimum on M. Therefore, each compact set is an existence set.

With a modified argument of F. Riesz (1918), which was earlier applied to polynomials by Kirchberger (1903), it follows that each finite dimensional linear subspace M of E is an existence set: Here given $f \in E$, let u_0 be a nearest point to

f from the compact subset $M \cap B_{2\|f\|}(0)$. Then we have even $u_0 \in P_M f$, because $\|u\| > 2\|f\|$ implies $\|f - u\| \geq \|u\| - \|f\| > \|f\| = \|f - 0\| \geq \|f - u_0\|$.

In the next chapter, more relaxed compactness assumptions will be considered for the treatment of nonlinear approximation problems.

If M is a convex set, then $P_M f$ is always convex. For, given $u_0, u_1 \in P_M f$, we obtain for $u_t = (1-t)u_0 + tu_1$, $0 < t < 1$:

$$\begin{aligned} \|f - u_t\| &= \|(1-t)(f - u_0) + t(f - u_1)\| \\ &\leq (1-t)\|f - u_0\| + t\|f - u_1\| = d(f, M). \end{aligned} \tag{1.2}$$

If, moreover, M is a convex set in a strictly convex space, then M is a uniqueness set. To verify this, let u_0 and u_1 be two distinct nearest points. Since $f - u_0 \neq f - u_1$, the strict inequality holds in (1.2). The element $u_{1/2}$ would be a better approximation than either u_0 or u_1.

The terms *Chebyshev approximation* or *uniform approximation* are commonly used for the approximation problem for $C(X)$ endowed with the uniform norm.

The set of polynomials (with one variable) of degree at most n will be denoted by Π_n, and ∂p will denote the degree of the polynomial p. As usual, $E_n(f)$ will be written instead of $d(f, \Pi_n)$ in the case of uniform approximation on $[-1, +1]$.

Finally, an element $u_0 \in M$ is called a *local best approximation* (LBA) to f from M, if u_0 is a nearest point to f from $M \cap U$ for some open set U containing u_0.

C. An Invariance Principle

In many situations there is an isomorphism $T: E \to E$ which maps the set of approximating functions M onto itself. For example, when functions on the symmetrical interval $[-1, +1]$ are considered with the uniform norm or a symmetric L_p-norm, the distance of two elements is invariant under the involution $(Tf)(x) = \pm f(-x)$. This involution maps the sets of polynomials of degree $\leq n$, the rational function of degree (m, n) and other interesting families onto themselves.

1.1 Invariance Principle (Meinardus 1963). *Let T be a linear mapping from the normed linear space E onto itself such that $\|Tf\| = \|f\|$ for each $f \in E$. Moreover, assume that $T(M) = M$. Then the element Tu_0 is a best approximation to Tf from M, whenever u_0 is a best approximation to f.*

If f is invariant, i.e., if $Tf = f$, and $u_0 \in P_M f$, then also $Tu_0 \in P_M f$.

If f is invariant and its best approximation u_0 is unique, then u_0 is invariant under T.

If f is invariant and no invariant element from M can be a best approximation, then f has either no best approximation or at least two.

Proof. It is sufficient to prove the first statement. Let $u_0 \in P_M f$ and $u_1 = T(u_2) \in M$. Since T does not change distances, it follows that

$$\|Tf - u_1\| = \|Tf - Tu_2\| = \|f - u_2\| \geq \|f - u_0\| = \|Tf - Tu_0\|.$$

Hence, Tu_0 is a nearest point. □

D. Divided Differences

Let h be a function defined on a real interval that is sufficiently differentiable. *Divided differences of order k* are defined iteratively for $k \geq 1$ by the relations

$$h(x_0, x_1, \ldots, x_k) = \begin{cases} \dfrac{h(x_1, x_2, \ldots, x_k) - h(x_0, x_1, \ldots, x_{k-1})}{x_k - x_0} & \text{if } x_k \neq x_0, \\ \dfrac{1}{k!} h^{(k)}(x_0) & \text{if } x_0 = x_1 = \cdots = x_k. \end{cases} \tag{1.3}$$

The relations (1.3) are sufficient for the definition, since divided differences of order k are symmetric functions of their $k + 1$ arguments. This symmetry is due to the fact that a divided difference equals the coefficient of a polynomial from a Hermite interpolation problem, which by definition has the symmetry property.

We will need the following properties which can be found for example, in Werner (1966).

1.2 Properties of Divided Differences. *Let X be a real interval and $h \in C^k(X)$, $k \geq 0$.*

(1) *There is a $z \in X$ such that*

$$h(x_0, x_1, \ldots, x_k) = \frac{1}{k!} h^{(k)}(z). \tag{1.4}$$

(2) *$h(x_0, x_1, \ldots, x_k)$ is a continuous function of its arguments.*

(3) *If $h \in C^{k+1}(X)$, then*

$$\frac{\partial}{\partial x_i} h(x_0, x_1, \ldots, x_k) = h(x_0, x_1, \ldots, x_k, x_i) \qquad \text{for } 0 \leq i \leq k \tag{1.5}$$

is the divided difference of order $k + 1$ with the argument x_i repeated.

(4) *If $g, h \in C^k(X)$, then*

$$g \cdot h(x_0, x_1, \ldots, x_k) = \sum_{i=0}^{k} g(x_0, x_1, \ldots, x_i) h(x_i, x_{i+1}, \ldots, x_k).$$

Exercises

1.3. Assume that $\|f - u\| < \|f - u_0\|$ and define $u_t = (1 - t)u_0 + tu$ for $t \in [0, 1]$. For which number $c > 0$ does $\|f - u_t\| \leq \|f - u_0\| - ct$ hold for $0 \leq t \leq 1$? Apply this often used inequality to the statement that each local best approximation from a convex set is a (global) best approximation.

1.4. Let M be a closed subset of a normed linear space E. Show that $P_M f$ is closed for all $f \in E$.

1.5. The best uniform approximation on $[1/2, 2]$ by rational functions of the form p/q, where $p, q \in \Pi_{2n}$ is known to be unique. Assume that $f(x) = f(1/x)$. Prove that the nearest point to f in the set of the rational functions above has the form

$$P(x + x^{-1})/Q(x + x^{-1})$$

where $P, Q \in \Pi_n$.

§ 2. A Review of the Characterization of Nearest Points in Linear and Convex Sets

A characterization of the solution of a linear or convex approximation problem can be obtained from the Hahn-Banach theorem. In many cases, the criteria are reformulated, and it is no longer apparent that a characterization in terms of peak functionals is given implicitly. Often, it is also easier to present a direct proof for a criterion, than to derive it from the general characterization theorem.

In particular, in non-smooth Banach spaces the characterizing functional is not unique. The possibility of choosing a practicable one leads to the Kolmogorov criterion.

A. Characterization via the Hahn-Banach Theorem and the Kolmogorov Criterion

The starting point for the characterization of nearest points is the geometric version of the Hahn-Banach theorem. Unless otherwise stated, E may be a real or complex space.

Separation Theorem for Convex Sets. *Let A and B be disjoint convex sets in a normed linear space E. Moreover, assume that A is open. Then, there is a bounded linear functional $l \in E'$ and a real number c such that*

$$\operatorname{Re} l(x) > c \quad \text{for } x \in A,$$

$$\operatorname{Re} l(x) \le c \quad \text{for } x \in B.$$

Henceforth, we will always assume that $f \notin \overline{M}$. Thus, we have $d(f, M) > 0$.

2.1 Characterization Theorem. *Let M be a convex set in a normed linear space E and let $f \notin \overline{M}$. Then $u_0 \in M$ is a best approximation to f from M if and only if there is a functional $l \in E'$ with the following properties:*

$$\|l\| = 1, \tag{2.1}$$

$$l(f - u_0) = \|f - u_0\|, \tag{2.2}$$

$$\operatorname{Re} l(u - u_0) \le 0 \quad \text{for } u \in M. \tag{2.3}$$

In its present formulation the characterization theorem has been given by Deutsch (1965) and independently by Rubinstein (1965). There are several predecessors, cf. the corollaries below and the historical notes by Singer (1974).

Proof. Assume that $u_0 \in P_M f$. Then M and $\mathring{B}_r(f)$, where $r = \|f - u_0\|$, are disjoint convex sets, and $\mathring{B}_r(f)$ is open. By the separation theorem, there is an $l_0 \in E'$ and $c \in \mathbb{R}$ such that

$$\begin{aligned} \operatorname{Re} l_0(u) &\le c, \qquad u \in M, \\ \operatorname{Re} l_0(g) &> c, \qquad g \in \mathring{B}_r(f). \end{aligned} \tag{2.4}$$

The continuity of l_0 implies that

$$\operatorname{Re} l_0(g) \ge c, \qquad g \in B_r(f). \tag{2.5}$$

Since $u_0 \in M \cap B_r(f)$, we obtain $\operatorname{Re} l_0(u_0) = c$. Also, since $f \in \mathring{B}_r(f)$, it follows that $\beta := \operatorname{Re} l_0(f) - c = \operatorname{Re} l_0(f - u_0) > 0$. Let

$$l = \beta^{-1} r l_0.$$

This normalization implies that

$$\operatorname{Re} l(f - u_0) = \|f - u_0\|, \tag{2.6}$$

and therefore $\|l\| \ge 1$.

Suppose that $\|l\| > 1$. Then there would exist an $h \in E$, with $\|h\| < 1$, such that $l(h)$ is real and $l(h) > 1$. For $g = f - rh$ we compute

$$\operatorname{Re} l_0(g) = \operatorname{Re}(l_0 f - r l_0 h) = (c + \beta) - \beta l(h) < c.$$

Since $g \in B_r(f)$, this contradicts (2.5). Hence $\|l\| = 1$.

As $\|l\| = 1$, it follows that $|l(f - u_0| \le \|f - u_0\|$. This and (2.6) imply (2.2). Finally, it is clear from (2.4) and (2.5) that $\operatorname{Re} l_0(u - u_0) \le 0$ for $u \in M$. Since l is a positive multiple of l_0, the last statement (2.3) has also been verified.

Therefore, the conditions (2.1) through (2.3) are necessary for a nearest point. The sufficiency follows from the next lemma, in which the convexity assumption is abandoned. □

2.2 Lemma. *Let M be a non-empty set in a normed linear space E and let $u_0 \in M$. Assume that there is an $l \in E'$ satisfying* (2.1), (2.2), *and* (2.3). *Then $u_0 \in P_M f$.*

Proof. Recalling that $|l(g)| \le \|l\| \cdot \|g\|$ for $g \in E$, we obtain from the given conditions for each $u \in M$,

$$\begin{aligned} \|f - u\| &\ge \|l\|^{-1} \operatorname{Re} l(f - u) \\ &= \operatorname{Re} l(f - u_0) - \operatorname{Re} l(u - u_0) \ge \|f - u_0\|. \end{aligned}$$

Hence, u_0 is a nearest point. □

If M is a linear subspace, then a simpler formulation is possible. We recall that $M^\perp := \{l \in E' : l(u) = 0 \text{ for } u \in M\}$.

2.3 Corollary (Singer 1956). *Let M be a linear subspace of a normed linear space E and let $f \notin \overline{M}$. Then $u_0 \in P_M f$ if and only if there is a functional $l \in M^{\perp}$ satisfying* (2.1) *and* (2.2).

A functional l which satisfies the conditions of the characterization theorem is a peak functional to the element $\varepsilon_0 := f - u_0$. Therefore, l is unique if E is a smooth space.

On the other hand, if there is more than one peak functional, one would like to choose a functional which is appropriate for the purpose at hand. Since the unit ball in E' is w^*-compact, by the Krein Milman theorem the set of admissible functionals is the closed convex hall of its extremal points. As was shown by Deutsch and Maserick (1967), one may always choose a functional for the characterization which is an extremal point of S'.

Specifically, the freedom in the choice of the characterizing functional is greater if an individual choice for each competing element u is admitted (Garkavi 1961).

2.4 Generalized Kolmogorov Criterion. *Let M be a convex set in a normed linear space E and let $f \in E$. Then, $u_0 \in M$ is a best approximation to f from M if and only if given $u \in M$, there is a functional $l \in E'$ with the following properties:*

$$\|l\| = 1,$$

$$l(f - u_0) = \|f - u_0\|,$$

$$\operatorname{Re} l(u - u_0) \leq 0.$$

In this case, u_0 is said to satisfy the *(generalized) Kolmogorov condition.* Its necessity follows from the characterization theorem 2.1, while the proof of sufficiency is similar to the proof of Lemma 2.2.

B. Special Function Spaces

The characterizations by the Hahn-Banach Theorem and the Kolmogorov Criterion are useful because, in the cases of actual interest, concrete functionals can be used.

The simplest case occurs with inner-product spaces.

2.5 Characterization Theorem for Inner-Product Spaces. *Let E be an inner-product space. Then u_0 is a nearest point to f from a linear subspace M if and only if $f - u_0$ is orthogonal to M. More generally, u_0 is a nearest point to f from a convex set M if and only if* $\operatorname{Re}(f - u_0, u - u_0) \leq 0$ *for any $u \in M$.*

Let $l \in E'$ be defined by $l(u) = (\varepsilon_0, u) \cdot \|\varepsilon_0\|^{-1}$, where $\varepsilon_0 = f - u_0$. Since l is the only peak functional to ε_0, Theorem 2.5 is obtained by concretization of the general theorems. Moreover, it may be verified independently by elementary calculations.

The characterization for Hilbert spaces covers the L_2-case. Similarly, the L_p-approximation is treated for $1 < p < \infty$.

2.6 Characterization of Best Approximations in L_p-spaces (Deutsch and Maserick 1967). *Let M be a linear subspace of $L_p(X, \mu)$, where $1 < p < \infty$. Then $u_0 \in M$ is a best approximation from M to $f \in L_p$ if and only if*

$$\int_X |\varepsilon_0(x)|^{p-2}\overline{\varepsilon_0(x)}u(x)\,d\mu(x) = 0 \quad \text{for } u \in M,$$

where $\varepsilon_0 = f - u_0$.

To verify this, we note that $l(u) = \|\varepsilon_0\|^{1-p}\int_X |\varepsilon_0|^{p-2}\bar{\varepsilon}_0 u\,d\mu$ describes the only peak functional to ε_0. From the Hölder inequality, it follows that the integral is well defined whenever $\varepsilon_0 \in L_p$, $u \in L_p$. The characterization for convex sets is analogous.

When considering L_1-approximation we encounter for the first time a nonsmooth Banach space. Usually the ambiguity in the choice of the peak functional is overcome by the use of a nonlinear characterization.

2.7 Characterization of Best Approximation in L_1-space (Kripke and Rivlin 1965). *Let M be a linear subspace of $L_1(X, \mu)$, where X is compact. Then $u_0 \in M$ is a best approximation to $f \in L_1(X, \mu)$ from M if and only if*

$$\operatorname{Re}\int_{X\backslash Z} \frac{\overline{\varepsilon_0(x)}}{|\varepsilon_0(x)|}u(x)\,d\mu \le \int_Z |u(x)|\,d\mu, \tag{2.7}$$

where $\varepsilon_0 = f - u_0$ and $Z := \{x \in X: \varepsilon_0(x) = 0\}$.

Proof. If (2.7) holds, then for any $u \in M$:

$$\begin{aligned}\|f - u_0 - u\| &= \int_{X\backslash Z} |f - u_0 - u|\,d\mu + \int_Z |u|\,d\mu \\ &\ge \operatorname{Re}\int_{X\backslash Z} (f - u_0 - u)\frac{\bar{\varepsilon}_0}{|\varepsilon_0|}\,d\mu + \int_Z |u|\,d\mu \\ &= \|f - u_0\| - \operatorname{Re}\int_{X\backslash Z} u\frac{\bar{\varepsilon}_0}{|\varepsilon_0|}\,d\mu + \int_Z |u|\,d\mu \ge \|f - u_0\|.\end{aligned}$$

Therefore, (2.7) implies optimality.

In order to avoid considerations from measure theory, the necessity of (2.7) will be verified only for f and M in $L_1(X, \mu) \cap C(X)$, and for real valued functions.

Assume that there is a $u_1 \in M$ such that

$$\int_{X\backslash Z'} u_1 \operatorname{sgn} \varepsilon_0\,d\mu > \int_Z |u_1|\,d\mu \tag{2.8}$$

where $Z' = Z$. Obviously we can choose an open neighborhood Z' of Z for which (2.8) remains true. By compactness, it follows that $\delta = \inf\{|\varepsilon_0(x)|: x \in X\backslash Z'\} > 0$.

After multiplying u_1 *by a positive factor, if necessary, we may assume that* $max\{|u_1|\} \le \delta$. Then $\operatorname{sgn}(f - u_0 - u_1) = \operatorname{sgn}(f - u_0)$ for $x \in X \backslash Z'$ and

$$\begin{aligned}\|f - u_0 - u_1\| &\le \int_{X\backslash Z'} (f - u_0 - u_1)\operatorname{sgn}\varepsilon_0\,d\mu + \int_{Z'} (|f - u_0| + |u_1|)\,d\mu \\ &= \int_X |f - u_0|\,d\mu - \int_{X\backslash Z'} u_1 \operatorname{sgn}\varepsilon_0\,d\mu + \int_{Z'} |u_1|\,d\mu \\ &< \|f - u_0\|.\end{aligned}$$

Therefore, u_0 is not optimal if (2.7) is violated. □

The simpler case where the set Z has measure zero and (2.7) becomes a linear equation has already been known to James (1947) and to Krein.

The advantage of the Kolmogorov criterion becomes apparent for uniform approximation, i.e., when $C(X)$ is endowed with the sup-norm.

2.8 The Kolmogorov Criterion for Uniform Approximation. *Let M be a convex set in $C(X)$, where X is compact. Then $u_0 \in M$ is a best uniform approximation to $f \in C(X)$ from M if and only if*

$$\inf_{x \in P[\varepsilon_0]} \operatorname{Re}\{\overline{\varepsilon_0(x)}[u(x) - u_0(x)]\} \le 0, \qquad u \in M, \tag{2.9}$$

where $\varepsilon_0 := f - u_0$ and $P[\varepsilon_0] := \{x \in X : |\varepsilon_0(x)| = \|\varepsilon_0\|\}$.

The sufficiency of (2.9) follows from the fact that $u \mapsto l(u) := \varepsilon_0(x_1)u(x_1)/\|\varepsilon_0\|$ is a peak functional whenever $x_1 \in P[\varepsilon_0]$. The condition (2.9) is also necessary. If (2.9) does not hold for $u_1 \in M$, then $u_t = u_0 + t(u_1 - u_0)$ is a better approximation than u_0 for sufficiently small $t > 0$. This may be verified by some elementary calculations, see e.g. Meinardus (1964). A derivation from the general Kolmogorov criterion on the other hand requires measure theoretical considerations.

The following criterion is obtained from 2.8 by an application of the Kuhn-Fourier theorem on linear inequalities (see e.g. Rivlin and Shapiro 1961):

2.9 Characterization of Best Uniform Approximation in Finite Dimensional Subspaces. *Let X be a compact set and M be an n-dimensional linear subspace of real valued functions (complex valued, resp.) of $C(X)$. If $f \notin M$, the following are equivalent:*

1°. *u_0 is a best uniform approximation to f from M.*

2°. *There are m points $x_i \in P[\varepsilon_0]$ and m positive numbers θ_i, $m \le n + 1$ ($m \le 2n + 1$, resp.) such that*

$$\sum_{i=1}^{m} \theta_i \overline{\varepsilon_0(x_i)} u(x_i) = 0, \qquad u \in M.$$

3°. *There are m points $x_i \in P[\varepsilon_0]$, $m \le n + 1$ ($m \le 2n + 1$, resp.) such that*

$$\min_{1 \le i \le m} \operatorname{Re} \overline{\varepsilon_0(x_i)} u(x_i) \le 0, \qquad u \in M.$$

Here ε_0 and $P[\varepsilon_0]$ are defined as in 2.8.

Noting that $u \mapsto l(u) = \sum_i \theta_i \overline{\varepsilon_0(x_i)} u(x_i) / \sum_i \theta_i |\varepsilon_0(x_i)|$, $x_i \in P[\varepsilon_0]$ is a peak functional, Condition 2° is seen to be the condition from the characterization via the Hahn-Banach theorem, while Condition 3° corresponds to the Kolmogorov criterion.

The characterization above may be extended to $C_0(X)$, where X is locally compact. Here, $C_0(X)$ is the completion of the set of functions with compact support.

Exercises

2.10. (Ascoli's lemma). Let $l \in E'$, $l \neq 0$, and let $M = \ker l$ be a (closed) hyperplane in E. Show that the distance is always given by

$$d(f, M) = \|l\|^{-1} |l(f)|.$$

Moreover, if there is a $g \in E$ such that $l(g) = \|l\| \cdot \|g\| = 1$, then $f - l(f) g \in P_M f$. Notice that in this case a multiple of g is a peak functional to l.

2.11. Characterize nearest points for the weighted Chebyshev approximation on X with the weighted norm

$$\|f\| = \|f\|_{\infty, w} := \sup_{x \in X} \{w(x) |f(x)|\},$$

where $w(x) > 0$ for $x \in X$.

2.12. Prove that $\|P_M f\| < 2\|f\|$ whenever M is a linear Chebyshev set, but not always $\|P_M f\| \leq \|f\|$.

2.13. Equip $C^1[\alpha, \beta]$ with the norm

$$\|f\| := \max\left\{|f(\alpha)|, \sup_{\alpha \leq x \leq \beta} |f'(x)\right\}$$

and formulate a characterization of nearest points from linear subspaces. Is there an analogue to Haar's uniqueness theorem (see below)?

2.14. Assume that u_1 and u_2 are two best L_1-approximations to f from a convex set M. Moreover assume that all functions are continuous. Show that

$$\operatorname{sgn}(f - u_1)(x) - \operatorname{sgn}(f - u_2)(x) \quad \text{for } x \in X!$$

Is there a generalization for the complex case?

§ 3. Linear and Convex Chebyshev Approximation

At the center of linear Chebyshev approximation is the theory of approximation in Haar subspaces. This theory is only briefly discussed here because the nonlinear families of functions which are varisolvent, or whose tangent sets satisfy the Haar condition, are considered as generalizations in Chapter III. A property which seems to be less important in linear problems, i.e. strong uniqueness of the best approximation, will turn out to be crucial in the nonlinear case. In order to be prepared for those cases in which the tangential sets are convex cones, the theory of Haar cones is included.

We will chiefly restrict ourselves to real valued functions. Then, the restriction to functions on an interval will be natural in the light of the Mairhuber-Curtis theorem.

A. Haar's Uniqueness Theorem, Alternants

3.1 Definition. An n-dimensional subspace M of $C(X)$ is a *Haar subspace* if one of the following equivalent conditions holds:

1°. Each $u \in M$, $u \neq 0$, has at most $n-1$ zeros in X.

2°. Given n distinct points $x_i \in X$ and n (real or complex) numbers y_i, the interpolation problem

$$u(x_i) = y_i, \qquad i = 1, 2, \ldots, n$$

has a solution $u \in M$.

3°. If $x_1, x_2, \ldots, x_n$ are distinct points in X and $\beta_1, \beta_2, \ldots, \beta_n$ are (real or complex) numbers such that

$$\sum_{i=1}^{n} \beta_i u(x_i) = 0 \quad \text{for } u \in M,$$

then $\beta_1 = \beta_2 = \cdots = \beta_n = 0$.

3.2 Haar's Uniqueness Theorem. *A finite dimensional subspace of $C(X)$, where X is locally compact and $C(X)$ is real or complex, is a Chebyshev set in $C(X)$ if and only if it is a Haar subspace.*

3.3 The Mairhuber-Curtis Theorem. *Let X be a compact set. If $C(X)$ contains a real Haar subspace of dimensional $n \geq 2$, then X is homeomorphic to a subset of the unit circle S^1.*

Since we shall consider only the approximation by real valued functions, in view of the Mairhuber-Curtis theorem the set X will be assumed to be an interval.

3.4 Definition. Assume that there are m points $x_1 < x_2 < \cdots < x_m$ such that for $\varepsilon \in C[\alpha, \beta]$:

$$|\varepsilon(x_i)| = \|\varepsilon\|, \qquad i = 1, 2, \ldots, m,$$

and

$$\varepsilon(x_{i-1}) = -\varepsilon(x_i), \qquad i = 2, 3, \ldots, m.$$

Then ε is said to have an *alternant of length m*. If moreover, $\varepsilon(x_m)$ is positive (or negative, resp.), then ε has a *positive* (or *negative*, resp.) *alternant* of length m.

The characterization of a best polynomial in terms of alternants was first proved by Borel (1905). The extension to arbitrary Haar subspaces was established by J.W. Young (1907).

3.5 Alternation Theorem. *An element u_0 in an n-dimensional Haar subspace M of $C[\alpha, \beta]$ is the best approximation to $f \in C[\alpha, \beta]$ if and only if there is an alternant of length $n+1$ to $f - u_0$.*

The alternation theorem is a consequence of the characterization theorem 2.9. An independent proof proceeds like the proof of the alternation theorem for varisolvent families (Theorem III, 3.8), from which this result follows as a special case.

Since the zero function is not a best approximation to the constant function $f(x) \equiv 1$, each Haar subspace contains a strictly positive function.

From a result of Newman and Shapiro the unique best approximation in a Haar subspace of $C[\alpha, \beta]$ is always strongly unique.

3.6 Definition (Newman and Shapiro 1963). An element u_0 is a *strongly unique best approximation* from M to a given f in a normed linear space E if there is a number $c > 0$ such that

$$\|f - u\| \geq \|f - u_0\| + c\|u - u_0\| \quad \text{for all } u \in M. \tag{3.1}$$

Because of the triangle inequality, the constant c in (3.1) cannot be larger than 1.

Strong uniqueness for the larger class of Haar cones will be proved in the next section.

B. Haar Cones

Many approximative properties of Haar subspaces are also found in the following type of cones, which, in the nonlinear theory, are encountered as tangent cones at boundary points of certain manifolds.

3.7 Definition (Braess 1973, Peisker 1983). Let $0 \leq m \leq n$ and let $v_1, v_2, \ldots, v_n \in C[\alpha, \beta]$. If the functions

$$\{v_j : j \in J\}$$

span a Haar subspace whenever $\{1, 2, \ldots, m\} \subset J \subset \{1, 2, \ldots, n\}$, then

$$M = \left\{ u = \sum_{j=1}^{n} a_j v_j \colon a_j \in \mathbb{R}, j = 1, 2, \ldots, m, a_j \geq 0, j = m+1, \ldots, n \right\} \tag{3.1}$$

is said to be a *Haar cone* of dimension n. Moreover, if $m < n$, then M is a *proper Haar cone*. To each Haar cone a *root number* is assigned:

$$r = r(M) := \inf\{k \in \mathbb{N} \colon \text{each } u \in M,\ u \neq 0, \text{ has at most } k-1 \text{ zeros}\}. \tag{3.2}$$

Let $u \in M$ be given as in (3.1) then

$$J(u) := \{1, 2, \ldots, m\} \cup \{j;\ m+1 \leq j \leq n,\ a_j > 0\}. \qquad \square$$

We note that multiplying v_n by (-1) leads to a different Haar cone if $m < n$.

3.8 Examples.

(1) The set of polynomials

$$\left\{ u(x) = \sum_{j=0}^{n} a_j x^j \colon a_j \in \mathbb{R},\ a_n \geq 0 \right\}$$

is a Haar cone with $r = n + 1$.

(2) Let $-1 < t_1 < t_2 < \cdots < t_n + 1$. Then

$$\left\{u(x) = \sum_{j=1}^{n} \frac{a_j}{1 - t_j x}: a_j \geq 0\right\}$$

is a Haar cone in $C[-1, +1]$ with $r = 1$. To verify this, observe that a sum with l nonzero coefficients a_j may be written in the form $p(x)/q(x)$ with p having degree at most $l - 1$.

3.9 Lemma. *Each Haar cone in $C[\alpha, \beta]$ is a Chebyshev set.*

Proof. Let $u^{(1)}$ and $u^{(2)}$ be best approximations to f in the Haar cone M. Then also $u = \frac{1}{2}(u^{(1)} + u^{(2)}) \in P_M f$. Notice that u is a local best approximation to f from the linear space $H := \operatorname{span}\{v_j : j \in J(u)\}$. From Problem 1.3 we know that each local best approximation from H is a best approximation from H. Since H is a Chebyshev set and $u^{(1)}, u^{(2)} \in H$, it follows that $u^{(1)} = u = u^{(2)}$. □

Obviously, u is a best approximation to f from the convex Haar cone M, if and only if 0 is a best approximation to $f - u$ from the set

$$C_u M = \left\{h = \sum_{j=1}^{n} \delta_j u_j : \delta_j \in \mathbb{R} \text{ for } j \in J(u) \text{ and } \delta_j \geq 0 \text{ otherwise}\right\}.$$

Since $C_u M$ is also a Haar cone, we may frequently reduce to the simpler case where 0 is a nearest point in a Haar cone.

The following theorem extends the results of Newman and Shapiro (1963) on strong uniqueness in Haar spaces. The proof makes use of an idea of Cheney (1966).

3.10 Strong Uniqueness Theorem (Braess 1973). *A best approximation from a Haar cone $M \subset C[\alpha, \beta]$ is a strongly unique best approximation.*

Proof. We may restrict ourselves to the case where 0 is the nearest point to f, $f \notin M$. Let $P[f] := \{x : |f(x)| = \|f\|\}$. We claim that

$$\min_{x \in P[f]} u(x) \operatorname{sgn} f(x) \leq -c\|u\|, \qquad u \in M, \tag{3.3}$$

for some $c > 0$. The function

$$\psi(u) := \min_{x \in P[f]} f(x) u(x)$$

attains its maximum on the compact set $M_1 = \{u \in M : \|u\| = 1\}$ for some element $u_0 \in M_1$. In view of the Kolmogorov criterion, we have $\psi(u_0) \leq 0$. If $\psi(u_0) < 0$ we are done. Suppose that $\psi(u_0) = 0$. We distinguish two cases concerning the optimality of u_0 in the linear space $H := \operatorname{span}\{v_j : j \in J(u_0)\}$.

Case 1. $0 \in P_H f$. By 2.9(2°) there are $k \leq \dim H + 1$ distinct points $x_1 < x_2 < \cdots < x_k$ in $P[f]$ and numbers $\theta_j > 0$ such that

$$\sum_{j=1}^{k} \theta_j f(x_j) u_0(x_j) = 0, \tag{3.4}$$

The Haar condition implies $k = \dim H + 1$. Since $\|u_0\| = 1$, u_0 has at most $\dim H - 1$ zeros and thus not all summands in (3.4) can vanish. Therefore, at least one summand is negative. This proves that $\psi(u_0) < 0$.

Case 2. $0 \notin P_H f$. From the Kolmogorov criterion it follows that

$$\min_{x \in P[f]} \{(f(x) - 0) u_1(x)\} > 0 \tag{3.5}$$

for some $u_1 \in H$. Note that $u_t = u_0 + t(u_1 - u_0) \in M$ for sufficiently small t. From $\psi(u_0) = \min f(x) u_0(x) = 0$ and (3.5) we conclude that

$$\min_{x \in P[f]} f(x) u_t(x) > 0.$$

By the Kolmogorov criterion, 0 is not a nearest point to f from the convex set M. This is a contradiction.

Therefore (3.3) holds at least for all $u \in M$ with $\|u\| = 1$ and some $c > 0$. A homogeneity argument yields (3.3) for all $u \in M$. From (3.3) we get, for any $u \in M$:

$$\begin{aligned} \|f - u\| &\geq \max_{x \in P[f]} |f(x) - u(x)| \\ &\geq \max_{x \in P[f]} [f(x) - u(x)] \cdot \operatorname{sgn} f(x) \\ &= \|f\| - \min_{x \in P[f]} u(x) \operatorname{sgn} f(x) \geq \|f\| + c\|u\|. \end{aligned}$$

As a consequence, 0 is a strongly unique best approximation. □

3.11 Remark. There is no strong uniqueness in complex Chebyshev approximation (Gutknecht 1978). Let $f(x) = x$ on $[-1 + 1]$. Then 0 is the best real or complex constant function. But when we consider the distance of f to the purely imaginary constants $u = ia$, $a \in \mathbb{R}$, we get

$$\|f - u\| = \sqrt{1 + a^2} \leq 1 + \tfrac{1}{2}a^2 = \|f - 0\| + \tfrac{1}{2}\|u - 0\|^2.$$

Also, the best approximation from a linear subspace of a real or complex Hilbert space is not strongly unique. For, if u_0 is the nearest point to f from a linear subspace M, then $(f - u_0, u) = 0$ for $u \in M$ and $\|f - u_0 - u\|^2 = \|f - u_0\|^2 + \|u\|^2$. Hence,

$$\begin{aligned} \|f - u_0 - u\| &= \|f - u_0\| + \frac{\|u\|^2}{\|f - u_0 - u\| + \|f - u_0\|} \\ &\leq \|f - u_0\| + \frac{\|u\|^2}{2d(f, M)}, \qquad u \in M. \end{aligned}$$

More generally, one cannot have strong uniqueness in linear subspaces of smooth Banach spaces.

C. Alternation Theorem for Haar Cones

The characterization of best approximations will be considered only for proper Haar cones, otherwise the results from linear Haar spaces can be applied. The solution from a cone in $C[\alpha, \beta]$ can be characterized in terms of signed alternants. Here the root number (3.1) is crucial. Obviously, $m + 1 \leq r(M) \leq n$.

3.12 Proposition. *Let $m + 1 \leq l \leq r(M)$. Then there exists an l-dimensional Haar cone $K_l \subset M$ and a nonzero function in K_l with exactly $l - 1$ zeros.*

Proof. The proof makes use of a Carathéodory type argument. By definition, $N_l := \{v \in M: v \neq 0, v \text{ has at least } l - 1 \text{ zeros in } [\alpha, \beta]\}$ is not empty. Choose a function $u_0 = \sum_j a_j v_j \in N_l$ for which the cardinality k of $J(u_0)$ is minimal. Denote the first $l - 1$ zeros of u_0 by $x_1 < x_2 < \cdots < x_{l-1}$.

We claim that $k = l$. Assume that $k > l$. By the definition of Haar cones, $\text{span}\{v_j: j \in J(u_0)\}$ contains an l-dimensional Haar subspace H. There is a nontrivial function $v \in H$ with zeros $x_1, x_2, \ldots, x_{l-1}$. Put $v = \sum_{j=1}^n b_j v_j$. After replacing v by $-v$ if necessary, we may assume that $b_j < 0$ for at least one $j > m$. Then $u_t = u_0 + tv, t \geq 0$, also vanishes at the given zeros. Set $t = \min\{|a_j/b_j|: j \geq m + 1, b_j < 0\}$. Then $u_t \in M$ but $|J(u_t)| \leq k - 1$, contradicting the minimality of $|J(u_0)|$.

By setting $K_l = \{u \in M: J(u) \subset J(u_0)\}$ the proof is complete. □

3.13 Proposition. *Assume that a proper l-dimensional Haar cone K_l contains a function with $l - 1$ zeros. Then there are nontrivial functions in K_l exhibiting $l - 1$ prescribed zeros. Moreover, each $v \in K_l$ with $l - 1$ zeros in (α, β) attains a fixed sign $\sigma = \sigma(K_l)$ on the right boundary point β.*

Proof. Denote the prescribed zeros by $x_1 < x_2 < \cdots < x_{l-1}$. Assume that $u_0 \in K_l$ has the zeros $z_1 < z_2 < \cdots < z_{l-1}$. For $0 \leq t \leq 1$ let $x_j(t) = z_j + t(x_j - z_j)$ and let u_t be the solution of the following interpolation problem in $H = \text{span}(K_l)$:

$$u_t(x_j(t)) = 0, \qquad j = 1, 2, \ldots, l - 1$$

$$u_t(\tfrac{1}{2}x_1(t) + \tfrac{1}{2}x_2(t)) = u_0(\tfrac{1}{2}[z_1 + z_2]).$$

Since each u_t has $l - 1$ zeros, none of its coefficients with index $j \geq m + 1, j \in J(u_0)$ can vanish. The mapping $t \mapsto u_t$ is continuous and the signs of the coefficients above are independent of t. Hence, $u_t \in K_l$, $0 \leq t \leq 1$, and u_1 is the solution of the interpolation problem.

The statement on the sign follows from the fact that $\text{sgn}\, u_t(\beta)$ is constant whenever $z_{l-1}, x_{l-1} \neq \beta$ and from the observation that $(-u_t) \notin K_l$. □

As a consequence of the above propositions, we obtain:

3.14 Lemma (Peisker 1983). *All functions in a proper Haar cone M having $r(M) - 1$ zeros in (α, β) have the same sign at $x = \beta$. (This sign will be denoted as $\sigma(M)$). Conversely given $m < l < r(M)$ and given $\sigma = +1$ or $\sigma = -1$, there is a cone K_l as stated in Proposition 3.12 with $\sigma(K_l) = \sigma$.*

Proof. Notice that the cones as spacified in Proposition 3.12 are not unique. Let K_r and $\tilde{K}_r$, $r = r(M)$, be two of them. We claim that $\sigma(K_r) = \sigma(\tilde{K}_r)$. Let

$v \subset K_r$ and $\tilde{v} \in \tilde{K}_r$ be functions with the same zeros $x_1 < x_2 < \cdots < x_{r-1} < \beta$. Moreover, we may assume that $|\tilde{v}(\beta)| = |v(\beta)|$. If v and $\tilde{v}$ attain the opposite sign at β, then $v + \tilde{v} \in M$ has an additional zero. Since $(-v) \notin M$ we have $v + \tilde{v} \neq 0$ and a contradiction to the maximality of r.

If $m < l < r$, then there is a function u_0 with l zeros $x_1 < x_2 < \cdots < x_{l-1} < x_l < \beta$ in K_{l+1}. Choose z_1 and z_2 such that $x_{l-1} < z_1 < x_l < z_2$. Then $\operatorname{sgn} u_0(z_1) = -\operatorname{sgn} u_0(z_2)$. Choose functions in the $(l + 1)$-dimensional Haar subspace H which contains K_{l+1} with the l zeros $x_1, x_2, \ldots, x_{l-1}, z_1$ and $x_1, x_2, \ldots, x_{l-1}, z_2$, resp. The reduction process from the proof of Proposition 3.12 yields two cones K_l and $\tilde{K}_l$, resp. such that $\sigma(K_l) = -\sigma(\tilde{K}_l)$. □

3.15 Alternation Theorem for Haar Cones (Peisker 1983). *The element 0 is the best approximation to f from a proper Haar cone M, if and only if $f - 0$ has an alternant of length $r(M)$ with the sign $-\sigma(M)$ to the right.*

Proof. Assume that there is an alternant $x_1 < x_2 < \cdots < x_r$ as stated. If $\|f - u\| < \|f\|$, then it follows from a de la Vallée-Poussin type argument that

$$(-1)^{r-i}\sigma(M)u(x_i) < 0, \qquad i = 1, 2, \ldots, r.$$

Hence, u has $r - 1$ zeros in (x_1, x_r) and $\operatorname{sgn} u(x_r) = -\sigma(M)$. This contradicts Lemma 3.14 and thus 0 is optimal.

Assume that there is an alternant $x_1 < x_2 < \cdots < x_l$ of exact length $l \leq r$. Moreover let the sign of the alternant be $\sigma(M)$, if $l = r$. Then we may choose a cone K_l as specified above, such that $\sigma(K_l)$ coincides with the sign of the alternant. Let H be the l-dimensional Haar subspace which contains K_l. Since there is no alternant of length $l + 1$, there is a better approximation $u_0 \in H$, $\|f - u_0\| < \|f\|$. It follows that

$$(-1)^{l-i}\sigma(K_l)u_0(x_i) > 0. \qquad i = 1, 2, \ldots, l.$$

Let $u_1 \in K_l$ be the function which has the same $l - 1$ distinct zeros as u_0. Moreover, since $\operatorname{sgn} u_0(x_l) = \sigma(K_l)$, after multiplying u_1 by a positive number, we get $u_1(x_l) = u_0(x_l)$. Hence, $u_0 = u_1 \in K_l \subset M$ and 0 is not a nearest point to f. □

Exercises

3.16. Determine the best uniform approximation to $f(x) = x^2$ on the interval $[-1, +1]$ from Π_1 and estimate the strong uniqueness constant.

3.17. Derive non-strong uniqueness for L_p-approximation, $1 < p < \infty$, from Clarkson's inequalities.

3.18. Determine the root number for the following Haar cones in $C[1, 2]$:

$$M_1 := \left\{u(x) = \sum_{k=0}^{n} a_k x^k \colon (-1)^k a_k \geq 0 \text{ for } k = 0, 1, \ldots, n\right\},$$

$$M_2 := \left\{u(x) = \sum_{k=0}^{n} a_k x^k \colon a_k \geq 0 \text{ if } k \text{ is odd}\right\}.$$

3.19. Let M be an n-dimensional subspace of $\mathbb{R}^m$. If $n = 1$ and $m = 2$, then there is a simple geometrical interpretation of the Haar condition: In this case M satisfies the Haar condition, if it is neither the x_1-axis nor the x_2-axis. How reads the analogous geometrical condition for arbitrary $m > n$?

3.20. Let M be an n-dimensional Haar subspace of $C[\alpha, \beta]$ and $\alpha \leq x_0 < x_1 < \cdots < x_n \leq \beta$. Show that the mapping φ: $M \to \mathbb{R}^n$ defined by

$$u \mapsto (u(x_0) - u(x_1), u(x_1) - u(x_2), \ldots, u(x_{n-1}) - u(x_n))$$

is a homeomorphism!

§4. L_1-Approximation and Gaussian Quadrature Formulas

The error curve of a solution of an L_1-approximation problem from an n-dimensional Haar subspace H has at least n zeros. In 1898, A.A. Markoff detected that the zeros are independent of a given f as long as there are exactly n zeros with sign changes. They are called the canonical points (with multiplicity 1) for the subspace H.

Similarly, when the best one-sided L_1-approximation is desired, an error function with $n/2$ double zeros or $(n-1)/2$ double zeros and one simple zero is found. The zeros are the canonical points of the classical Gaussian quadrature formulas.

The canonical points and their extremal properties will be briefly studied in a unified theory.

A. The Hobby-Rice Theorem

Let $d\mu(x) = w(x)\,dx$ be a nonnegative measure on $[\alpha, \beta]$ with $w \in C(\alpha, \beta)$. Assume that u^* is a best $L_{1,\mu}$-approximation to $f \in C[\alpha, \beta]$ from a real n-dimensional linear subspace $H \subset C[\alpha, \beta] \cap L_{1,\mu}[\alpha, \beta]$. Moreover, assume that $f - u^*$ changes its sign at exactly n points $\alpha = x_0 < x_1 < \cdots < x_n < x_{n+1} = \beta$. Then by the characterization theorem 2.7:

$$\sum_{i=0}^{n} (-1)^i \int_{x_i}^{x_{i+1}} v(x)\,d\mu(x) = 0 \quad \text{for } v \in H. \tag{4.1}$$

Note that (4.1) is independent of the function f. Therefore, under the conditions above, the best $L_{1,\mu}$-approximation to f is given by the interpolation at the points $\{x_i\}_{i=1}^n$. The latter are called the *canonical points* of H.

The existence of the canonical points can even be established for subspaces without the Haar condition (Hobby and Rice 1965).

4.1 Hobby-Rice-Theorem. *Let H be a real n-dimensional linear subspace of $L_1([\alpha, \beta], \mu)$, where μ is a finite, nonatomic, nonnegative measure. Then there exist $\{x_i\}_{i=1}^r$, $r \leq n$, $\alpha = x_0 < x_1 < \cdots < x_{r+1} = \beta$ such that*

$$\sum_{i=0}^{r} (-1)^i \int_{x_i}^{x_{i+1}} u(x)\,d\mu(x) = 0 \quad \text{for } u \in H. \tag{4.2}$$

The main tool for its proof will be a topological argument (Borsuk 1933, see also Amann 1974):

Borsuk Antipodality Theorem. *Let Ω be a bounded, open, symmetric neighborhood of 0 in $\mathbb{R}^{n+1}$ and $T\colon \partial\Omega \to \mathbb{R}^n$ be an odd, continuous mapping. Then there exists some $x \in \partial\Omega$ for which $T(x) = 0$.*

We note that a mapping T is said to be odd if $T(x) = T(-x)$.

Proof of Theorem 4.1 (Pinkus 1976). Let $S^n \subset \mathbb{R}^{n+1}$ be defined by

$$S^n = \left\{\mathbf{y} = (y_0, y_1, \ldots, y_n) \colon \sum_{i=0}^{n} |y_i| = \beta - \alpha\right\}.$$

Given $\mathbf{y} \in S^n$, we associate to $\mathbf{y}$ the vector $\mathbf{x} = \mathbf{x}(\mathbf{y})$ specified by $x_0 = \alpha$, $x_{i+1} = x_i + |y_i|$, $i = 0, 1, \ldots, n$. Let $\{v_j\}_{j=1}^{n}$ be a basis of H and define $T\colon S^n \to \mathbb{R}^n$ by

$$T_j(\mathbf{y}) = \sum_{i=0}^{n} (\operatorname{sgn} y_i) \int_{x_i(\mathbf{y})}^{x_{i+1}(\mathbf{y})} v_j(x)\, d\mu(x) \quad \text{for } j = 1, 2, \ldots, n. \tag{4.3}$$

Since $d\mu$ is a nonatomic measure, T is continuous. Moreover, T is odd. By the Borsuk Antipodality Theorem, we have $T(\mathbf{y}^*) = 0$ for some $\mathbf{y}^* \in S^n$, and the subset $\{x_i(\mathbf{y}^*);\ 1 \le i \le n,\ y_i y_{i+1} \le 0\}$ yields a set of canonical points. □

If H is a Haar subspace, then each set of canonical points must consist of exactly n points.

B. Existence of Generalized Gaussian Quadrature Formulas

A system $\{v_j\}_{j=1}^{N}$ of continuous functions on $[\alpha, \beta]$ is called an *extended Chebyshev system* on $[\alpha, \beta]$ or *ET-system**, if it is the basis of an N-dimensional space H such that the Hermite interpolation problem in H is always solvable.

Let $\{v_i\}_{i=0}^{l+1}$ be a set of integers, $v_0 \ge 0$, $v_{l+1} \ge 0$ and $v_i > 0$ for $i = 1, 2, \ldots, l$ such that

$$\sum_{i=0}^{l+1} v_i = N. \tag{4.4}$$

To each set of knots $\mathbf{x} = \{x_i\}_{i=1}^{l}$ with $\alpha = x_0 < x_1 < \cdots < x_{l+1} = \beta$ we associate a sign function

$$\sigma_{\mathbf{x},\mathbf{v}}(\xi) := \operatorname{sgn} \prod_{i=1}^{l} (x_i - \xi)^{v_i} = (-1)^{[\sum_{1 \le i \le l, (x_i \le \xi)} v_i]}. \tag{4.5}$$

The famous quadrature formulas of Gauss (1809) have been an attractive subject of investigation for many mathematicians (see e.g. Krein 1951, Karlin

* The abbreviation "*ET*-system" refers to the old transcription Tschebyscheff instead of Chebyshev.

and. Pinkus 1976a, b, Barrow 1978). With the notation above we can formulate the main result on generalized Gaussian quadrature formulas (GGQF).

4.2 Theorem on Generalized Gaussian Quadrature Formulas (Bojanov, Braess, and Dyn 1986). *Let H be a Haar subspace of $C[\alpha, \beta]$ spanned by an extended Chebyshev system $\{v_j\}_{j=1}^N$ and let* (4.4) *hold. Moreover, assume that $d\mu(x) = w(x)\,dx$ with a positive weight function $w \in C(\alpha, \beta)$. Then there is a unique set $\mathbf{x}$ of canonical points $\alpha = x_0 < x_1 < \cdots < x_l < x_{l+1} = \beta$ such that*

$$\int u\sigma_{\mathbf{x},\mathbf{v}}\,d\mu(x) = \sum_{i=0}^{l+1}\sum_{j=0}^{v_i-1} a_{ij}u^{(j)}(x_i) \quad \text{for } u \in H \tag{4.6}$$

holds with appropriate weight factors a_{ij} with

$$a_{i,v_i-1} = 0 \quad \text{for } i = 1, 2, \ldots, l. \tag{4.7}$$

Obviously, the special case $v_1 = v_2 = \cdots = v_l = 1$ and $v_0 = v_{l+1} = 0$ refers to the Hobby-Rice-Theorem. If $v_1 = v_2 = \cdots = v_l = 2$ and no boundary terms occur, then from (4.6) and (4.7), the classical Gaussian quadrature formulas are obtained. In that case, the sign function $\sigma_{\mathbf{x},\mathbf{v}}$ is always positive and may be abandoned in the formula. The same holds for Gaussian quadrature formulas with even multiplicities v_i $(1 \le i \le l)$.

It is obvious from the special cases above that the sign function (4.5) heavily depends on the parity of the multiplicities of the knots. The next lemma shows that $+\sigma_{\mathbf{v}}$ and $-\sigma_{\mathbf{v}}$ are the only sign functions which admit quadrature formulas of Gaussian type.

4.3 Lemma. *Let the conditions of Theorem* 4.2 *hold and let $\alpha = x_0 < x_1 < \cdots < x_l < x_{l+1} = \beta$. Assume that $|s(x)| \equiv 1$ and that s is constant in the subintervals (x_i, x_{i+1}) for $i = 0, 1, \ldots, l$. Moreover, let $s(x) = +1$ for $x \le x_1$. If there is a formula of the form*

$$\int u(x)s(x)\sigma_{\mathbf{x},\mathbf{v}}(x)\,d\mu = \sum_{i=0}^{l+1}\sum_{j=0}^{v_i-1} a_{ij}\cdot u^{(j)}(x_i) \quad \text{for } u \in H, \tag{4.8}$$

such that (4.7) *holds, then $s(x) = +1$ almost everywhere.*

Proof. Define an auxiliary multiplicity vector $\boldsymbol{\omega}$ by

$$\omega_i = \begin{cases} v_i - 1 & \text{if } s \text{ changes its sign at } x_i, \\ v_i & \text{otherwise.} \end{cases}$$

Moreover, set $\omega_0 = v_0$ and choose ω_{l+1} such that $\sum_i (\omega_i - v_i) = 0$. Then by construction, we have $\sigma_{\boldsymbol{\omega}} = s \cdot \sigma_{\mathbf{v}}$ almost everywhere. Consider the function $u \in H$ which solves the interpolation problem

$$u^{(j)}(x_i) = \delta_{i,l+1}\delta_{j,\omega_l-1} \quad \text{for } j = 0, 1, \ldots, \omega_i - 1, \qquad i = 0, 1, \ldots, l+1. \tag{4.9}$$

Since $\{v_j\}_{j=1}^N$ is an ET-system, u has not more zeros than is specified by (4.9). Therefore $u \cdot \sigma_{\boldsymbol{\omega}}$ does not change its sign, and the integral on the left hand side of

(4.8) does not vanish. If $\mathbf{v} \neq \boldsymbol{\omega}$, then $\omega_{l+1} > v_{l+1}$, and the right hand side of (4.8) equals zero. Consequently, $s \neq +1$ leads to a contradiction. □

We will only verify the existence of canonical points. We note that for the simplest cases, i.e. $v_i = 1$ or $v_i = 2$ $(1 \leq i \leq l)$, uniqueness follows immediately by elementary arguments from the extremal properties below.

Proof of the Existence of Canonical Points (as stated in Theorem 4.2). Let $S^l \subset \mathbb{R}^{l+1}$ be defined by

$$S^l = \left\{ \mathbf{y} = (y_0, y_1, \ldots, y_l) \colon \sum_{i=0}^{l} |y_i| = \beta - \alpha \right\} \tag{4.10}$$

Given $\mathbf{y} \in S^l$, we associate with $\mathbf{y}$ the vector $\mathbf{x} = \mathbf{x}(\mathbf{y})$ specified by $x_0 = \alpha$, $x_{i+1} = x_i + |y_i|$, $i = 0, 1, \ldots, l$. Next, we define $\{z_j\}_{j=1}^N$ such that this set contains each knot x_i with its multiplicity v_i for $i = 0, 1, \ldots, l+1$. Furthermore, we number the z_j's such that

$$z_{N+1-i} = x_i, \qquad i = 1, 2, \ldots, l.$$

Since we will verify only à posteriori that the knots x_i and x_{i+1} do not coalesce, divided differences will be used when functionals of the following form are introduced:

$$L_{\mathbf{y}}(u) := \sum_{i=0}^{l} (\operatorname{sgn} y_i) \int_{x_i(\mathbf{y})}^{x_{i+1}(\mathbf{y})} u(x)\sigma_{\mathbf{v}}(x)\,d\mu - \sum_{j=1}^{N} b_j u(x_1, x_2, \ldots, x_j). \tag{4.11}$$

Here $b_j = b_j(\mathbf{y})$ for $j = 1, 2, \ldots, N$ is to be chosen such that

$$L_{\mathbf{y}}(v_m) = 0, \qquad m = 1, 2, \ldots, N. \tag{4.12}$$

For this purpose, (4.11) is to be understood as a linear system of N equations in the N coefficients $\{b_j\}_{j=1}^N$. The matrix of this system is nonsingular, since it is the adjoint of a matrix for a Hermite interpolation problem and $\{v_m\}_{m=1}^N$ is an ET-system. Moreover, the divided differences are continuous functions of the arguments. Therefore the matrix and the right hand side of the linear system above are continuous functions of $\mathbf{y}$. The mapping $T\colon S^l \to \mathbb{R}^l$

$$T(\mathbf{y}) = (b_N(\mathbf{y}), b_{N-1}(\mathbf{y}), \ldots, b_{N+1-l}(\mathbf{y}))$$

is continuous and odd. By the Borsuk Antipodality Theorem there is a $\mathbf{y}^* \in S$ such that $T(\mathbf{y}^*) = T(-\mathbf{y}^*) = 0$.

The numbering of the z_j's implies that a quadrature formula of type (4.6) is defined by $\mathbf{y}^*$, into which $x_1, x_2, \ldots, x_l$ enter with reduced multiplicities. Thus $T(\mathbf{y}^*) = 0$ yields (4.7).

To complete the proof, one has to verify that $y_i^* \neq 0$ for $i = 0, 1, \ldots, l$. This follows with the same type of argument as in the proof of Lemma 4.3. Finally, it follows from that same lemma that $\operatorname{sgn} y_0^* = \operatorname{sgn} y_1^* = \cdots = \operatorname{sgn} y_l^*$. Hence, there is a formula as postulated. □

The weight factors in the classical Gaussian quadrature formulas are positive. The signs of the leading terms in the generalized formulas are also known.

4.4 Lemma. *If there is a formula of type* (4.6) *with* $a_{m,\nu_m-1} = 0$ *for some* $m \in \{1, 2, \dots, l\}$, *then* $a_{m,\nu_m-2} \neq 0$ *and*

$$\operatorname{sgn} a_{m,\nu_m-2} = \sigma_m \tag{4.13}$$

holds where the abbreviations $\sigma_m := \sigma_{\mathbf{x},\mathbf{v}}(x_m + 0)$ *and*

$$a_{m,-1} = [\sigma_{\mathbf{x},\mathbf{v}}(x_m + 0) - \sigma_{\mathbf{x},\mathbf{v}}(x_m - 0)]\, w(x_m) \tag{4.14}$$

are used.

Proof. We only need to consider the case where $\nu_m \geq 2$. Choose $u \in H$ according to the interpolation conditions:

$$\begin{aligned} u^{(j)}(x_m) &= \delta_{j,\nu_m-2}, && j = 0, 1, \dots, \nu_m - 2, \\ u^{(j)}(x_i) &= 0, && j = 0, 1, \dots, \nu_i - 1, 0 \leq i \leq l + 1, i \neq m, \\ & && \text{and } j = \nu_{l+1}, i = l + 1. \end{aligned}$$

Again, we deduce that u has no more zeros than the $N - 1$ specified ones. Specifically, u has a zero of multiplicity $\nu_m - 2$ at x_m, and ν_i at x_i, $i \neq l$, $1 \leq i \leq m$. Hence $u(x)\sigma_{\mathbf{x},\mathbf{v}}(x)$ does not change its sign. Moreover, $u(x) > 0$ holds in (x_m, x_{m+1}). Hence $\operatorname{sgn} \int u\sigma_{\mathbf{x},\mathbf{v}}\, d\mu = \sigma_m$. This proves (4.13). □

The lemma shows that in *GGQF*'s at least one of the leading terms a_{m,ν_m-1} and a_{m,ν_m-2} for any knot does not vanish.

C. Extremal Properties

Assume that f is in the *convexity cone* of $\{v_j\}_{j=1}^N$. This means that $\{v_1, v_2, \dots, v_N, f\}$ is again an *ET*-system. Moreover, the sign of f is fixed by $\operatorname{sgn}(f - u) = \sigma_{\mathbf{v}}$ whenever $u \in H = \operatorname{span}\{v_j\}_{j=1}^N$, and $f - u$ has N zeros with the multiplicities given by $\mathbf{v}$.

The best $L_{1,\mu}$-approximation to f from H is the function u^* which interpolates at the N canonical points with multiplicity one. Indeed, set $\mathbf{v} = \mathbf{1} := (1, 1, \dots, 1)$. Then for any $u \in H$,

$$\begin{aligned} \|f - u\|_{1,\mu} &\geq \int (f - u)\sigma_{\mathbf{1}}\, d\mu \\ &= \int (f - u^*)\sigma_{\mathbf{1}}\, d\mu + \int (u^* - u)\sigma_{\mathbf{1}}\, d\mu \\ &= \|(f - u^*)\|_{1,\mu}. \end{aligned} \tag{4.15}$$

Here we have made use of $\int v\sigma_{\mathbf{1}}\, d\mu = 0$ for $v = u^* - u \in H$.

Let N be even. Then the one-sided $L_{1,\mu}$-approximation to f from H is the function u^* which interpolates f at the $l = N/2$ canonical points with multiplicity

2, i.e. at the classical Gaussian points. Then $\sigma_{\mathbf{v}} \geq 0$, and for any $u \in H$ with $u \leq f$ we obtain with the classical *GQF*

$$\begin{aligned}
\|f - u\|_{1,\mu} &= \int (f - u)\, d\mu \\
&= \int (f - u^*)\, d\mu + \int (u^* - u)\sigma_2\, d\mu \\
&= \|f - u^*\|_{1,\mu} + \sum_{i=1}^{l} a_i(u^* - u)(x_i) \\
&\geq \|f - u^*\|_{1,\mu}.
\end{aligned} \tag{4.16}$$

Since the weights a_i are positive, (4.16) is a consequence of the restrictions $u(x_i) \leq f(x_i) = u^*(x_i)$, $i = 1, 2, \ldots, l$. As a byproduct, we obtain that $f - u^*$ has $N/2$ distinct double zeros.

In both cases equality holds only for $u = u^*$, and u^* is the unique best approximation. Hence, we get uniqueness of the canonical points.

In the general case, the extremal properties of the canonical points are given in terms of an L_1-approximation problem for a submanifold of H. Given a multiplicity vector $\mathbf{v}$ such that (4.4) holds and given f in the convexity cone, consider the nonlinear family

$$\begin{aligned}
H(f, \mathbf{v}) := \{u \in H\colon \exists \xi \in \Delta_l \text{ with } \xi_0 < \xi_1 < \cdots < \xi_{l+1} \text{ such that} \\
(f - u)^{(j)}(\xi_i) = 0, j = 0, 1, \ldots, v_i - 1, i = 0, 1, \ldots, l + 1\}.
\end{aligned} \tag{4.17}$$

Here

$$\Delta_l := \Delta_l[\alpha, \beta] := \{\mathbf{x} = (x_1, x_2, \ldots, x_l)\colon \alpha < x_1 < \cdots < x_l < \beta\}.$$

The set $H(f, \mathbf{v})$ is an l-dimensional submanifold of the linear space H, (and the zeros $\xi_1, \xi_2, \ldots, \xi_l$ may be chosen as the coordinates of the elements).

4.5 Theorem (Bojanov, Braess and Dyn 1985). *Let the conditions of Theorem 4.2 hold. Assume that f is in the convexity cone of H. Then u is the unique best $L_{1,\mu}$-approximation to f from $H(f, \mathbf{v})$ if and only if u interpolates f at the canonical points for H associated with the multiplicities $v_0, v_1, \ldots, v_{l+1}$.*

Chapter II. Nonlinear Approximation: The Functional Analytic Approach

In this chapter, existence and uniqueness of nearest points are considered in a general setting in which only properties from functional analysis such as the different kinds of compactness, convexity, and characterizations of best approximation in terms of linear functionals are used. In this framework the influence of uniform convexity, smoothness or non-smoothness of the underlying normed linear space is considered.

The most famous result in this context is the theorem that each approximatively compact Chebyshev set in a Hilbert space must be convex. This means that given a non-convex set in a Hilbert space, under very weak assumptions an element with at least two nearest points or with none can be found. In this context, solar properties turn out to be crucial. In contrast to the nonuniqueness result mentioned, one has uniqueness for *almost all* elements in a uniformly convex space.

§ 1. Approximative Properties of Arbitrary Sets

A. Existence

As mentioned in the last chapter, compact sets are existence sets but compactness assumptions have to be relaxed in order to treat the problems of interest.

Firstly, we note that weakly (sequentially) compact sets are also existence sets. More generally, the extension of existence results to the weak topology can be done with the following argument: If $u \in M$ is the weak limit of a minimizing sequence (u_n) in M for f, then $u \in P_M(f)$. Indeed, let $l \in S'$ be a peak functional for $f - u$, i.e. $l(f - u) = \|f - u\|$. Then

$$\|f - u\| = l(f - u) = \lim_{n\to\infty} l(f - u_n) \leq \lim_{n\to\infty} \|l\| \cdot \|f - u_n\| = d(f, M). \tag{1.1}$$

Hence, u is a nearest point.

Next, the arguments used in Chapter I for the existence of nearest points in finite dimensional linear spaces are considered in a more general framework. The concept of approximative compactness was introduced by Efimov and Stechkin (1961), extended by Breckner (1968) to the weak topology, and further generalized by Deutsch (1980) for non-topological existence proofs.

1.1 Definition. A set $M \subset E$ is called *boundedly compact* (*boundedly weakly compact*, resp.) if the intersection of M with each closed ball is compact (weakly compact, resp.). A set $M \subset E$ is called *approximatively compact* (*approximatively weakly compact*) if each minimizing sequence contains a subsequence which converges (converges weakly, resp.) to an element in M.

Since each minimizing sequence is bounded, one has: Each boundedly compact (boundedly weakly compact) set is approximatively compact (approximatively weakly compact). Each approximatively compact (approximatively weakly compact) set is an existence set.

In the following we will repeatedly see that, in uniformly convex spaces, existence is established more easily than in other normed linear spaces.

1.2 Theorem (Sz.-Nagy 1942). *A closed convex set M in a uniformly convex Banach space is an approximatively compact Chebyshev set.*

Proof. Let M be such a set. When considering the approximation to a given $f \in E$, we may perform a translation such that $f = 0$. Let $d = d(0, M)$. Obviously, we may restrict ourselves to the case $d > 0$. Let (u_n) be a minimizing sequence. Set $\lambda_n = \|u_n\|^{-1}$ for $n = 1, 2, \ldots$. Given $\varepsilon > 0$, let δ be taken as in the definition of uniform convexity. Then for n, m sufficiently large, we have $\|u_n\|, \|u_m\| < d(1 + \delta/2)$. Thus:

$$\begin{aligned}\left\|\frac{1}{2}\lambda_n u_n + \frac{1}{2}\lambda_m u_m\right\| &= \frac{1}{2d}\|u_n + u_m - (1 - d\lambda_n)u_n - (1 - d\lambda_m)u_m\| \\ &\geq \frac{1}{d}\left\|\frac{1}{2}(u_n + u_m)\right\| - \frac{1}{2d}(\lambda_n^{-1} - d)\|\lambda_n u_n\| \\ &\quad - \frac{1}{2d}(\lambda_m^{-1} - d)\|\lambda_m u_m\| > 1 - \delta/2,\end{aligned}$$

since $\frac{1}{2}(u_n + u_m) \in M$ has a distance $\geq d$ from the origin. By uniform convexity, $\|\lambda_n u_n - \lambda_m u_m\| < \varepsilon$ and $\|u_n - u_m\| < \lambda_n^{-1}\varepsilon + |1 - \lambda_m/\lambda_n| \cdot \|u_m\|$. Consequently, (u_n) is a Cauchy-sequence and converges to some u in the Banach space. Since M is closed, we have $u \in M$ and M is approximatively compact.

The uniqueness of the best approximation is a consequence of uniform (strict) convexity. □

Almost completely opposite is the situation in c_0, i.e., in the space of those sequences $x = (x_1, x_2, \ldots)$ with $\lim x_k = 0$.

1.3 Example. Let c_0 be endowed with the sup-norm, and let M be the hyperplane

$$M = \{x \in c_0 : l(x) = 0\}, \quad \text{where} \quad l(x) := \sum_{k=1}^{\infty} 2^{-k} x_k. \tag{1.2}$$

Then $P_M f = \emptyset$ for each $f \in c_0 \setminus M$.

To prove this, write $f = (f_1, f_2, \ldots)$. By definition, $f \notin M$ means

$$\lambda := \sum_k 2^{-k} f_k \neq 0.$$

To verify that $d(f, M) \leq |\lambda|$, consider the sequence (u_n), defined by

$$u_n = f - \lambda(1 - 2^{-n})^{-1} \cdot (1, 1, \ldots, 1, 0, 0, \ldots).$$

Specifically, the first n components of the second vector are nonzero. A simple calculation yields $l(u_n) = 0$ and $u_n \in M$. From $\|f - u_n\| = (1 - 2^{-n})^{-1}|\lambda|$ we get $d(f, M) \leq \lim \|f - u_n\| = |\lambda|$.

Suppose that there is a best approximation $u^* = (y_1, y_2, \ldots)$. Since $(f - u^*)$ is a sequence which converges to zero, we have

$$|f_k - y_k| < \tfrac{1}{2}|\lambda|, \quad \text{for all } k \geq n$$

and some $n \in \mathbb{N}$. From this, $l(u^*) = 0$, and $\|f - u^*\| \leq |\lambda|$, we obtain

$$\begin{aligned} |\lambda| = |l(f - u^*)| &= \left| \sum_{k=1}^{n-1} 2^{-k}(f_k - y_k) + \sum_{k=n}^{\infty} 2^{-k}(f_k - y_k) \right| \\ &\leq \sum_{k=1}^{n-1} 2^{-k}|\lambda| + \sum_{k=n}^{\infty} 2^{-k} \cdot \frac{1}{2}|\lambda| < |\lambda|. \end{aligned}$$

From this contradiction, it follows that no nearest point exists. □

Similarly, the set $\{x \in c_0 : |l(x)| \leq 1\}$ has the same property. By taking the intersection of sets of this type, Edelstein and Thompson (1972) have constructed a closed bounded convex symmetric set M in c_0 such that no element outside M has a nearest point.

A crucial point in the above example is the non-reflexivity of the space c_0. There is no $g \in c_0$ to which the functional in (1.2) is a peak functional. More generally, we have the following: If $g \in S$ and $l \in S'$ satisfy $l(g) = \|g\| = 1$, then $u = f - l(f)g$ is a best approximation to f from the hyperplane $\{u \in E : l(u) = 0\}$.

Let E be a reflexive space. Then each bounded sequence has a weakly converging subsequence. Since any minimizing sequence is bounded, each weakly closed set in a reflexive normed linear space is an existence set.

However, most of the interesting nonlinear families in nonlinear Chebyshev approximation are not boundedly weakly compact. Typical of the families of rational functions, exponential sums, and spline functions with free knots is a behaviour which is illustrated for a 2-parameter family of rational functions in $C[0, 1]$:

$$M = \left\{ u(x) = \frac{a}{1 + tx} : a \in \mathbb{R}, t \geq 0 \right\}. \tag{1.3}$$

If (u_n), where $u_n = a_n/(1 + t_n x)$, is a bounded sequence, then $|a_n| = |u_n(0)| \leq \|u_n\|$ implies that (a_n) is bounded. If (t_n) is also bounded, an accumulation point in M is easily obtained from a subsequence with convergent parameters. If, on the other hand, $t_n \to \infty$, we get

$$\lim u_n(x) = \begin{cases} \lim a_n & \text{for } x = 0, \\ 0 & \text{for } x > 0. \end{cases}$$

The sequence converges to $u_0 = 0 \in M$ pointwise with $x = 0$ excluded. Since $\lim l(u_n) \neq l(u_0)$ for the functional $l(f) := f(0)$, M is not weakly compact.

Nevertheless, a general result asserts that u_0 is a nearest point to f whenever (u_n) is a minimizing sequence in (1.3) for the approximation of f. One has $u_n \to u_0$ if *convergence* is understood in a non-topological framework, see de Boor (1969) and Deutsch (1980).

It turns out that from the beginning of rigorous proofs for nonlinear approximation problems, the following lemma has been applied more or less hidden.

1.4 Lemma. *If each bounded sequence in $M \subset C(X)$ contains a subsequence which converges pointwise on a dense subset of X to an element in M, then M is* an existence set.

Proof. Given $f \in C(X)$, let (u_n) be a minimizing sequence in M. The sequence is bounded. After passing to a subsequence if necessary, we may assume that (u_n) converges pointwise to some $u^* \in M$ on a dense set $X_1 \subset X$.

Given $\varepsilon > 0$, by the definition of the sup-norm we have

$$\|f - u^*\| \leq |(f - u^*)(x_0)| + \varepsilon$$

for some $x_0 \in X$. Since $f - u^*$ is continuous and X_1 is dense in X, one gets

$$|(f - u^*)(x_0)| < |(f - u^*)(x_1)| + \varepsilon$$

for some $x_1 \in X_1$. The convergence of the sequence (u_n) at x_1 implies that

$$|(f - u^*)(x_1)| < |(f - u_n)(x_1)| + \varepsilon \leq \|f - u_n\| + \varepsilon$$

for all n sufficiently large. Summing up we have $\|f - u^*\| \leq \|f - u_n\| + 3\varepsilon$. Since (u_n) is minimizing and ε may be arbitrarily small, it follows that $\|f - u^*\| \leq d(f, M)$. □

We emphasize that the families to which thc above lemma applies are not necessarily approximatively compact. Moreover, often the assumptions of Lemma 1.4 are established in two steps. In a first step convergence on a dense set to functions in an augmented set (as rational functions with poles or spline functions with jump discontinuities) are established. In the second step the elements which are only in the augmented set and not in the original one are excluded.

An analogous result for L_p-approximation is treated by Deutsch (1980), see also Theorem V.1.4.

B. Uniqueness from the Generic Viewpoint

If there are two best approximations to some f from M, then the metric function $u \mapsto \|f - u\|$ has two global minima in M. One would expect that small perturbations will induce one of the global minima to become a local one. In this sense it is an exceptional case that the distance to f is the same at the two local minima and it will turn out that one does indeed encounter uniqueness of the nearest points for "most" of the f's in a uniformly convex Banach space.

The naive definition of "most" would be that the statement refers to "a dense set". However, this notion is inadequate because in that concept *most* numbers would be rational but *most* numbers would be irrational as well. Therefore we are led to consider the notion of a generic property.

1.5 Definiton. Let E be a topological space, M a subset of E. M is called *residual* in E if it contains a countable intersection of dense, open subsets of E. A property is called *generic* if it is true for all elements of a residual set.

Equivalently, M is a residual set if its complement is a set of first category in E, i.e., if $E \backslash M$ is the countable union of nowhere dense subsets of E.

Note that the set of irrational numbers is the countable intersection of open dense subsets $\bigcap_{n>1} \{x \in \mathbb{R}: nx \notin \mathbb{Z}\}$. Another example which will be important when counting the solutions of nonlinear problems, will be given in Exercise 1.14. But first we will be concerned with uniqueness from the generic viewpoint.

For a given non-empty set M in a normed linear space E, Stechkin (1963) introduced the sets:

$$E_M = \{f \in E: P_M f \neq 0\},$$

$$U_M = \{f \in E: P_M f \text{ is empty or a singleton}\},$$

$$T_M = \{f \in E: P_M f \text{ is a singleton}\}.$$

Obviously, U_M is dense in strictly convex spaces. Indeed, if f has more than one nearest point, select $u \in P_M f$. Then U_M contains $g = tf + (1-t)u$ whenever $0 < t < 1$. In order to determine T_M, a structurally more stable set T'_M is considered:

$$P_\delta f := P_{\delta,M} f = M \cap B_{d(f,M)+\delta}(f), \quad \delta > 0,$$

$$D_M f = \lim_{\delta \to +0} \operatorname{diam}(P_{\delta,M} f) \quad \text{where } \operatorname{diam}(U) = \sup\{\|u - v\|: u, v \in U\}, \tag{1.4}$$

$$T'_M = \{f \in E: D_M f = 0\}.$$

Here, $\operatorname{diam}(M)$ is the diameter of the set M and $P_\delta f$ is called the set of *δ-nearest points* to f. We note that later the sets of δ-nearest points will be rediscovered in a different setting as the level sets of the metric function.

Let M be a closed set in a Banach space. Obviously $D_M f \geq \operatorname{diam}(P_M f)$. Therefore $D_M f = 0$ implies $f \in U_M$. Moreover, in this case each minimizing sequence is a Cauchy sequence and $P_M f \neq \varnothing$. Combining these arguments, we get $T'_M \subset T_M$.

If M is approximatively compact, then

$$T'_M = T_M = U_M. \tag{1.5}$$

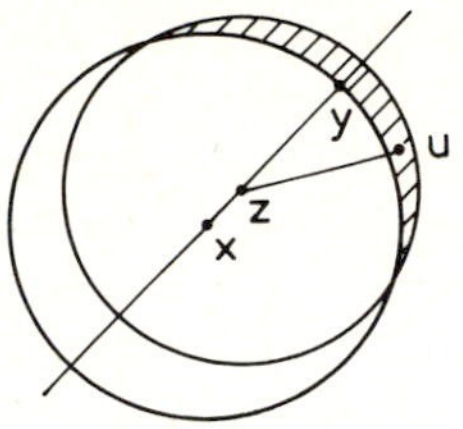

Fig. 2

Indeed, let $f \in T_M = U_M$ and $u \in P_M f$. Then a minimizing sequence (u_n) satisfying $\|u_n - u\| \geq (1/2) D_M f$, $n = 1, 2, \ldots$ exists. Since a subsequence converges to the unique nearest point, it follows that $D_M f = 0$.

1.6 Proposition (Stechkin 1963). *Let E be a uniformly convex space, and let $0 < r < R$. Then*

$$\operatorname{diam}[B_{r+\delta}(z) \backslash \mathring{B}_R(x)]$$

tends to zero as $\delta \to +0$, uniformly for all $x, z \in E$ satisfying $\|x - z\| = R - r$.

Proof. Since a similarity transformation does not change the statement, we may assume that $x = 0$ and $R = 1$. Moreover, we may assume $r > \frac{1}{2}$, since the bound for diameters is obviously an increasing function of r. Let $\|y\| = 1$ and $z = (1 - r)y$. If $u \in B_{r+\delta}(z)$, then $u = (1 - r)y + \alpha v$ for $\alpha \leq r + \delta$ and some $\|v\| = 1$. It is sufficient to restrict ourselves to $\alpha = r + \delta_1$ with $0 \leq \delta_1 \leq \delta$. Then

$$\begin{aligned} \|u\| &= \|(1 - r)(y + v) + (2r - 1 + \delta_1)v\| \\ &\leq (1 - r)\|y + v\| + (2r - 1 + \delta) \\ &\leq (1 - r) \cdot 2(1 - \tilde{\delta}(\varepsilon)) + (2r - 1) + \delta = 1 + \delta - 2(1 - r)\tilde{\delta}(\varepsilon), \end{aligned}$$

if $\|y - v\| > \varepsilon$ and $\tilde{\delta}(\varepsilon)$ is taken as in the definition of uniform convexity. Hence, we have $\|u\| \geq 1$ only if $\tilde{\delta}(\varepsilon) \leq \delta/(2 - 2r)$. On the other hand, $\|u - y\| = \|r(v - y) + \delta_1 v\| \leq r\varepsilon + \delta$. Hence, the diameter in question is bounded by $2(\varepsilon + \delta)$ and tends to 0 as $\delta \to 0$. □

The following theorem contains two results of Stechkin, one of them being given in an improved form.

1.7 Theorem. *Let M be a closed subset of a strictly convex Banach space E. Moreover assume that M is approximatively compact or that E is a uniformly convex space. Then T_M and T'_M are residual sets.*

Proof. (1) Let $a > 0$. We claim that $G_a := \{f \in E, D_M f < a\}$ is open.

Given $f \in G_a$, there is a $\delta > 0$ such that $\operatorname{diam}(P_{2\delta} f) < a$. If $\|g - f\| < \delta$ then $P_\delta g \subset P_{2\delta} f$ and $D_M g \leq \operatorname{diam}(P_{2\delta} f) < a$. Hence G_a is open.

(2) Assume that M is approximatively compact. Since E is strictly convex, U_M is dense in E. Recalling (1.5) and noting that $G_a \supset T'_M$, we get the density of G_a.

(3) We claim that G_a is also dense, if E is uniformly convex.

Let $f \in E \backslash M$. Without loss of generality we may assume that $f = 0$ and $d(f, M) = 1$. Given $r < 1$, by the preceding proposition there is a $\delta > 0$ such that $\operatorname{diam}[B_{r+2\delta}(g) \backslash \mathring{B}_1(0)] < a$ whenever $\|g\| = (1 - r)$. Since $d(f, M) = 1$, there is some $u_1 \in M$ with $\|u_1\| < 1 + \delta$. Let $h = u_1/\|u_1\|$ and $g = (1 - r)h$. Then $d(g, M) \leq \|g - u_1\| \leq r + \delta$. Hence, $P_\delta g \subset B_{r+2\delta}(g) \backslash \mathring{B}_1(0)$, and from the choice of δ we know that $\operatorname{diam}(P_\delta g) < a$. This means that $g \in G_a$. Since we may choose $1 - r$ arbitrarily small, G_a is dense in E.

(4) Under the assumptions of the theorem

$$T'_M = \bigcap_{n>1} G_{1/n}$$

is represented as a countable intersection of dense open sets. □

In Theorem 1.7, the assumption on the strict convexity cannot be abandoned. Let the n-space be equipped with the sup-norm and let $M = B_1(0)$ be the unit ball. Then $f = (f_1, f_2, \ldots, f_n)$ has only a unique nearest point if $f \in M$ or $|f_1| = |f_2| = \cdots = |f_n|$. In this case, even non-uniqueness is a generic property in $E \backslash M$.

Exercises

1.8. Determine E_M, U_M, T_M and T'_M for the following sets, where $\mathbb{R}^2$ is endowed with the Euclidean norm and with the supremum norm, resp.:
a) $M = S^1 = \{(x, y): x^2 + y^2 = 1\}$,
b) $M = \{(x, 0): |x| < 1\}$,
c) $M = \{(x, y): |x|^{1/2} + |y|^{1/2} \leq 1\}$.

1.9. Let M be an existence set (an approximatively compact set, resp.) in $C[-1, +1]$. Is the induced set of symmetrized functions

$$M_1 = \{u(x) = \tfrac{1}{2}[v(x) + v(-x)]: v \in M\}$$

an existence set?

1.10. Assume that $M \subset \mathbb{R}^n$ is not an existence set. Show that the complement of E_M contains infinitely many points.

1.11. Are the union and the intersection of two existence sets also existence sets?

1.12. Consider the set (1.3) and verify that a closed subset of an existence set is not necessarily an existance set. Is a closed subset of an approximatively compact set also approximatively compact?

1.13. Derive from Jackson's theorems that $C[0, 1] \backslash C^1[0, 1]$ is a residual set in $C[0, 1]$.

1.14. Let f be a mapping from a metric space X into the finite set $\{1, 2, \ldots, N\}$. Assume that, to each $x_0 \in X$ there is a neighborhood U such that

$$f(x) \geq f(x_0), \quad \text{for } x \in U.$$

Show that f is continuous on a residual set. Is this still true if the range of f is $\mathbb{N}$?

1.15. For $n = 1, 2, \ldots$ let $u_n(x) = \max\{1 - |1 - nx|, 0\}$. Prove that $M = \{0\} \cup \{u_n: n = 1, 2, \ldots\}$ is an existence set in $C[0, 1]$. Is $M \backslash \{0\}$ still an existence set?

1.16. Suppose that M is not an existence set in E. Does there always exist an existence set $M_1 \supset M$ such that

$$d(f, M_1) = d(f, M) \quad \text{for } f \in E?$$

1.17. Show that $M = \{u \in C[0,1]: u(0) = 0\}$ is an existence set in $C[0,1]$ which is not approximatively weakly compact.
Hint: By $u_m(x) = x^{1/n}$, a minimizing sequence is given for $f \equiv 1$.

1.18. Denote a function $f \in C[0,1]$ as non-exceptional if $f - p_n$ has not an alternant of length $n + 3$ whenever $p_n \in \Pi_n$. Show that the set of these functions is a residual set in $C[0,1]$.

§2. Solar Properties of Sets

A. Suns. The Kolmogorov Criterion

In nonlinear approximation theory, geometrical properties which are weaker than convexity turn out to be important. Certain properties which admit a separation (such as the separation of convex sets by hyperplanes) are called *solar* properties.

The starting point is the following observation: Let $u_0 \in P_M f$, $f \notin M$. Then u_0 is also a nearest point to

$$f_t = u_0 + t(f - u_0) \tag{2.1}$$

whenever $0 < t < 1$, (see Fig. 3). Moreover, if E is a strictly convex space, then u_0 is the unique nearest point.

Indeed, we have $\|f - u_0\| = \|f - f_t\| + \|f_t - u_0\|$. From this and $u_0 \in P_M f$ we obtain for any $u \in M$:

$$\|f_t - u\| \geq \|f - u\| - \|f - f_t\| \geq \|f - u_0\| - \|f - f_t\| = \|f_t - u_0\|, \tag{2.2}$$

with the first inequality being strict whenever E is a strictly convex space.

On the other hand, u_0 is possibly not a nearest point to other elements on the ray $\{f_t: t > 1\}$, see Fig. 4.

2.1 Definition. An existence set M in a normed linear space is a *sun* if, given $f \notin M$, there is a $u_0 \in P_M f$ such that u_0 is also a nearest point to

$$f_t = u_0 + t(f - u_0) \quad \text{for each } t > 1.$$

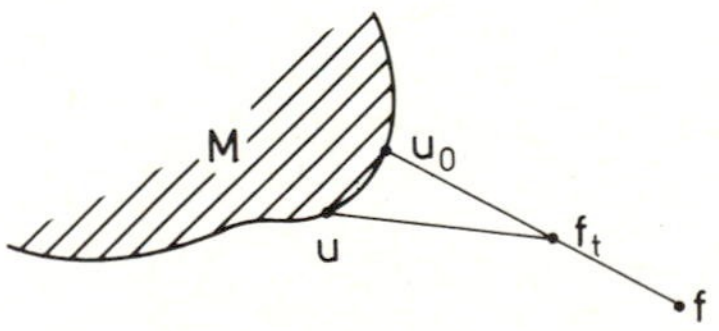

Fig. 3

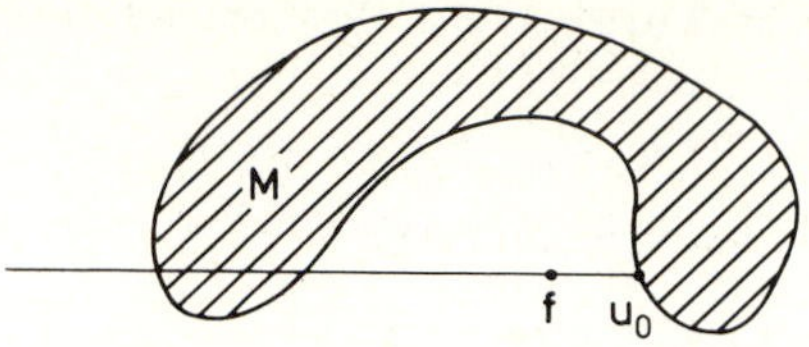

Fig. 4. A set which is not a sun

A nearest point with this property is said to be a *solar point* to f. If each $u_0 \in P_M f$ is a solar point to f, then M is said to be a *strict sun*.

The definitions above are due to Efimov and Stechkin (1958a). The underlying concept with the subsequent considerations was previously used by Klee (1949). (We note that in the literature sometimes suns but not strict suns are assumed to be existence sets.)

Obviously, each convex existence set is a strict sun, and a strict sun is a sun. In the class of Chebyshev sets, suns and strict suns coincide. An example of a sun which is not a strict sun is $\{(x_1, x_2) \in \mathbb{R}^2 \colon x_1 \geq 0 \text{ or } x_2 \geq 0\}$ in 2-space equipped with the supremum-norm (see Fig. 5).

2.2 Remark. A sun in a strictly convex normed linear space is a Chebyshev set.

Proof. Let $u_0 \in P_M f$ be a solar point. If u_1 is another nearest point to f, by strict convexity, we have $\|f_2 - u_1\| < \|f_2 - u_0\|$ for $f_2 = 2f - u_0$. This is a contradiction. □

2.3 Remark (Brosowski and Deutsch 1974a). Each local best approximation in a strict sun is a (global) best approximation.

Proof. Assume that u_0 is a local best approximation to f from a strict sun M. This means that $u_0 \in P_{M \cap B_r(u_0)} f$ for some $r > 0$. Recalling the notation (2.1), we note that $u_0 \in P_M f_t$ where $t = r/3\|f - u_0\|$. Since M is a strict sun, we have $u_0 \in P_M f$. □

The converse is not true in general. The complement of the open unit ball in a Hilbert space is not a sun, but each local solution is a global one. We will return to this problem later in connection with the discussion of moons (cf. Exercise 2.13).

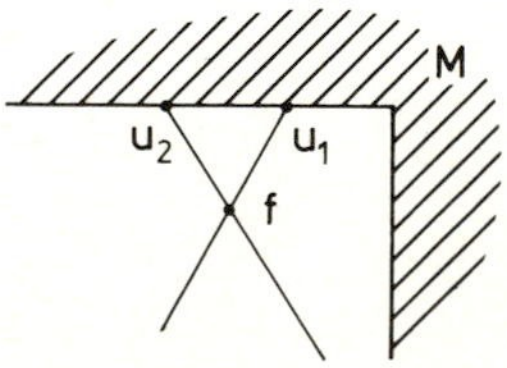

Fig. 5. A sun which is not a strict sun: u_1 is a solar point to f, but u_2 is not

The first hint to the separation property of suns is the connection with a convex problem and the Kolmogorov criterion.

2.4 Lemma (Brosowski 1969, Breckner 1970). *Let M be an existence set in a normed linear space E. Then the following are equivalent*:
1°. *M is a strict sun.*
2°. *For any $f \in E$, $u_0 \in M$ we have $u_0 \in P_M f$ if and only if u_0 is a nearest point to f from $[u_0, u]$ whenever $u \in M$.*
3°. *Each nearest point can be characterized by the (generalized) Kolmogorov condition: $u_0 \in P_M f$ if and only if given $u \in M$ there is a peak functional l such that*

$$\|l\| = 1,$$
$$l(f - u_0) = \|f - u_0\|,$$
$$\operatorname{Re} l(u - u_0) \leq 0.$$

Proof. 1° ⇔ 2°. M is a strict sun if, for any $u_0 \in P_M f$ and $u \in M$:

$$\|f_t - u\| \geq \|f_t - u_0\| \quad \text{for } t \geq 1. \tag{2.3}$$

Here we have used the notation (2.1) again. Therefore (2.3) means

$$\|u_0 - u + t(f - u_0)\| \geq t\|f - u_0\| \quad \text{for } 1 \leq t < \infty.$$

After dividing by t we get the equivalent relation:

$$\|f - u_0 - t^{-1}(u - u_0)\| \geq \|f - u_0\| \quad \text{for } 0 < t^{-1} \leq 1.$$

This is the optimality of u_0 in $[u_0, u]$, (see Fig. 6). The converse is proven in the same manner.

2° ⇒ 3°. Assume that $u_0 \in P_M f$. Given $u \in M$, u_0 is also a nearest point to the convex set $[u_0, u]$. By applying the characterization via the Hahn-Banach theorem to the auxiliary approximation problem, we get a functional as required by the Kolmogorov criterion.

3° ⇒ 2°. Let $u_0 \in P_M f$. Assume that given $u \neq u_0$, $u \in M$ there is a functional l as stated in the Kolmogorov condition. Then obviously $\operatorname{Re} l(v - u_0) \leq 0$ holds for any $v \in [u_0, u]$. From the characterization theorem I.2.1 it follows that u_0 is a nearest point to f from the convex set $[u_0, u]$. □

In a uniformly convex space E, we get extra information on suns. Let $f \in E \backslash M$ and u_0 be a solar point to f. Let $r = d(f, M)$. Given $\delta > 0$, we know that $P_\delta f \in B_{r+\delta}(f) \backslash \mathring{B}_{2r}(f_2)$. By Proposition 1.5, diam$(P_\delta f)$ tends to zero as $\delta \to 0$.

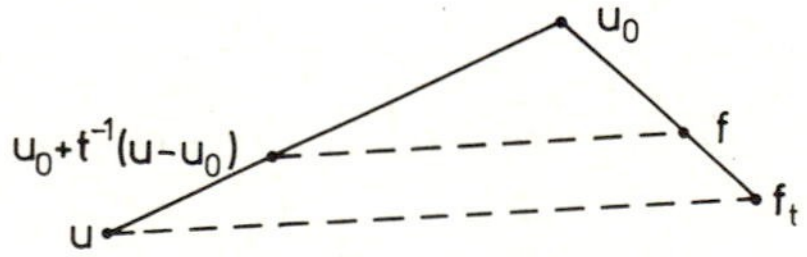

Fig. 6. Equivalence of 1° and 2° in Lemma 2.2

Hence $f \in T'_M$ and $T'_M = E$. Therefore, each sun in a uniformly convex set is an approximatively compact Chebyshev set.

B. The Convexity of Suns

2.5 Theorem. *The following conditions on a normed linear space E are equivalent*:
1°. *E is smooth.*
2°. *Each sun in E is convex.*

Proof. (1) Assume that M is a sun in the smooth space E. Let $u_1, u_2 \in M$, $0 < \alpha < 1$ and suppose that $f := \alpha u_1 + (1-\alpha)u_2 \notin M$. Let $u_0 \in P_M f$ be a solar point to f. With the same argument as in the proof of Lemma 2.4 (cf. Exercise 2.15), it follows that u_0 can be characterized as a nearest point by the Kolmogorov condition. Specifically, there exist linear functionals l_1 and $l_2 \in E'$ such that

$$\left.\begin{aligned} \|l_i\| &= 1, \\ l_i(f - u_0) &= \|f - u_0\|, \\ \operatorname{Re} l_i(u_i - u_0) &\le 0, \end{aligned}\right\} \quad i = 1, 2. \tag{2.4}$$

From the first and the second condition we conclude that $\|\frac{1}{2}(l_1 + l_2)\| = 1$. Since E is smooth, we get $l_1 = l_2$. By using the third relation, we estimate $l_1(f - u_0)$:

$$\begin{aligned} \operatorname{Re} l_1(f - u_0) &= \operatorname{Re} l_1[\alpha u_1 + (1-\alpha)u_2 - u_0] \\ &= \alpha \operatorname{Re} l_1(u_1 - u_0) + (1-\alpha)\operatorname{Re} l_1(u_2 - u_0) \le 0. \end{aligned}$$

This contradicts the second relation from (2.4). Therefore $f \in M$ and M is convex.

(2) Assume that E is not a smooth space. We may restrict ourselves to the real case, because a complex space may be identified with a sum of real spaces. To construct a non-convex sun, let $g \in S$ and $l_1, l_2 \in S'$, $l_1 \ne l_2$ be such that $l_1(g) = l_2(g) = \|g\| = 1$. Define

$$M_i = \{u \in E : l_i(u) \ge 0\} \quad \text{for } i = 1, 2, \text{ and } M = M_1 \cup M_2.$$

(Note that in the special case where $E \in \mathbb{R}^2$ is equipped with the uniform norm, the choice $l_i(x) = x_i$, $i = 1, 2$ and $g = (1, 1)$ leads to the set M depicted in Fig. 4).

Let $f \in E \backslash M$. Then for $i = 1, 2$ we have $l_i(f) < 0$ and from $d(f, M_i) \ge |l_i(f)|$, $f - l_i(f)g \in M_i$ we conclude that $d(f, M_i) = -l_i(f)$. Without loss of generality we may assume that $c := d(f, M_1) \le d(f, M_2)$ and one has $u_0 = f + cg \in P_M f$.

Recalling (2.1) we notice that $l_1(f_t) \ge l_2(f_t)$. Hence, $d(f_t, M) = d(f_t, M_1) \le d(f_t, M_2)$. Since M_1 is convex, u_0 is also a best approximation to f_t from M_1 and thus $u_0 \in P_M f_t$. Therefore M is a non-convex sun. □

Notice that the proof of the second part mainly proceeds in the 2-dimensional space spanned by f and g.

The theorem above means that the characterization of nearest points via the

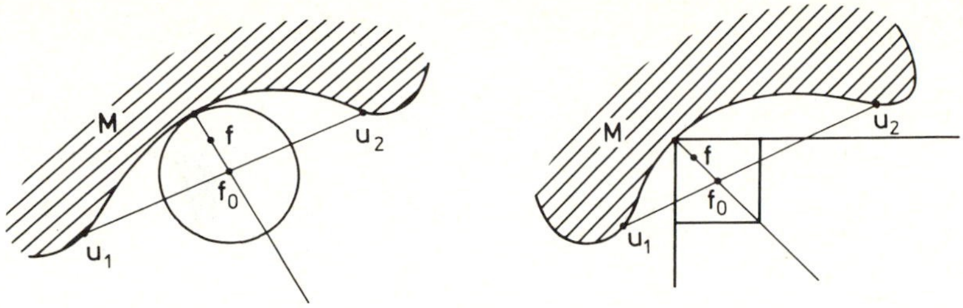

Fig. 7. Impossibility and possibility for the separation of $f \notin M$ by an arbitrary large ball from M

Hahn-Banach theorem and the Kolmogorov criterion coincide for nonlinear sets in smooth spaces as well.

A geometrical interpretation for the theorem above is given in Fig. 7. According to Efimov and Stechkin (1959) a set is called *a-convex* if each $f \notin M$ is contained in a ball of radius a which is disjoint from M. Suns are a-convex for each $a > 0$, because $f \in B_a(f_t)$, where $t = \max\{1, (a/d(f, M) + 2)^{-1}\}$. This means that f can be separated from a sun by an arbitrarily large sphere. If E is a smooth space, then the separation by an arbitrarily large sphere is equivalent to the separation by a hyperplane.

C. Suns and Moons in $C(X)$

Of special interest are characterizations of suns by inner properties* which refer only to the elements of the sets and not to the approximation problem. Such a property, namely convexity, has been established for smooth spaces in Theorem 2.5. Another property of this kind will be the closed sign property of sets in $C(X)$.

We note that both properties state that there are sufficiently many elements in each neighborhood of u_0 in M. More precisely, given u, $u_0 \in M$ we have information on elements in a neighborhood of u_0 in M which depends on u though u may be far away from u_0. This is consistent with the fact that a local best approximation in a strict sun is a global solution.

For reasons which become apparent later, we restrict ourselves to real spaces in this section.

2.6 Characterization of Suns in $C(X)$ (Brosowski 1968). *Let X be a compact set and let $C(X)$* be equipped with the uniform norm. For an existence set, *the following are equivalent*:

1°. *M is a strict sun.*

2°. *Each local best approximation from M to any $f \in E$ is a nearest point to f.*

3°. *M has the closed sign property, i.e., given a pair u_0, $u_1 \in M$ and a closed set $A \subset X$ with*

---

* in German: *innere Eigenschaften*

$$\inf_{x \in A} |u_1(x) - u_0(x)| > 0 \tag{2.5}$$

the element u_0 is contained in the closure of the set

$$\{v \in M \colon [v(x) - u_0(x)][u_1(x) - u_0(x)] > 0, x \in A\}. \tag{2.6}$$

The sets satisfying condition 2° were called *regular by* Brosowski (1968) and were independently introduced by Dunham (1969a) with the notation given here.

Proof. 2° ⇒ 3°. Assume that Condition 2° holds and that u_0, $u_1 \in M$, $\|u_1 - u_0\| = r > 0$. Let $A \subset X$ and $a = \inf_{x \in A} |u_1(x) - u_0(x)| > 0$. Put $\rho(x) = \max\{a, |u_1(x) - u_0(x)|\}$,

$$h(x) = \rho^{-1}(x)[u_1(x) - u_0(x)], \tag{2.7}$$

and $f = u_0 + rh$. Obviously, $\|h\| = 1$, $h \cdot (u_1 - u_0) = |h| \cdot |u_1 - u_0|$, and $\|f - u_0\| = r$. For $x \in X$:

$$\begin{aligned} |f - u_1|^2 &= |u_1 - u_0 - rh|^2 \\ &= |u_1 - u_0|^2 - 2rh(u_1 - u_0) + r^2|h|^2 \\ &= |u_1 - u_0|^2 - 2r|h| \cdot |(u_1 - u_0)| + r^2|h|^2 \\ &= [|u_1 - u_0| - r|h|]^2 < r^2 = \|f - u_0\|^2. \end{aligned}$$

Hence $u_0 \notin P_M f$. By Condition 2°, u_0 is not a local best approximation to f from M. In each neighborhood of u_0 there is a v such that $\|f - v\| < \|f - u_0\|$. By construction v, belongs to the set (2.6). Therefore it has the closed sign property.

3° ⇒ 1°. Let M have the closed sign property. Suppose that u_0, $u_1 \in M$ and that $\|f - u_1\| < \|f - u_0\|$. Then $\delta = \frac{1}{2}(\|f - u_0\| - \|f - u_1\|) > 0$. Put

$$A = \{x \in X \colon |(f - u_0)(x)| \geq \|f - u_0\| - \delta\}.$$

Obviously $(u_1 - u_0)(f - u_0)(x) > 0$ whenever $x \in A$. Given $t \in (0, 1)$, by the closed sign property there is a $v \in M \cap B_{t\delta}(u_0)$ such that

$$(v - u_0)(f - u_0)(x) > 0 \quad \text{for } x \in A.$$

Therefore, for $x \in A$ we have $\operatorname{sgn}(v - u_0) = \operatorname{sgn}(f - u_0)$ and

$$\begin{aligned} |f_t(x) - v(x)| &= |t(f - u_0) - (v - u_0)| \\ &= t|f - u_0| - |v - u_0| < t\|f - u_0\|. \end{aligned}$$

On the other hand, for $x \in X \backslash A$ it follows from the definition of A that

$$\begin{aligned} |f_t(x) - v(x)| &\leq t|f - u_0| + \|v - u_0\| \\ &\leq t[\|f - u_0\| - 2\delta] + t\delta < t\|f - u_0\|. \end{aligned}$$

Consequently, $\|f_t - v\| < t\|f - u_0\| = \|f_t - u_0\|$ and $u_0 \notin P_M f_t$, $0 < t \leq 1$. This means that u_0 is only a nearest point to f_t, $0 < t \leq 1$, if u_0 is also a nearest point to f. By definition, M is a strict sun.

1° ⇒ 2°. By Remark 2.3 each local best approximation in a strict sun is a global solution and the proof is complete. □

An analogue to the closed sign property can be given in any normed linear space, if one abandons the postulate that the properties must be specified by conditions on the set M alone without referring to other points of the linear space into which M is embedded.

Given $h \in E$ let $P[h] := \{l \in S' \colon l(h) = \|h\|\}$. For $u_0, h \in E$ define the cone

$$K(u_0, h) := \{u_0 + g \in E \colon l(g) > 0 \text{ for all } l \in P[h]\}. \tag{2.8}$$

2.7 Definition (Amir and Deutsch 1972). A set M in a normed linear space is a *moon* if $u_0 \in M$, $h \in E$ and $M \cap K(u_0, h) \neq \emptyset$ implies $u_0 \in \overline{M \cap K(u_0, h)}$.

A set in $C(X)$ is a moon if and only if it satisfies the closed sign property. Let $u_0, u_1 \in M$ and $A \subset X$ satisfy (2.5). Recall the definition of h from (2.7) and note that $l(g) > 0$ for all $l \in P[h]$ is equivalent to $l(g) > 0$ for all extreme elements l in $P[h]$. Therefore, the set (2.6) is $K(u_0, h)$ and we have $u_1 \in K(u_0, h)$. The Condition 3° asserts that $u_0 \in \overline{M \cap K(u_0, h)}$.

Obviously, M intersects $K(u_0, f - u_0)$ if u_0 is not a nearest point. In the same spirit, the Kolmogorov condition 2.4 (3°) may be rephrased: $u_0 \in P_M f$ if and only if $M \cap K(u_0, f - u_0) = \emptyset$. Furthermore, the reformulation of the statement that u_0 is not a local best approximant in a strict sun if it is not a nearest point, yields the lunar property from Definition 2.7. Finally, the second part of the proof of Theorem 2.6 may be generalized. Summarizing we have some of the results on the connection between suns and moons obtained by Amir and Deutsch (1972), Brosowski and Deutsch (1974).

2.8 Proposition.

(1) *Each sun in a normed linear space is a moon.*

(2) *A set in $C(X)$, X compact, is a sun if and only if it is a moon.*

(3) *If each local best approximation from a set M in a normed linear space E is a nearest point, then M is a moon.*

The example after Remark 2.3 shows that not each moon in a strictly convex space is a sun. In particular, Euclidean 2-space may be identified with the set of complex valued functions whose domain is a single point. Consequently, Theorem 2.6 cannot be extended to the complex case.

Geometric properties of suns and strict suns in finite dimensional spaces are described by Braess (1974) and Berens and Hetzelt (1982).

Exercises

2.9. Construct an existence set M in 2-space such that for each $f \notin M$ there is no solar point in $P_M f$.

2.10. A set $M \subset C(X)$ is said to have the betweenness property if, given $u_1, u_2 \in M$ there is a $u \in M$ such that

$$u(x) = (1 - t(x))u_1(x) + t(x)u_2(x),$$

where $0 < t(x) < 1$ for $x \in X$ (Dunham 1969). Prove that each set M of this kind is a strict sun for uniform approximation.

2.11. Why can a finite set which consists of more than one point, never be a sun?

2.12. Verify that the unit sphere in a strictly convex space is a moon. (What is known about local best approximations?)

2.13. Consider $M = \{(x, y) \in \mathbb{R}^2;\ x^2 + 4y^2 \geq 1\}$ in Euclidean 2-space. Show that M is a moon but not each local best approximation is a global one.

2.14. Are the union and the intersection of two suns always suns?

2.15. Let M be an existence set in a normed linear space which is not necessarily a strict sun. Modify Lemma 2.4 such that it provides equivalent properties to the statement that an individual element in M is a solar point.

§ 3. Properties of Chebyshev Sets

The convexity of Chebyshev sets in smooth spaces under certain additional assumptions is the main result of this section, e.g. each approximatively compact Chebyshev set in a Hilbert space must be convex. It is an open question whether this statement is true without the compactness condition. From the preceding section we know that it is sufficient to show the solarity of a set, provided that the space is smooth. On the other hand, there are examples of Chebyshev sets in $C(X)$ and in pre-Hilbert spaces which are not suns.

As before, we chiefly derive approximative properties from the features of the underlying spaces. For the converse, refer to the survey article by Vlasov (1973).

A. Approximative Compactness

The most important consequence of approximative compactness is the fact that it implies continuity of the metric projection.

The metric projection onto a Chebyshev set will be understood as a mapping $P = P_M\colon E \to M$.

3.1 Continuity Theorem. *The metric projection onto an approximatively compact Chebyshev set is continuous.*

Proof. Assume that M is an approximatively compact Chebyshev set and that $f = \lim f_n$. Since $\|f_n - u\|$ attains its minimum in M at $u = Pf_n$, it follows that

$$\begin{aligned} \|f - Pf_n\| &\leq \|f - f_n\| + \|f_n - Pf_n\| \leq \|f - f_n\| + \|f_n - Pf\| \\ &\leq 2\|f - f_n\| + \|f - Pf\|. \end{aligned} \tag{3.1}$$

Hence (Pf_n) is a minimizing sequence to f. Since each subsequence contains an accumulation point in M and each accumulation point is a best approximation to f, it follows from uniqueness that $\lim Pf_n = Pf$. □

Similarly, if a Chebyshev set is approximatively weakly compact, then the metric projection is continuous when the image is endowed with the weak topology. The proof of this statement, which may also be extended to more general topologies of the topological linear space E, proceeds as the proof of the above theorem.

For completeness, an analogous result for sets which are not uniqueness sets, is mentioned. The set-valued metric projection P onto an approximatively compact set is upper semi-continuous, i.e., $\{f \in E: P_M f \subset U\}$ is open in E for each open set $U \subset E$. On the other hand, there exists only a continuous mapping ϕ from E onto an approximatively compact subset M such that $\phi f \in P_M f$ for any $f \in E$ if M is a Chebyshev set (Nürnberger 1977).

A consequence of the continuity theorem is

3.2 Corollary. *Each local best approximation in an approximatively compact Chebyshev set is a best approximation.*

Proof. Assume that $u_0 \neq P_M f$. Define f_t as in (2.1) and consider the curve $\{u_t = P_M f_t: 0 \le t \le 1\}$. By the preceding theorem the curve is continuous. By assumption $u_1 \neq u_0$. Given $\varepsilon > 0$, there is a $t \in (0, 1)$ such that $u_t \neq u_0$ and $\|u_t - u_0\| < \varepsilon$. Obviously,

$$\|f - u_0\| = \|f - f_t\| + \|f_t - u_0\| > \|f - f_t\| + \|f_t - u_t\| \ge \|f - u_t\|.$$

Therefore, u_0 is not a local best approximation to f. □

The following observation is often useful. Let u be the unique nearest point to f from an approximatively compact set M. Then we have $\lim_{\delta \to +0} \operatorname{diam}(P_\delta f) = 0$. This is true because any minimizing sequence must converge to u (see p. 29).

A first application is found in the

3.3 Lemma on the Removal of a Ball (Efimov and Stechkin 1958b). *Suppose that $B_r(f)$ intersects the approximatively compact set M at the unique point u_0. For any $g \in \mathring{B}_r(f)$, there is a $t_0 > 0$ such that for all $t \in (0, t_0)$ the displaced ball $t(g - u_0) + B_r(f)$ is disjoint from M.*

Proof. Let $0 < t \le 1$. The shift $v \to v_t := v + t(g - u_0)$ sends u_0 to the interior point g, when $t = 1$. By continuity we have $v_1 \in \mathring{B}_r(f)$ whenever $v \in B_\varepsilon(u_0)$, $\varepsilon < r - \|f - g\|$. Since each orbit $\{v_t: 0 \le t \le 1\}$ is a segment, we even have $v_t \in \mathring{B}_r(f)$, whenever $0 < t \le 1$, $v \in B_\varepsilon(u_0) \cap B_r(f)$. From the preceding discussion

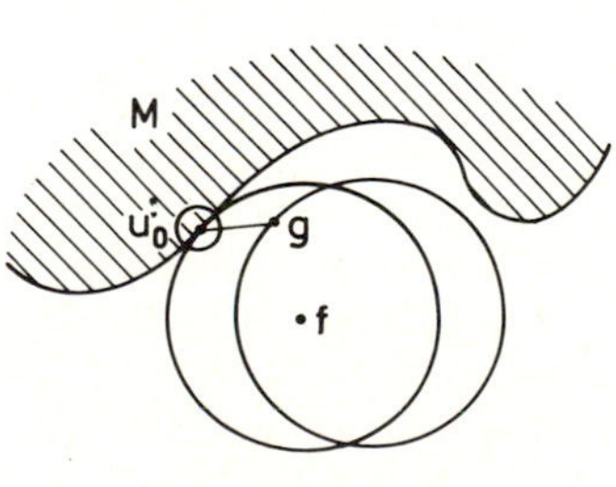

Fig. 8. Removal of a ball

we know that $\operatorname{diam}(P_\delta f) < \varepsilon/2$ whenever δ is sufficiently small. Moreover, after reducing δ if necessary we may assume that $\delta < \varepsilon/2$. Let $t_0 = \delta\|g - u_0\|^{-1}$.

Now consider $v_t = v + t(g - u_0)$ for $v \in B_r(f)$ and $0 < t < t_0$. Then $v_t \in B_{r+\delta}(f)$. If $v \in B_\varepsilon(u_0)$, then from the choice of ε we know that $v_t \in B_r(f)$. Hence in this case $v_t \notin M$. If, on the other hand $v \notin B_\varepsilon(u_0)$, then $\|v_t - u_0\| \geq \|v - u_0\| - \|t(g - u_0)\| > \varepsilon - \delta > \varepsilon/2$. From this and $M \cap B_{r+\delta}(f) = P_\delta f \subset B_{\varepsilon/2}(u_0)$ it follows that $v_t \notin M \cap B_{r+\delta}(f)$. Therefore, $v_t \notin M$. □

B. Convexity and Solarity of Chebyshev Sets

3.4 Theorem. *Each boundedly compact Chebyshev set in a Banach space is a sun.*

Remark. It is sufficient to prove that each locally compact Chebyshev set with a continuous metric projection is a sun. Moreover, in view of Theorem 3.1 the continuity of the metric projection may be replaced by approximative compactness. Similarly, local compactness may be replaced by the assumption that the set $P_{\delta,M} f$ is compact for some $\delta = \delta(f) > 0$, or that P_M is continuous in the weak topology. There are other variants.

The theorem and the corollaries below have a long history. The first proofs for finite dimensional (Euclidean) spaces are due to Bunt (1934), Motzkin (1935), and Kritikos (1938). The first variants in infinite dimensional spaces were formulated by Efimov and Stechkin (1959) and proved by Fikken (cf. Klee (1961)). Other proofs were established by Klee (1961) and Vlasov (1961). They make use of the

Schauder Fixed Point Theorem. *Let A be a closed convex set in a Banach space. Every map $f: A \to A$ such that $f(A)$ is relatively compact, has a fixed point.*

Proof of Theorem 3.4. (1) If M is not a sun, then by definition there is an $f \in E \backslash M$ and a $t_0 > 0$ such that $Pf_{t_0} \neq Pf =: u_0$. Here we have used the abbreviation (2.1) again. Let $t_1 = \sup\{t > 0: Pf_t = u_0\}$. Since $Pf_t = u_0$ for each $t \in (0.t_1)$, from the continuity of the metric projection it follows that $Pf_{t_1} = \lim_{t \to t_1 - 0} Pf_t = u_0$.

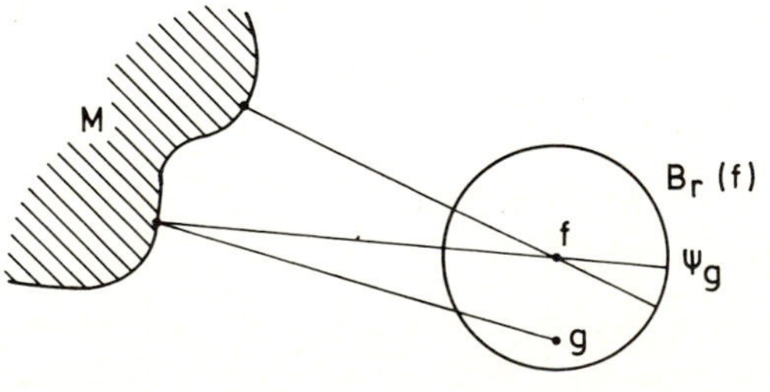

Fig. 9. Construction of the image for the mapping in the proof of Theorem 3.3

For convenience, we may assume that f has already been chosen such that $f_{t_1} = f$.

(2) Since M is locally compact, $M \cap B_\delta(u_0)$ is compact for some $\delta > 0$. By the continuity of P we may choose $0 < r < d(f, M)$, such that $P(B_r(f)) \subset B_\delta(u_0)$. Define Φ: $M \to B := B_r(f)$ by

$$\Phi(u) = f + r\frac{f - u}{\|f - u\|}. \tag{3.2}$$

Since Φ and P are continuous, the composition $\Psi = \Phi \circ P$: $B \to B$ is also continuous. The set $\Phi(M \cap B_\delta(u_0))$ is the image of a compact set and contains $\Psi(B)$. Therefore Ψ maps B onto a relatively compact subset of B.

(3) The mapping Ψ satisfies the hypothesis of the Schauder fixed point theorem. There is a $g \in B$ such that $\Psi(g) = g$. By construction, $f \in [g, Pg]$. Recalling that $u_0 = Pf$, we have

$$\begin{aligned} \|g - Pg\| &= \|g - f\| + \|f - Pg\| \\ &\geq \|g - f\| + \|f - u_0\| \geq \|g - u_0\|. \end{aligned}$$

Now, uniqueness of the nearest point implies that $u_0 = Pg$ and $g = f_{t_2}$, where $t_2 = 1 + r/\|f - u_0\| > 1$. This contradicts the fact that $Pf_t \neq Pf$ for $t > 1$. □

From the theorems 2.5 and 3.4, we get

3.5 Corollary. *Each boundedly compact Chebyshev set in a smooth Banach space is convex.*

Moreover, in a finite dimensional space each closed set is boundedly compact.

3.6 Corollary. (1) *Each Chebyshev set in a finite dimensional normed linear space is a sun.*

(2) *In a smooth and strictly convex finite dimensional space, the following conditions are equivalent:*

1°. *M is closed and convex.*

2°. *M is a Chebyshev set.*

3°. *M is a sun.*

For uniform approximation, the solarity of Chebyshev sets is more easily established. By Corollary 3.2 approximative compactness is sufficient to guarantee that each local best approximation is a global solution. From Theorem 2.6 we have the following result for $C(X)$.

3.7 Theorem. *Each approximatively compact Chebyshev set of real valued functions in $C(X)$ is a sun.*

In the proofs of the preceding theorems, the continuity of the metric projection has been always crucial. This property does not hold for the most famous Chebyshev set in nonlinear uniform approximation, i.e. the set of rational functions of given degree (see V. § 2). Nevertheless, the rational functions form a sun. But the following example provides a Chebyshev set which is not even a sun. Thus—at least in $C(X)$—one cannot completely abandon the assumption on the continuity.

3.8 Example (Dunham 1975). Let $\varphi: \mathbb{R}_+ \to \mathbb{R}$ be a strictly monotonic function such that $\varphi(0) = 1$ and $\lim_{x\to\infty} \varphi(x) = 0$, e.g., $\varphi(x) = (1 + x)^{-1}$. Put

$$v_a(x) = \begin{cases} (2 + a)\varphi(x/a), & a > 0 \\ 0, & a = 0. \end{cases}$$

and

$$M = \{v_a\colon a \geq 0\} \subset C[0, 1]. \tag{3.3}$$

Let (u_n) be a bounded sequence. If we write $u_n = v_{a_n}$, from $\|v_a\| = |v_a(0)| \geq a$ we conclude that (a_n) is bounded and has an accumulation point a^*. If $a^* > 0$, a subsequence of (u_n) converges to v_{a*}. If $a^* = 0$, then the subsequence converges pointwise to $v_0 = 0$ on the open interval $(0, 1)$. By Lemma 1.4, M is an existence set.

Moreover, M is a uniqueness set. Suppose that $v_a, v_b \in P_M f$, where $a < b$. The pointwise inclusion $v_a(x) < v_{(a+b)/2}(x) < v_b(x)$ for $x \in [0, 1]$ implies that $v_{(a+b)/2}$ would be a better approximation.

M is not a sun. The best approximation to the constant function $f_1(x) \equiv 1$ is $v_0 = 0$. But, for each $t > 1$ we have $\|f_t - v_{t-1}\| < \|f_t - 0\|$ where $f_t = tf_1$. Thus M is not a sun, and the metric projection is not continuous at f_1. □

We note that the Chebyshev set (3.3) is not weakly compact. Let $l(f) := f(0)$. The bounded sequence $(v_{1/n})$ has no accumulation point u such that $l(v_{1,n}) \to l(u)$.

Moreover we note that the sets of δ-nearest points in the Chebyshev set (3.3) are not always connected, see Exercise 3.17. This is in contrast to the result for uniformly convex spaces given in Proposition IV.1.6.

The set (3.3) is a one-dimensional example. Similarly n-dimensional examples may be constructed (Braess 1975b). Let C be the Haar cone of nonnegative functions from Example I.3.8(2). The set of transformed functions $\{v_h\colon h \in C\}$; where

$$v_h(x) = \begin{cases} [2 + h(x)]\varphi(x/h(x)), & \text{if } h(x) > 0, \\ 0, & \text{otherwise,} \end{cases}$$

exhibits the same properties as the set (3.3). To verify uniqueness, the theory of varisolvent functions from Chapter III combined with the technique from the cones can be used. □

Another counterexample refers to an incomplete inner-product space.

Example (Johnson 1986). Let E be the inner-product space of real number sequences having at most finitely many nonzeros terms, $\|x\|^2 = (x, x) = \sum_i |x_i|^2$. For each $n > 0$ let

$$E_n^- = \{x \in E\colon x_n \leq 0, \text{ and } x_i = 0 \text{ if } i > n\}.$$

A non-convex Chebyshev set $M = \bigcup_{n\geq 1} S_n$ with $S_n \subset E_{n+1}^-$ is defined by induction. Let $1 = A_0 > A_1 > A_2 > \cdots > 0$, $L_0 = F_0 \equiv 1$ and

$$a_n(x) = 1 + A_n L_n(x),$$
$$L_{n+1}(x) = a_n(x)F_n^2(x) + [a_n(x) - 1]F_n(x)x_{n+1} - x_{n+1}^2,$$
$$F_{n+1}^2(x) = 2L_{n+1}(x)/[a_n(x) + 1].$$

Note that a_n, L_n, and F_n are functions of $x_1, x_2, \ldots, x_n$. Set

$$S_n = \{x \in E_{n+1}^- : x_{n+1} = -F_n(x) \text{ and } L_j(x) \geq 0 \text{ for } j = 1, 2, \ldots, n\}.$$

Since $x_{n+1} = -F_n(x)$ implies that $F_{n+1}(x) = 0$, it follows that $S_n \subset S_{n+1}$. The sets S_n are very similar to the semi-ellipsoids considered in Exercise 3.19. In particular, to each $f \in E_{n+1}^-$ there is a unique best approximation from S_n. As was shown by Johnson (1986), the best approximation to $f \in E_{n-1}^-$ from any S_m where $m > n$, belongs to S_n, provided that the sequence (A_n) has been chosen appropriately. By this observation, M is an existence set. Therefore, M is a Chebyshev set which is not a sun. □

In the preceding discussion, the *metric* properties were always central. The following shows that *topological* properties also play a certain role.

3.9 Remark. Let M be a compact Chebyshev set in a normed linear space. Then each continuous mapping Φ from M into itself has a fixed point.

Proof. M is bounded and is contained in some ball $B = B_r(0)$. Since the metric projection is continuous, $\Phi \circ P: B \to B$ is a continuous mapping from a convex set onto a convex subset. Hence, it has a fixed point u_0. Obviously $u_0 \in M$, and u_0 is a fixed point of Φ. □

In finite dimensional spaces, the topology is independent of the norm and the fixed-point property is a topological feature. In particular, the sphere $S^{n-1} = \{x \in \mathbb{R}^n : \sum_k x_k^2 = 1\}$, or any closed curve (continuous image of S^1), cannot be a Chebyshev set, independently of the chosen norm.

C. An Alternative Proof

In certain Banach spaces, the convexity of Chebyshev sets can be proved under the weaker assumption that M is only approximatively compact. We have already seen this for the uniform approximation in Section 2. The central idea of Vlasov's concept for the treatment of the problem in uniformly convex spaces is the *almost convexity* of a set. This is a separation property; therefore it is not surprising that this property (as the existence of separating hyperplanes for convex sets) is established via Zorn's lemma.

Although in Vlasov's theory δ-suns are considered in a more general framework, we restrict ourselves to their discussion in Chebyshev sets.

3.10 Proposition. *Let M be an approximatively compact Chebyshev set in a Banach space. Then M is a δ-sun, i.e., for every $f \in E \backslash M$ there is a sequence (g_n) for which $g_n \neq f$, $g_n \to f$ and*

$$\frac{d(g_n, M) - d(f, M)}{\|g_n - f\|} \to 1. \tag{3.4}$$

Proof. Let $u = Pf$, $g = u + t(f - u)$ for $t > 1$, and $v = Pg$. Obviously, $\|g - v\| \leq \|g - u\| \leq \|g - f\| + \|f - u\|$. Hence,

$$I := 1 - \frac{d(g, M) - d(f, M)}{\|g - f\|} = 1 - \frac{\|g - v\| - \|f - u\|}{\|g - f\|} \geq 0.$$

In order to estimate I from above, let $l \in S'$. Then

$$I \leq 1 + \frac{\|f - u\| - l(g - v)}{\|g - f\|} = 1 - \frac{l(g - f)}{\|g - f\|} + \frac{\|f - u\| - l(f - v)}{\|g - f\|}.$$

Now we fix l such that $l(f - v) = \|f - v\|$. Since $u = Pf$ implies $\|f - v\| \geq \|f - u\|$, the last quotient is not positive. Note that $g - f$ is a multiple of $f - u$ and $l(g - f)/\|g - f\| = l(f - u)/\|f - u\|$. By recalling the choice of l once more, we obtain

$$\begin{aligned} I \leq 1 - \frac{l(f - u)}{\|f - u\|} &= \frac{\|f - u\| - l(f - v)}{\|f - u\|} + \frac{l(v - u)}{\|f - u\|} \\ &\leq \frac{\|v - u\|}{\|f - u\|} = \frac{\|Pf - Pg\|}{d(f, M)}. \end{aligned} \tag{3.5}$$

Since the metric projection is continuous, $\|Pf - Pg\|$ is arbitrarily small if g is sufficiently close to f. This yields (3.4) if $g_n = f + \frac{1}{n}(f - Pf)$. □

3.11 Proposition. *Each δ-sun M in a Banach space E is almost convex, i.e., given a ball $B_r(f)$ with a positive distance from M and $r' > r$, there is a ball $B_{r'}(f')$ which contains $B_r(f)$ and is also disjoint from M.*

Proof. (1) Let $\sigma > 1$ and $f \notin M$. We claim that the set

$$K(\sigma, f) := \{g \in E \colon \|g - f\| \leq \sigma[d(g, M) - d(f, M)]\}$$

is not bounded.

To prove this, given $R > 0$, we order the ball $B_R(f)$. Specifically, we put $g \leq g'$ if $\|g - g'\| \leq \sigma[d(g', M) - d(g, M)]$. The ordering is antisymmetric and transitive. Let $\{g_\alpha\}$ be a chain in $B_R(f)$, i.e., for $\alpha \leq \alpha'$ we have $g_\alpha \leq g_{\alpha'}$. The net of numbers $\{d(g_\alpha, M)\}$ converges, since it is bounded. Hence, the net $\{g_\alpha\}$ is a Cauchy net. Since E is a Banach space, $g_\alpha \to g \in B_R(f)$. Obviously $g_\alpha \leq g$ for every α, and the chain has the upper bound g. By Zorn's lemma there is a maximal element g_0 such that $f \leq g_0$. Suppose that g_0 is an interior point of $B_R(f)$. Then, by the preceding proposition there is a $g_1 = g_0 + t(g_0 - Pg_0) \in B_R(f)$ with $t > 0$ such that $[d(g_1, M) - d(g_0, M)]/\|g_1 - g_0\| > 1/\sigma$, which contradicts the maximality of g_0. Therefore g_0 lies on the boundary of $B_R(f)$. We have $\|g_0 - f\| = R$ and $g_0 \in K(\sigma, f)$.

(2) Let $B_r(f)$ be disjoint from M. Then $d(f, M) = r + \varepsilon$ where $\varepsilon > 0$. Given $r' > r + \varepsilon$, let $\sigma = (r' - r)/(r' - r - \varepsilon/2)$. Choose $f' \in K(\sigma, f)$, $\|f' - f\| = r' - r$. Then $d(f', M) \geq r' + \varepsilon/2$ and $B_{r'}(f')$ is a ball with the required properties. □

Now we are in a position to prove

3.12 Theorem (Vlasov 1967). *Each approximatively compact Chebyshev set in a uniformly convex space is a sun.*

Proof. It is sufficient to show that an almost convex set M in a uniformly convex Banach space is a sun.

Without loss of generality, we assume that $f = 0$ and $d(f, M) = 1$. Let $u_0 = P_M f$. By Proposition 3.11 there is a ball $B_2(g_n) \supset B_{1-1/n}(0)$ with $B_2(g_n) \cap M = \varnothing$. The inclusion implies that $\|g_n\| \leq 1 + 1/n$. The distance of $\tilde{g}_n = g_n/\|g_n\|$ to M is easily estimated: $d(\tilde{g}_n, M) \geq d(g_n, M) - \|g_n - \tilde{g}_n\| \geq 2 - \frac{1}{n}$. Hence,

$$\|\tilde{g}_n - u_0\| \geq 2 - \frac{1}{n}, \qquad \|\tilde{g}_n\| = \|u_0\| = 1.$$

Now, uniform convexity yields $\tilde{g}_n \to -u_0$. The continuity of the distance function yields $d(-u_0, M) = \lim d(g_n, M) = 2$, and u_0 is a nearest point to $-u_0 = u_0 + 2(f - u_0)$. Therefore M is a sun. □

The results for Hilbert spaces are summarized by:

3.13 Corollary. *For an approximatively compact set M in a Hilbert space the following conditions are equivalent:*
1°. *M is convex.*
2°. *M is a sun.*
3°. *M is a Chebyshev set.*

Exercises

3.14. Why is a strict sun always connected?

3.15. Let $M \in E$, $l \in E'$ and assume that

$$M_1 = \{u \in M : l(u) \leq 0\}.$$

If M is an existence set, a uniqueness set, approximatively compact, or a sun, resp, does M_1 inherit the property?

3.16. Let $M = \{(x, y) \in \mathbb{R}^2 : |x| = |y|\}$ and $\mathbb{R}^2$ be endowed with the Euclidean norm or the supremum norm. Prove that $0 \in M$ is not a nearest point to any $f \neq 0$. Why does this property imply that M is not a Chebyshev set?

3.17. Consider the Chebyshev set from Example 3.8 and construct an $f \in C[0, 1]$ such that for each $\delta > 0$ the set of δ-nearest points is not connected.

3.18. Assume that $M \subset C(X)$ has the betweenness property (see Exercise 2.10). Show that $P_M f$ is connected for each $f \in C(X)$.

3.19. Let $n \geq 2$ and $a_1 > a_2 > \cdots > a_n > 0$. Set $S_{n-1}^- := \{x \in \mathbb{R}^n : \sum_{i=1}^n x_i^2/a_i^2 = 1,\ x_n \leq 0\}$. Show that there is a unique best approximation to $f \in \mathbb{R}^n$ from S_{n-1}^- provided that the n-th coordinate of f is nonpositive.
Hint: If the n-th coordinate equals zero, there is a simple formula for the best approximation.

The following exercise is concerned with approximation from a linear family. Nevertheless, it is more easily treated by methods from the nonlinear theory.

3.20. Let M be an n-dimensional linear subspace of a normed linear space E. Show that each $(n+1)$-dimensional subspace E_{n+1} of E contains a nonzero element f for which 0 is a best approximation from M (Krein, Krasnosel'ski, and Milman 1948).
Hint: Assume first that M is a Chebyshev set. Consider the metric projection on $S^n = \{f \in E_{n+1} : \|f\| = 1\}$ and apply the Borsuk Antipodality Theorem.

3.21. Show that each $(n+1)$-dimensional subspace of $C[\alpha, \beta]$ contains a function $f \neq 0$ such that f has an alternant of length $\geq n+1$.
Hint: Use Exercise 3.20.

3.22. Give an existence set M such that no best approximation to f from M satisfies the Kolmogorov condition whenever $f \notin M$.

Chapter III. Methods of Local Analysis

In the local theory, we shall study the characterization and other properties of local best approximations. Local solutions will be determined to be global solutions only if this is possible by simple methods, i.e. without using topological methods. Specifically, we shall use results from the linear theory and from convex approximation, where the characterization of local solutions is performed via approximation on tangent sets. This involves replacing the nonlinear problem by a linearized one.

As in other variational problems, the concept of critical points plays a central role in this context. The implications will become apparent later, when methods of global analysis will be incorporated.

The approximating families which are of most interest are generally given in a parametric form. This is the case for the family of rational functions on a real interval X:

$$R_{m,n} := \left\{ u(x) = \sum_{k=0}^{m} a_k x^k \Big/ \sum_{k=0}^{n} b_k x^k : \right.$$

$$\left. a_k, b_k \in \mathbb{R}, \sum_{k=0}^{n} b_k x^k > 0 \text{ for } x \in X \right\}, \quad \text{where } m, n \geq 0.$$

We note that special functions are often represented in computers by means of rational functions. We also have the family of (proper) exponential sums which are met in the analysis of decay processes:

$$E_n^0 = \left\{ u(x) = \sum_{\nu=1}^{n} \alpha_\nu e^{t_\nu x} : \alpha_\nu, t_\nu \in \mathbb{R} \right\}, \quad \text{where } n \geq 1.$$

These families are parametrized in a natural way.

If only one parametrization is used in the study of the family, the analysis may often be impossible. One must abandon the representation given in the definition of the family, and choose one which is better suited to the element under consideration. Thus, we shall (as much as possible) state the definitions and results in a form independent of the chosen parametrization.

The families described above are not manifolds (but varifolds) and the presence of "singular" points seems to be a characteristic of actual problems of nonlinear approximation. Nevertheless, knowledge of approximation from manifolds may be of direct use in the treatment of those families. Finally, we may use the

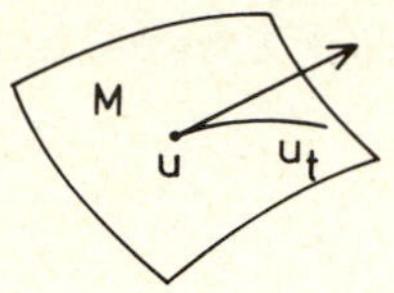

Fig. 10. Tangent ray and associated curve (u_t)

language of the elementary part of the theory of manifolds to formulate information in a less technical fashion.

§ 1. Critical Points

A. Tangent Cones and Critical Points

We will introduce tangent cones as cones which are "close" to a given family in a normed linear space. In many cases, the tangent cones are linear spaces and may be viewed as a linearization of the set.

1.1 Definition. Let M be a non-empty set in a normed linear space E. An element $h \in E$ is a *tangent ray* at $u \in M$, if there is a continuous mapping $[0,1] \to M$, $t \mapsto u_t$ such that

$$\|u_t - u - th\| = o(t) \quad \text{as } t \to 0. \tag{1.1}$$

The set of all tangent rays at u is the *tangent cone* $C_u M$. If $C_u M$ is a linear subspace of E, then it is said to be a *tangent space*, and we write $T_u M$ instead of $C_u M$.

We note that the tangent cones as the set M itself are considered as embedded into the linear space E. In some branches of mathematics, tangent spaces are introduced in a more abstract setting.

A slightly different concept of cones which are close to the set M has been investigated by Dubovitskii and Miljutin (1965), Brosowski (1968) and Laurent (1972): An element $h \in E$ is said to be an *admissible direction* at $u \in M$, if there is a mapping from a subset $T \subset (0,1]$ with $0 \in \bar{T}$, to M, $t \mapsto u_t$, such that (1.1) holds. The *cone of admissible directions* at $u \in M$ will be denoted by $K_u M$.

Obviously, $C_u M \subset K_u M$. There are cases in which $C_u M$ may be strictly contained in $K_u M$. But we are mainly interested in those nonpathological cases in which both definitions yield the same cone. Then the concept of tangent cones is more suitable for critical point theory.

1.2 Examples. (1) Let M be the curve $\{(t, t \cdot \sin(\pi/t)\colon t \in \mathbb{R}\}$ in 2-space. Then at the origin, we have

$$C_0 M = \{0\}, \qquad K_0 M = \{(x, y) \in \mathbb{R}^2, |y| \leq |x|\}.$$

Thus $C_0 M \neq K_0 M$.

(2) Let $M = \{(x, y) \in \mathbb{R}^2 \colon x \geq 0, |y| \leq x^2\}$. Then

$$C_0 M = \{(x, 0) \in \mathbb{R}^2 \colon x \geq 0\}.$$

This elucidates that a tangent cone of a two-dimensional space may collapse to a one dimensional set, though $C_0 M = K_0 M$.

(3) A similar situation in an infinite dimensional space is found with $M = \{u \colon u(x) = ax + \sin bx, a, b \in [0, 2]\} \subset C[0, 1]$. Then the tangent cone at the corner

$$C_0 M = \{h \colon h(x) = \alpha x, \alpha \geq 0\}$$

also collapses.

(4) Let $M = B_1 = \{(x, y) \in \mathbb{R}^2 \colon x^2 + y^2 \leq 1\}$. Then

$$C_{(-1,0)} M = \mathbb{R}_+ \times \mathbb{R}, \quad \text{and} \quad C_{(x,y)} M = T_{(x,y)} M = \mathbb{R}^2 \quad \text{if } (x, y) \in \mathring{B}_1.$$

The search for a best approximation in a set is equivalent to the determination of the minimum of the metric function $\|f - .\|$ on M. As in other nonlinear variational problems it is natural to introduce the concept of critical points. Since the metric function is not always differentiable, critical points cannot be defined via conditions on its derivative.

Example. Let $E = C[-1, +1]$ be endowed with the sup-norm and let $M = \{u_a(x) = a \colon a \in \mathbb{R}\}$ be the set of constant functions. The natural metric function for the approximation of $f(x) = x$ is

$$\rho(a) := \|f - u_a\| = \sup_{-1 \leq x \leq +1} |x - a| = 1 + |a|.$$

Unfortunately, this function is non-differentiable at the point of interest.

1.3 Definition (Braess 1973a). Let M be a non-empty set in a normed linear space E and $f \in E$. Then $u \in M$ is a *critical point* for the metric function $M \to \mathbb{R}$, $v \mapsto \|f - v\|$, if 0 is a best approximation to $(f - u)$ from the tangent cone $C_u M$. For brevity u is then said to be a critical point to f in M.

The motivation for the definition becomes obvious from the following lemma.

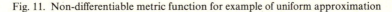

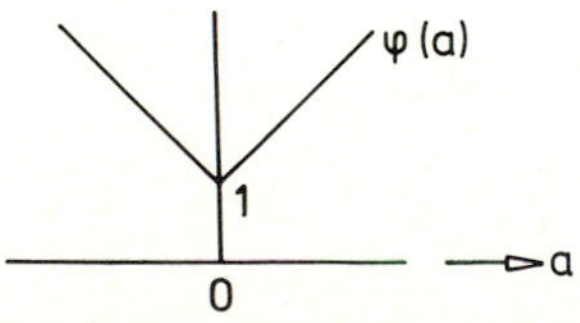

Fig. 11. Non-differentiable metric function for example of uniform approximation

1.4 Lemma of First Variation. *Each local best approximation to f from a subset M of a normed linear space E is a critical point to f in M.*

Proof. Assume that u is not a critical point. Then by definition

$$\|f - u - h\| < \|f - u\|$$

for some $h \in C_u M$. Let (u_t) be a curve in M corresponding to h. Then $c := \|f - u\| - \|f - u - h\| > 0$ and

$$\begin{aligned} \|f - u_t\| &\leq \|f - u - th\| + \|u_t - u - th\| \\ &= \|(1 - t)(f - u) + t(f - u - h)\| + o(t) \\ &\leq \|(f - u)\| - tc + o(t) < \|f - u\| \end{aligned}$$

for sufficiently small positive t. Consequently, u is not an LBA (local best approximation). □

The lemma of first variation as given above was stated in 1973 by Braess, but there are predecessors for manifolds without boundaries by Wulbert (1971a), for families with parametric representations by Meinardus and Schwedt (1965) and in some other investigations of other special sets (see also Brosowski (1968), Laurent (1972), Collatz and Krabs (1973)).

This lemma is useful since the tangent cones are often easier to handle than the original sets. Often the cones (or at least subsets of them), are linear spaces or convex sets, and results from linear and convex theory are applicable.

To get more insight into the lemma, we consider two examples in 2-space. First, let M be a smooth curve in Euclidean 2-space as shown in Fig. 12a. Then u is an LBA to f only if $f - u$ is orthogonal to the tangent line, since u must be the closest point on the tangent line. This is also the case in Fig. 12b, where the situation for the supremum-norm is illustrated.

From Fig. 12 we obtain our first knowledge on the validity of the converse of Lemma 1.4. In the Euclidean case the critical point u is an LBA to f_1 but not to f_2. On the other hand, in the supremum-norm case, u is an LBA whenever it is a critical point. We will see in §4 that this fact is not restricted to 2-space, but is true more generally whenever the tangent cones satisfy the Haar condition.

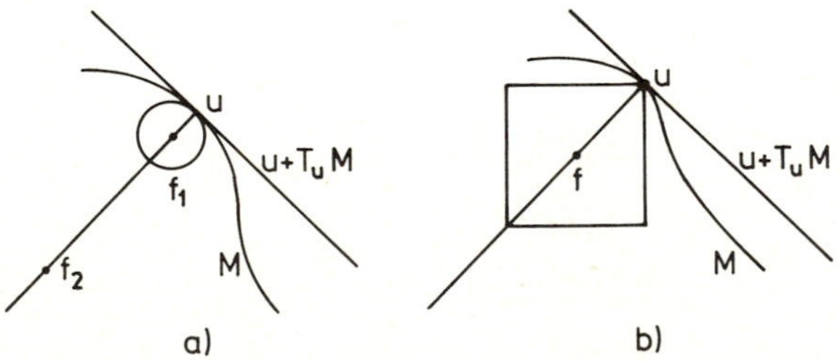

Fig. 12. Critical points in 2-Space. a) Euclidean norm b) Sup-norm

B. Parametrizations and C^1-Manifolds

Usually, the tangent cones are determined from the parametric representations of the families of interest. Obviously the natural parametrizations of the rational functions at the beginning of this chapter, specifically their numerators and denominators are not one-one. Therefore in nonlinear approximation theory any continuous map from a subset $A \subset \mathbb{R}^n$ into a family M in a normed linear space E is called a *parametrization*, and A is called the parameter set.

On the other hand, a *chart* is a set $A \subset \mathbb{R}^n$ together with a one-one mapping F onto some open subset $F(A) \subset M$ such that both F and F^{-1} are continuous.

As a preparation let A be an open set in n-space and $F\colon A \to M \subset E$. If the Fréchet derivative $d_a F$ exists at $a \in A$, then

$$d_a F(\mathbb{R}^n) \subset C_u M, \quad \text{where } u = F(a). \tag{1.2}$$

Indeed, let $h = d_a F \cdot b$, $b \in \mathbb{R}^n$. It follows from the definition of differentiability that

$$\|F(a + tb) - u - th\| = o(t) \quad \text{as } t \to 0.$$

Thus $t \mapsto F(a + tb)$ describes a curve for the tangent ray h in the sense of Definition 1.1.

However, $C_u M$ may strictly contain $d_a F(\mathbb{R}^n)$, even if F is surjective.

1.5 Examples. (1) Let $M = \mathbb{R}_+ \subset E = \mathbb{R}$,

$$F\colon \mathbb{R} \to \mathbb{R},$$
$$F(a) = a^2.$$

Obviously, $d_0 F(\mathbb{R}) = \{0\}$, but $C_0 M = \mathbb{R}_+$.

(2) Let $M \subset E_2 = \mathbb{R}^2$ be the image of the mapping

$$F\colon \mathbb{R} \to \mathbb{R}^2,$$
$$a \mapsto \left(\frac{a}{1 + a^4}, \frac{a^3}{1 + a^4}\right).$$

F is one-one and continuous, but F^{-1} is not continuous at $u_0 = F(0) = 0$. Therefore, $C_0 M$ cannot be determined from $d_0 F$.

Recall that we are considering approximation on the set $F(A)$. The function

Fig. 13. Example 1.5(2)

$a \mapsto \|f - F(a)\|$ shall only be considered an auxiliary tool. In this spirit the parameter set is often translated in n-space, such that the zero vector is the parameter for the element under consideration.

The examples above are characteristic of all pathological cases. They are not found in manifolds.

1.6 Definition. A topological space M with a countable basis is an *n-dimensional manifold* (*with boundary*) if, given $u \in M$, there is an open neighborhood U of u and a chart

$$F: V \cap C \to U \subset M,$$

such that the parameter set is the intersection of an open set V and a closed convex body C in n-space. A set of charts $(F_\alpha, V_\alpha \cap C_\alpha)$ is an *atlas* if $M = \bigcup_\alpha F_\alpha(V_\alpha \cap C_\alpha)$.

If, in particular the choice $C = \mathbb{R}^n$ is always possible, then M is a manifold without boundary or short a manifold.* This means that boundaries are admitted but not necessarily present in manifolds with boundaries. Moreover, in most cases we will restrict ourselves to simple forms of boundaries (see below). Manifolds are locally compact and locally path connected. Hence a connected manifold is path connected.

We will not deal with abstract manifolds, but with those which are embedded into a normed linear space E. In particular, we will often encounter C^1-manifolds without boundary. Then, given $u_0 \in M$, there is a chart $F: V \to U$ from an open set $V \subset \mathbb{R}^n$ onto an open neighborhood u of U_0. Moreover, F is differentiable and the continuous derivative

$$d_aF: \mathbb{R}^n \to E$$

is one-one for $a \in V$.

More generally, the differentiable structure is strongly related to an approximation property of the tangent cones.

1.7 Definition. A manifold $M \subset E$ is a C^1-*manifold* (with boundary) in E, if given $u_0 \in M$ there is a chart $F: V \cap C \to U$ in the atlas (in the sense of Definition 1.6) such that the following holds:

(i) (Differentiability property) F is Fréchet-differentiable in $C \cap V$ and the map $a \mapsto d_aF$ is continuous.

(ii) (Approximation property) Without loss of generality let $F^{-1}(u_0) = 0$. There is a continuous mapping

$$\kappa: U \to d_0F\left(\overline{\bigcup_{t>0} tC}\right)$$

---

*Another definition: M is a manifold without boundary of dimension n, if each point in M has a neighborhood which is homeomorphic to an open set in n-space.

such that

$$\|u - u_0 - \kappa(u)\| = o(\|u - u_0\|) \quad \text{as } u \to u_0. \tag{1.3}$$

If moreover, all parametrizations of the charts are C^k-functions for $k \geq 1$, then M is a C^k-manifold.

Remark. In particular, one may assume that $\|u - u_0 - \kappa(u)\| < \frac{1}{3}\|u - u_0\|$. By using the triangle inequality we conclude that

$$\tfrac{2}{3}\|u - u_0\| \leq \|\kappa(u)\| \leq \tfrac{3}{2}\|u - u_0\|. \tag{1.4}$$

Hence, (1.3) is equivalent to

$$\begin{aligned} \kappa(u_0) &= 0, \\ \|u - u_0 - \kappa(u)\| &= o(\|\kappa(u)\|). \end{aligned} \tag{1.5}$$

We consider two cases in which the approximation property may be determined from a simple condition on the derivative of the parametrization.

Let M be a C^1-manifold without boundary as previously defined. Let $u_0 = F(a_0)$ and assume that $d_{a_0}F$ is one-one. Then $\|d_{a_0}Fb\| \geq c\|b\|$ holds for all $b \in \mathbb{R}^n$ and some $c > 0$. The approximation property follows from

$$\|F(a) - u_0 - d_{a_0}F(a - a_0)\| = o(\|a - a_0\|) = o(c^{-1}\|d_{a_0}F(a - a_0)\|). \tag{1.6}$$

Hence, by setting $\kappa(u) = d_{a_0}F(F^{-1}u - a_0)$, we gain the estimate (1.5).

Next, let M be a ∂-manifold, i.e., boundaries and corners admit the parameter sets to be open in convex cones of the form

$$C = \mathbb{R}^m \times R_+^{n-m} \quad \text{where } 0 \leq m \leq n.$$

More generally, if the domain is a closed convex cone with vertex at $a_0 = F^{-1}(u_0)$, then Condition 1.7(ii) may be replaced by

1.7(ii)′. Without loss of generality, let $F^{-1}(u_0) = 0$. Then

$$d_0Fb \neq 0 \quad \text{for all } b \in C,\, b \neq 0. \tag{1.7}$$

Indeed, compactness implies $c = \min\{\|d_0Fb\| : b \in C \subset \mathbb{R}^n, \|b\| = 1\} > 0$. Letting $u_0 = 0$ we have again (1.6) and κ as above satisfies (1.5). □

The following example shows the standard behaviour for a boundary with $m = n - 1$, while Example 1.2(2) elucidated a situation where the tangent cone collapsed to a lower dimensional set.

1.8 Example. Let $M \subset C[0, 1]$ be the family of those rational functions with at most one pole (on the negative real line) which are defined by the parametrization

$$F: \mathbb{R} \times \mathbb{R}_+ \to C[0, 1],$$

$$F(a, t)(x) = \frac{a}{1 + tx}.$$

If $a \neq 0, t = 0$ then $\frac{\partial}{\partial a}F = 1, \frac{\partial}{\partial t}F = -ax$, and the tangent cone contains the set of polynomials

$$\{h(x) = \delta_1 + \delta_2 x; \delta_1, \delta_2 \in \mathbb{R}, \delta_2 \cdot \operatorname{sgn} a \leq 0\}. \tag{1.8}$$

We may conclude from the discussion below that indeed (1.8) is exactly the tangent cone.

From arguments similar to those used in deriving (1.2), it is obvious that $d_0 F$ maps the cone $\overline{\bigcup_{t>0}(tC)}$ into the tangent cone. The following lemma shows the equality

$$C_{F(0)}M = d_0 F\left[\overline{\bigcup_{t>0}(tC)}\right] \tag{1.9}$$

when the approximation property 1.7(ii) holds. In particular, in a manifold without boundary, ($d_a F$ nonsingular), one has

$$T_{F(a)}M = d_a F(\mathbb{R}^n). \tag{1.10}$$

1.9 Remark (Braess 1973a). Let $M \subset E$ and let $N \subset C_u M$ be a closed cone with vertex at 0. If there is a neighborhood U of u in M and a continuous mapping $\kappa: U \to N$, such that

$$\|v - u - \kappa(v)\| = o(\|v - u\|),$$

then $N = C_u M = K_u M$.

Proof. It is sufficient to verify that $K_u M \subset N$. Assume that $h \in K_u M$. By the definition of $K_u M$, given $m > 0$, there is a point $v_m \in U \subset M$ and $t_m \in \left(0, \frac{1}{m}\right)$ such that $\|(v_m - u)/t_m - h\| < 1/m$. Consequently, $\|v_m - u\| < t_m(1 + \|h\|)$ and

$$\begin{aligned} \|\kappa(v_m) - t_m h\| &\leq \|\kappa(v_m) - v_m + u\| + \|v_m - u - t_m h\| \\ &\leq o(\|v_m - u\|) + \frac{1}{m}t_m = o(t_m). \end{aligned}$$

Hence, $\lim_{m\to\infty} \kappa(v_m)/t_m = h$. Since N is a closed set, we have $h \in N$ and the proof is complete. □

It is clear that results should not depend on the choice of parametrizations. In this sense, the lemma above assures independence for the "computed tangent cone" given by the right hand side of (1.9).

C. Local Strong Uniqueness

As we know, critical points are not always *LBA*'s. In 1971, Wulbert observed that in connection with strong uniqueness critical points can be identified as local solutions of the approximation problem. Though uniqueness is met only in nonsmooth normed linear spaces, specifically in the case of uniform approximation or l_1-approximation, this phenomenon is crucial in nonlinear Chebyshev approximation, cf. Definition I.3.6.

1.10 Definition. An element $u \in M$ is a *strongly unique best approximation* to f from M, if there is a number $c > 0$ such that

$$\|f - v\| \geq \|f - u\| + c\|v - u\| \tag{1.11}$$

for all $v \in M$. An element $u \in M$ is a *strongly unique local best approximation* to f from M, if u is the strongly unique best approximation from some open neighborhood of u in M.

Clearly, the triangle inequality implies that $c \leq 1$.

Firstly, we observe that for the vertex of a cone (or a point in a linear space), strong uniqueness and local strong uniqueness are equivalent. Indeed, let 0 be the strongly unique LBA to f from the cone M, whose vertex is the origin. Then by definition given $h \in M$, there exists a $t \in (0, 1)$ such that

$$\|f - th\| \geq \|f - 0\| + c\|th\| = \|f\| + ct\|h\|.$$

Moreover we have

$$\|f - th\| \leq t\|f - h\| + (1 - t)\|f\|.$$

From these inequalities we obtain $\|f - h\| \geq \|f\| + c\|h\|$ by simple manipulations. □

1.11 Theorem (Wulbert 1971a, Braess 1973a). *In a C^1-manifold M with boundary in a normed linear space E, the following are equivalent:*

1°. *u is a strongly unique local best approximation to f from M.*

2°. *0 is the strongly unique best approximation to $(f - u)$ from the tangent cone $C_u M$.*

Further, in any non-empty set M, property 1° *implies* 2°.

Proof. (1) Assume that 1° holds, i.e., (1.11) holds for all v in some neighborhood $V \subset M$. Given $h \in C_u M$ for the elements on the associated curve (u_t) and sufficiently small t, we have

$$\|u_t - u - th\| \leq \frac{c}{3}\|th\|,$$

and $u_t \in V$. Since $c \leq 1$, we have $\|u_t - u\| \geq \frac{2}{3}t\|h\|$. Now this and (1.11) imply that

$$\begin{aligned}\|f - u - th\| &\geq \|f - u_t\| - \|u_t - u - th\| \\ &\geq \|f - u\| + c\|u_t - u\| - \frac{c}{3}t\|h\| \\ &\geq \|f - u\| + \frac{c}{3}\|th\|.\end{aligned}$$

From the remark preceding the theorem, we now conclude that strong uniqueness holds in the tangent cone with the constant $\frac{1}{3}c$.

(2) Assume that M is a C^1-manifold and that 2° holds, i.e.,

$$\|f - u - h\| \geq \|f - u\| + c\|h\|, \qquad h \in C_u M.$$

For all v in some neighborhood U of u we have

$$\|v - u - \kappa(v)\| \leq \frac{c}{3}\|\kappa(v)\|.$$

Recalling (1.4), by combining the last two inequalities, we obtain for any $v \in U$

$$\begin{aligned}\|f - v\| &\geq \|f - u - \kappa(v)\| - \|v - u - \kappa(v)\| \\ &\geq \|f - u\| + c\|\kappa(v)\| - \frac{1}{3}c\|\kappa(v)\| \geq \|f - u\| + \frac{c}{3}\|v - u\|.\end{aligned}$$

Hence, u is a strongly unique *LBA*. □

Due to Theorem 1.11, strong uniqueness of local solutions in a nonlinear family can only be expected in those spaces in which strong uniqueness is found in the linear sets. Consequently, this phenomenon is restricted to the uniform and to l_1-approximation. On the other hand, in those cases the converse of the lemma of first variation also holds:

1.12 Corollary. *Let M be a C^1-manifold with boundary. Assume that each best approximation in any tangent cone is a strongly unique best approximation. Then the following are equivalent:*
1°. *u is a critical point to f in M.*
2°. *u is a local best approximation to f from M.*
3°. *u is a strongly unique local best approximation to f from M.*

At the beginning of this chapter, we observed the nondifferentiability of the metric function (cf. Fig. 2). We recall that due to this fact, a definition of critical points was necessary which differs slightly from the definition that one would expect. Now we see that the non-differentiability is related to strong uniqueness, which in turn makes the analysis simpler (compare Corollary 1.12 and Fig. 12).

We conclude this section with a useful tool which requires only strict local uniqueness.

1.13 Local Continuity Lemma. *Assume that u is a strict local best approximation to f from a locally compact family M. Then given $\varepsilon > 0$ there is a $\delta > 0$ such that to each $g \in B_\delta(f)$ there is a local best approximation v to g from M with $\|v - u\| < \varepsilon$.*

Proof. By assumption, u is the unique nearest point to f from some compact neighborhood U of u in M. Let $\varepsilon > 0$. After reducing ε if necessary, we may assume that $M \cap B_\varepsilon(u) \subset \mathring{U}$. From the uniqueness and the compactness, we conclude that

$$\delta = \tfrac{1}{3}[\inf\{\|f - v\|: v \in U \backslash \mathring{B}_\varepsilon(u)\} - \|f - u\|] > 0. \tag{1.12}$$

Let $g \in B_\delta(f)$. Then $d(g, M \cap B_\varepsilon(u)) \leq \|g - u\| \leq \|f - u\| + \delta$. There is a best approximation v to g from the compact set $M \cap B_\varepsilon(u)$, which by (1.12) cannot be located on the boundary of this set. Hence v is a local best approximation from M. □

Exercises

1.14. Let C be a cone with vertex u_0 which needs not to be convex. Show that the Kolmogorov criterion applies to the vertex, i.e., $u_0 \in P_C f$, if and only if for each $u \in C$ there is a functional $l \in S'$ such that $l(f - u_0) = \|f - u_0\|$ and $\operatorname{Re} l(u - u_0) \geq 0$.

1.15. Meinardus and Schwedt (1964) introduced the concept of asymptotic convexity which may be rephrased as follows: A set $M \subset C(X)$ is said to be *asymptotically convex* if, given $u, u_1 \in M$ there is an $h \in C_u M$ and a positive function $g \in C(X)$ such that

$$u_1(x) - u(x) = g(x) \cdot h(x) \quad \text{for } x \in X.$$

Show that each asymptotically convex set is a sun, has the closed sign property (II.2.6), and each critical point in an asymptotically convex set is a nearest point.
Hint: Use Lemma II.2.4 (cf. Krabs 1967).

1.16. Show that the set of rational functions $R_{m,n}$ where $m, n \geq 0$, is asymptotically convex.
Hint: If p/q and $p_1/q_1 \in R_{m,n}$, then $p_1 q^{-1} - p q_1 q^{-2} \in C_{p/q} R_{m,n}$.

1.17. Consider the Chebyshev set $M = \{x \in \mathbb{R}^3: x_1 \geq 0, x_2 \geq 0, x_3 = 0\}$ in Euclidean 3-space, and define the function $M \to \mathbb{Z}$ by

$$m(x) = \begin{cases} 0 & \text{if } x = 0, \\ 1 & \text{if } x_1 + x_2 > 0 \text{ and } x_1 \cdot x_2 = 0, \\ 2 & \text{if } x_1 > 0,\ x_2 > 0. \end{cases}$$

Obviously $T_x M$ contains a linear space of dimension $m(x)$. Show that the function: $m \circ P_M: \mathbb{R}^3 \to \mathbb{Z}$ is continuous on a residual set.

1.18. A point u is said to be a farthest point to u from M, if $\|f - u_1\| \leq \|f - u\|$ for all $u_1 \in M$. Prove that each locally farthest point in a C^1-manifold without boundary is also a critical point for the problem of best approximation. Why is the same not true for manifolds with boundaries?

1.19. Show that the tangent cones for a set M in n-space are independent of the underlying norm. On the other hand investigate the tangent spaces for the set of spline functions

$$M = \{u: u(x) = \max\{x - a, 0\}, 0 \leq a \leq 1\}$$

in $C[0,1]$ and in $L_\infty[0,1]$, resp.

1.20. Formulate a necessary condition for a best L_1-approximation from the set $R_{m,n}$ of rational functions.

1.21. Consider the set M from Example 1.5(2) with $\mathbb{R}^2$ equipped with the L_∞-norm. There are bounds for the maximal number of best approximations from M and of critical points, respectively. Do the two bounds coincide? How is the situation with the Euclidean metric?

1.22. Let $m \geq 1$, $\omega \geq 2$, and $N = m\omega$. Given $f \in C^\omega[\alpha, \beta]$ let M be the family of those polynomials u from Π_{N-1} for which $f - u$ has m distinct zeros in (α, β) the multiplicity being precisely ω. Prove that the tangent space $T_u M$ consists of the polynomials of the form

$$h(x) = p(x)\prod_{i=1}^{m}(x - x_i)^{\omega-1} \quad \text{where } p \in \Pi_{m-1},$$

if $\{x_i\}_{i=1}^m$ are the zeros of $f - u$.

1.23. Assume that $C_u M$ contains an n-dimensional Haar subspace of $C[\alpha, \beta]$. Show that there is an alternant of length $n + 1$ to $f - u$, if u is a best uniform approximation from M.

§2. Nonlinear Approximation in Hilbert Spaces

In approximation theory, two types of Hilbert spaces are of interest. Firstly, there are the L_2-spaces $E = L_2(X)$ of (real-valued) square integrable functions on a domain $X \subset \mathbb{R}^m$, with the norm induced by the inner product

$$(f, g) = \int_X w(x) f(x) g(x)\, dx.$$

Here, w is a positive weight function. Secondly, the Hardy spaces are encountered in the theory of optimal quadrature formulae. Specifically, the space of functions which are analytic in the unit disc of the complex plane, and square-integrable on its boundary, may be endowed with the inner product

$$(f, g) = \frac{1}{2\pi}\int_{|z|=1} \bar{f}(z) g(z)\, dz = \frac{1}{2\pi}\int_0^{2\pi} \bar{f}(e^{i\varphi}) g(e^{i\varphi})\, d\varphi.$$

General results, which go beyond the basic theory of critical points given in the last section, are only obtained for manifolds without boundaries.

These results do, however, enable us to construct functions which have an arbitrarily large number of local best approximations in the family of rational functions.

A. Nonlinear Approximation in Smooth Banach Spaces

If the underlying space E is a Hilbert space, or more generally, a smooth normed linear space, then critical points may be characterized as solutions of convex approximation problems.

Let 0 be a best approximation to f from some (not necessarily convex) cone C whose vertex is the origin. Given $u \in C$, zero is also the best approximation to f from the segment $[0, u]$. From the approximation in convex sets, it is known that for some functional $l \in E'$:

$$l(f) = \|f\|, \qquad l(u) \leq 0, \qquad \|l\| = 1. \tag{2.1}$$

Since E is smooth, there is only one functional $l \in E'$ for which the first and the third conditions hold, l is independent of u and $l(u) \leq 0$ holds for all $u \in C$. Consequently, (2.1) is satisfied even for all u in the convex hull co(C), and 0 is a best approximation to f from $\mathrm{co}(C)$. The consequences are formulated in the following remarks.

2.1 Remarks. Let E be a smooth Banach space.

(1) u is a critical point to f in a set $M \subset E$, if and only if 0 is a best approximation to $f - u$ from $\mathrm{co}(C_u M)$.

(2) If $\mathrm{co}(C_u M)$ is dense in E, then u cannot be a (local) best approximation to any $f \in E \backslash M$.

From the second remark, we conclude that in rational L_p-approximation either the numerator or the denominator has the maximal degree.

2.2 Example. Let $1 < p < \infty$. Any local best approximation to $f \in L_p[0,1] \backslash R_{m,n}$ from $R_{m,n}$, $n \geq 1$ is not contained in $R_{m-1,n-1}$.

Here, we define $R_{-1,n} := \{0\}$.

First we show that $\mathrm{co}(R_{0,1})$ is dense in $L_p[0,1]$. By the Weierstraß approximation theorem, given $f \in L_p$ there is a polynomial p which is close to f. Let $n = \partial p + 1$. Then we may choose n small numbers $t_1, t_2, \ldots, t_n$ such that p/q with $q(x) = \prod_\nu (1 + t_\nu x)$ is also close to f. From the decomposition into partial fractions $p/q = \Sigma_\nu a_\nu (1 + t_\nu x)^{-1}$, we see that $p/q \in \mathrm{co}(R_{0,1})$.

To continue the proof, let $u_0 = p_0/q_0 \in R_{m-1,n-1}$. Then $[p_0 + a(1 + tx)^{-1}]/q_0 \in R_{m,n}$ for any $a \in \mathbb{R}$. Hence, $\pm[q_0(1 + tx)]^{-1} \in C_{u_0} R_{m,n}$ and $\mathrm{co}(q_0^{-1} R_{0,1}) \subset \mathrm{co}(C_{u_0} R_{m,n})$. From the discussion above, we conclude that the latter is dense in $L_p[0,1]$. □

B. A Classification of Critical Points

Let M be a non-empty set in a real Hilbert space E with inner product $(.,.)$. Moreover, let $F: A \to M$ be a mapping whose domain A is open in n-space. Rather than the metric function $\|f - F(a)\|$, one usually considers its square:

$$\rho(a) := \|f - F(a)\|^2 = (f - F(a), f - F(a)). \tag{2.2}$$

If F is Fréchet-differentiable at $a \in A$, i.e.,

$$F(a + b) = F(a) + d_a F b + o(\|b\|) \quad \text{for } b \to 0,$$

then we have $\rho(a + b) = \rho(a) - 2(f - F(a), d_a F b) + o(\|b\|)$. Hence, ρ is differentiable and

$$d_a \rho b = -2(f - F(a), d_a F b). \tag{2.3}$$

2.3 Lemma. *Let M be a C^1-manifold in a Hilbert space E, and let F be a parametrization for a neighborhood of $u = F(a) \in M$. Then u is a critical point to f in M, if and only if $d_a \rho = 0$.*

Proof. From $T_u M = d_a F(\mathbb{R}^n)$ and (2.3) it follows that $d_a \rho = 0$ if and only if $(f - u) \in (T_u M)^\perp$, i.e., if $f - u$ is orthogonal to $T_u M$. This in turn characterizes 0 as the best approximation to $f - u$ from the linear subspace T_uM in a Hilbert space. □

The lemma shows once again that the approximation-theoretic definition 1.3 is consistent with the definition of critical points for the distance function from the viewpoint of differential topology.

For the specification of critical points assume that $\rho\colon A \to \mathbb{R}$ is a C^2-function. It is well known that ρ has a strict local minimum at a critical point, if $d_a^2\rho$ is a positive definite bilinear form.

Let F be twice differentiable at the critical parameter $a_0 = 0$ and $u = F(a_0)$. Then $F(a) = F(0) + d_0Fa + \frac{1}{2}d_0^2F(a, a) + o(\|a\|^2)$ implies that

$$\rho(a) = \rho(0) - (f - u, d_0^2F(a, a)) + (d_0Fa, d_0Fa) + o(\|a\|^2).$$

Hence,

$$d_0^2\rho(a, b) = -(f - u, d_0^2F(a, b)) + (d_0Fa, d_0Fb). \tag{2.4}$$

The second term belongs to a positive semi-definite bilinear form. If d_0F is injective, it is even positive definite.

2.4 Definition. (1) Let B be a Hermitean matrix. The *nullity* of B is the dimension of its kernel. The *index* of B is the dimension of the largest linear space on which B is negative definite.

(2) The nullity and the index of a critical point in a C^2-manifold are the nullity and the index of the second derivative $d^2\rho$. A critical point is *degenerate* if its nullity is positive.

The nullity and the index of a critical point are independent of the parametrization chosen (Milnor 1963).

Now the characterization of a local minimum (which is well known from classical analysis), may be reformulated for the metric function (cf. Jongen, Jonker, and Twilt 1983):

A critical point is a strict local best approximation if its index and nullity are zero. A critical point is a local best approximation only if its index is zero.

The role of the nullity is elucidated by returning to arguments from the previous chapter in the investigation of suns. A point $u \in M$ which is critical for the approximation of f, is also critical to all points $f^{(\lambda)}$ on the ray through f with center u:

$$f^{(\lambda)} = u + \lambda(f - u), \qquad \lambda > 0.$$

The derivatives are obtained from (2.4) when f is replaced by $f^{(\lambda)}$. Hence,

$$d_0^2\rho^{(\lambda)}(a, b) = -\lambda(f - u, d_0^2F(a, b)) + (d_0F(a), d_0F(b)). \tag{2.5}$$

This form is positive definite for sufficiently small λ. Because of the splitting we have the following: If $d_0^2\rho^{(\lambda_0)}$ is negative definite on some subspace $H \subset \mathbb{R}^n$, then $d_0^2\rho^{(\lambda)}$ is also negative definite on H for all $\lambda > \lambda_0$. Consequently, the index is a

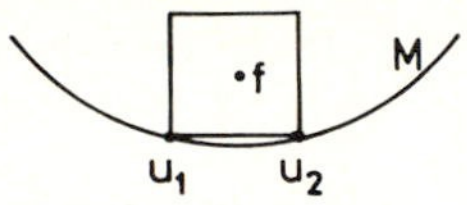

Fig. 14. Example of a set M such that there are elements close to M which have 2 solutions

nondecreasing function of λ. Specifically we have (see Milnor 1963, Jongen, Jonker and Twilt 1983).

2.5 Morse Index Theorem. *Let u be a non-degenerate critical point to f in a C^2-manifold (without boundary). Then the index is equal to the number of points $f^{(\lambda)}$ lying on the segment from u to f, for which u is a degenerate critical point, each such point being counted with its nullity.*

By this theorem, the index of a critical point in M is zero, provided that the distance to f is sufficiently small. Each f in some neighborhood of M has exactly one best approximation from M (cf. Fig. 12a). This is in contrast to the situation elucidated in Fig. 14, where $M \subset \mathbb{R}^2$ is the graph of a symmetric convex C^∞-function, and $\mathbb{R}^2$ is endowed with the uniform norm.

C. Continuity

From a general continuity theorem for locally compact sets, (Lemma 1.13), one knows that a strict local best approximation on a manifold does not vanish under small perturbations. In Hilbert spaces, this result can be improved and extended to all non-degenerate critical points.

2.6 Continuity Theorem for Non-Degenerate Critical Points. *Let u_0 be a non-degenerate critical point to f_0 from a C^2-manifold M in a Hilbert space E. Then there is a neighborhood U of u_0 in M and a $\delta > 0$ such that, for each $f \in E$ satisfying $\|f - f_0\| < \delta$, there is exactly one critical point in U.*

To prove this, we recall the

Lipschitz-Inverse-Function-Theorem. *Let L: $\mathbb{R}^n \to \mathbb{R}^n$ be a one-one linear mapping and ψ: $U \to \mathbb{R}^n$ for $U \subset \mathbb{R}^n$ be a Lipschitz-continuous mapping with Lipschitz constant $c > 0$ which preserves the origin, so that*

$$\psi(0) = 0,$$

$$\|\psi(x) - \psi(y)\| \le c\|x - y\| \quad \text{for } x, y \in U.$$

If $c < \|L^{-1}\|^{-1}$, then $(L + \psi)$: $U \to \mathbb{R}^n$ is one-one, and for all sufficiently small y we have exactly one $x = (L + \psi)^{-1} y \in U$ with

$$\|x\| \le \frac{\|y\|}{\|L^{-1}\|^{-1} - c}.$$

Proof of Theorem 2.6. Let $F: A \to M$ be a C^2-parametrization, such that $u_0 = F(0)$ and d_0F is injective. Consider the nonlinear equation $2(f - F(a), d_aF) = 0$ or

$$d_a\rho - 2(f - f_0, d_aF - d_0F) = 2(f - f_0, d_0F). \tag{2.6}$$

Here $\rho = \|f_0 - F(a)\|^2$. Since $d^2\rho$ is continuous and u_0 is a nondegenerate critical point, $d_a\rho$ only differs from the linear injective map $d_0^2\rho(a, .)$ by a Lipschitz-continuous function where the Lipschitz constant is arbitrarily small. Since the Lipschitz constant of the second term in (2.6) and a bound for the right hand side can be made arbitrarily small by choosing $\|f - f_0\|$ small, the statement of the theorem is an immediate consequence of the Lipschitz-Inverse-Function-Theorem. □

Moreover, we note that the solutions of (2.6) are unique:

2.7 Corollary. *Non-degenerate critical points in a C^2-manifold are isolated.*

The following example elucidates that degenerate critical points must be indeed excluded in Theorem 2.6 and its corollary.

2.8 Example. Let $M = \left\{(\cos a, \sin a) \in \mathbb{R}^2;\ -\frac{\pi}{8} < a < +\frac{\pi}{8}\right\}$ be an arc of the circle in Euclidean 2-space. The point $u_0 = (1, 0)$ is a degenerate critical point in M for the approximation of $f_0 = 0$. Obviously u_0 is not an isolated critical point. Since all normal lines pass through the origin, there is no critical point in M to $f = (0, \delta)$ for $\delta > 0$, even if δ is arbitrarily small.

By a perturbation, u_0 can be made an isolated critical point with destroying the non-existence for $f = (0, \delta)$. To this end, consider the perturbed arc

$$M' = \left\{(\cos a - \sin^3 a, \sin a) \in \mathbb{R}^2;\ -\frac{\pi}{8} < a < \frac{\pi}{8}\right\}$$

which is also shown in Fig. 15.

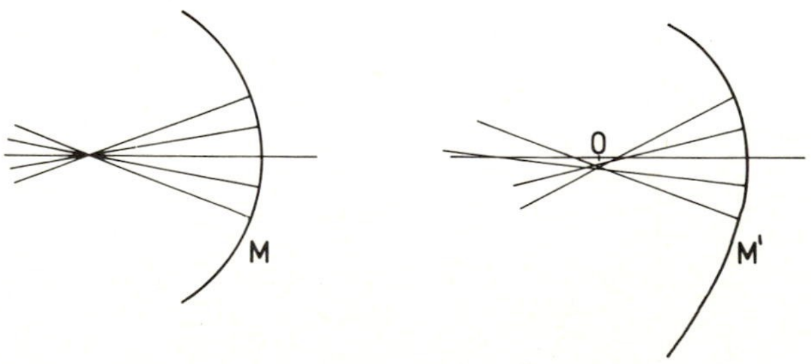

Fig. 15. Degenerate critical points in M and M′

D. Functions with Many Local Best Approximations

By applying the ideas from the preceding sections it is possible to construct functions with many local solutions in the family of rational functions. To this end the considerations from the local analysis are completed by simple functional analytic arguments. The negative result is in contrast to rational approximation in the sense of Chebyshev, where uniqueness does occur.

For convenience, the central result will be formulated and proved only for the Hilbert space $L_2(X)$, X being a compact real interval, although with slightly more care it may be established for any reasonable Hilbert space.

2.9 Theorem (Wolfe 1974). *Let* $0 \leq m < n$. *Assume that* $u_i = p_i/q_i \in R_{m,n} \setminus R_{m-1,n-1}$, $\partial q_i = n$, $i = 1, 2, \ldots, N$, *are such that* q_i *and* q_j *have no common factors for* $i \neq j$. *Then, there is an* $f \in L_2(X)$ *to which* $u_1, u_2, \ldots, u_N$ *are strict local best approximations from* $R_{m,n}$.

In order to prove the theorem, we first evaluate the tangent space at $u = p/q \in R_{m,n} \setminus R_{m-1,n-1}$. To this end consider the parametrization defined by

$$(a_0, a_1, \ldots, a_m, b_0, \ldots, b_n) \mapsto \frac{\sum_{k=0}^m a_k x^k}{\sum_{k=0}^n b_k x^k} =: \frac{p}{q}.$$

This mapping is a homeomorphism, if we identify each vector (a, b) with its non-zero multiples and restrict the domain, such that $\partial q = n$ and p and q have no common divisors for any p/q in the image. Obviously,

$$\frac{\partial F}{\partial a_k} = \frac{x^k}{q}, \qquad k = 0, 1, \ldots, m,$$

$$\frac{\partial F}{\partial b_k} = \frac{x^k p}{q^2}, \qquad k = 0, 1, \ldots, n-1.$$

From (1.10) it follows that the partial derivatives span the tangent space if they are independent. We claim that

$$T_{p/q} R_{m,n} = \left\{ h = \frac{Pq - Qp}{q^2}; P \in \Pi_m, Q \in \Pi_{n-1} \right\}. \tag{2.7}$$

Here, q is normalized by choosing $b_n = 1$. Then $h = 0$ implies $Q = 0$, since otherwise h would not vanish at the n zeros of q. This implies $P = Q = 0$ and $\{Pq - Qp \colon P \in \Pi_m, Q \in \Pi_{n-1}\}$ is an $(m + n - 1)$-dimensional subset of Π_{m+n}. It thus coincides with Π_{m+n}, and so

$$T_{p/q} R_{m,n} = q^{-2} \Pi_{m+n} := \left\{ h = \frac{v}{q^2}; v \in \Pi_{m+n} \right\}. \tag{2.8}$$

Moreover, by the explicit calculation of second derivatives, one verifies that the image of $d^2_{(a,b)} F$ is contained in $q^{-3} \Pi_{m+2n} \subset q^{-3} \Pi_{3n-1}$.

Now the main idea of the proof of Theorem 2.9 is to show the existence of an element f in the Hilbert space such that

$$\left.\begin{aligned}(f - u_i, d_{a_i}F) &= 0 \\ (f - u_i, d_{a_i}^2F) &= 0,\end{aligned}\right\} \quad \text{for } i = 1, 2, \ldots, N, \qquad \begin{aligned}&(2.9)\\&(2.10)\end{aligned}$$

where dF and d^2F are to be evaluated at $F^{-1}u_i$. Then (2.9) is the condition for u_i to be a critical point. Furthermore, (2.10) implies that, in the formula (2.4) for $d^2\rho$, only the positive definite term survives. Note that both (2.9) and (2.10) may be reformulated as

$$f - u_i \in (q_i^{-3}\Pi_{3n-1})^{\perp} \qquad i = 1, 2, \ldots, N,$$

which in turn is

$$(f, q_i^{-3}v) = (u_i, q_i^{-3}v) \quad \text{for } v \in \Pi_{3n-1}, i = 1, 2, \ldots, N. \tag{2.11}$$

This is a linear system of $3nN$ equations. It is a linear system for the same number of unknowns, if f is restricted to a $3nN$-dimensional subspace $E_1 \subset E$. Now the restricted system always has a solution, whenever the corresponding homogeneous system

$$(g, q^{-3}v) = 0 \quad \text{for } v \in \Pi_{3n-1}, i = 1, 2, \ldots, N, \tag{2.12}$$

has only the trivial solution $g = 0$ in E_1. Indeed,

$$E_1 = \operatorname{span}\{q_i^{-3}\Pi_{3n-1}: i = 1, 2, \ldots, N\} \tag{2.13}$$

is a $3nN$-dimensional space. Otherwise, there would exist polynomials $v_1, v_2, \ldots, v_N \in \Pi_{3n-1}$, not all zero, such that

$$\sum_{i=1}^{N} \frac{v_i}{q_i^3} = 0.$$

Then $v_j = -q_j^3 \sum_{i \neq j}(v_i/q_i^3)$ vanishes at the zeros of q_j^3 and has $3n$ zeros counting multiplicities. Since $v_j \in \Pi_{3n-1}$, we have $v_j = 0$. Hence, from (2.12) we get $(g, g) = 0$, i.e. $g = 0$ and (2.11) has a solution f in the subspace (2.13). This completes the proof of the theorem. □

The extension to rational functions in $R_{m,n}$ where $m \geq n$, is possible with slight modifications (Braess 1976). The nonuniqueness is a consequence of nonlinearity and is disturbed by the fact that $R_{m,n}$, $m \geq n$, contains the linear subspace Π_{m-n}. Therefore, if we want $u_1, u_2, \ldots, u_N$ to be critical points, we must have

$$f - u_i \in \Pi_{m-n}^{\perp}, \qquad i = 1, 2, \ldots, N.$$

This implies that $u_i - u_j \in \Pi_{m-n}^{\perp}$ for $i, j = 1, 2, \ldots, N$.

To overcome this difficulty, let $u_i = p_i/q_i$, $i = 1, 2, \ldots, N$ be given functions whose denominators satisfy the assumptions of Theorem 2.9. Then choose $v_i \in \Pi_{m-n}$ such that

$$\tilde{u}_i = u_i + v_i \in \Pi_{m-n}^{\perp}, \quad \text{for } i = 1, 2, \ldots, N.$$

In a similar fashion, f may constructed in a $3nN$-dimensional subspace of the reduced space $\Pi_{m-n}^{\perp} \subset L_2(X)$, having the *LBA*'s $\tilde{u}_1, \tilde{u}_2, \ldots, \tilde{u}_N$. The rest of the construction proceeds as in the proof of Theorem 2.9 and is omitted. □

We will draw two conclusions from Theorem 2.9 and its extension to $R_{m,n}$, $m \geq n$.

The construction of the function f guaranteed that the given functions u_1, $u_2, \ldots, u_N$ were non-degenerate critical points. From the continuity theorem, we know that for each $g \in L_2(X)$, with $\|f - g\| < \delta$, there are N distinct strict *LBA*'s in neighborhoods of u_1, u_2, ..., and u_N, respectively, provided that $\delta > 0$ is arbitrarily small. Hence, the set of f's for which there are at most $N - 1$ *LBA*'s is not dense in $L_2(X)$. We reformulate this observation (cf. Theorem II.1.7).

2.10 Corollary. *Let N be a natural number, $n \geq 1$, and $m \geq 0$. It is not a generic property for f in $L_2(X)$ that f has at most N local best approximations from $R_{m,n}$.*

If u is the best approximation to f from a linear subspace of a Hilbert space, then $\|f - u\|^2 = \|f\|^2 - \|u\|^2$. If u is a local best approximation from $R_{m,n}$, u must be the best approximation in the one dimensional space spanned by u. Therefore, all the solutions from Theorem 2.9 lie on the same level whenever $\|u_1\| = \|u_2\| = \cdots = \|u_N\|$.

2.11 Corollary. *Let N be a natural number, $n \geq 1$, and $m \geq 0$. Then there is an $f \in L_2(X)$ which has N equally good strict local best approximations from $R_{m,n}$.*

Exercises

2.12. Let $f \in L_p[\alpha, \beta] \cap C[\alpha, \beta]$ where $1 < p < \infty$. Assume that $C_u M$ contains an m-dimensional Haar subspace. Show that $f - u$ has at least m sign changes if u is a nearest point to f from M. How many sign changes has the error function of a best rational L_p-approximation?

2.13. Assume that f is given on a discrete set which contains M points and consider the l_2-approximation from $R_{0,1}$. Given $b \in \mathbb{R}$, there is only a critical point of the form $a/(x + b)$, if the determinant with the scalar products vanishes:

$$\det \begin{vmatrix} \left(\dfrac{1}{x+b}, \dfrac{1}{x+b}\right) & \left(f, \dfrac{1}{x+b}\right) \\ \left(\dfrac{1}{x+b}, \dfrac{1}{(x+b)^2}\right) & \left(f, \dfrac{1}{(x+b)^2}\right) \end{vmatrix} = 0.$$

Show that the determinant is a rational function of b with the degree depending on M. Derive a bound for the number of critical points.

2.14. Prove that each $f \in C[\alpha, \beta]$ may be arbitrarily well approximated by exponential sums

$$\sum_{\nu} \alpha_\nu e^{t_\nu x}, \quad \text{where } \alpha_\nu, t_\nu \in \mathbb{R}.$$

Hint: Note that $x^n = (\partial^n/\partial t^n) e^{tx}|_{t=0}$.

2.15. Show that 0 is not the nearest L_2-point to $f \neq 0$ in the family $\{u(x) = \alpha e^{tx}: \alpha, t \in \mathbb{R}\}$. Hint: Recall Exercise 2.14.

§ 3. Varisolvency

In the theory of nonlinear uniform approximation, the generalization of the Haar condition plays a central role. In the linear case, the Haar condition may be defined in terms of the number of zeros of the underlying functions, or equivalently, by the existence of interpolating functions. In 1961, when introducing the concept of varisolvency, Rice observed that, in the nonlinear case, both properties have to be postulated separately. Specifically the interpolation property is only required locally, while the property on the zeros must be stated in a global way.

There is one more difference to the linear families. It is possible (and even necessary for a treatment of the actual and interesting examples) that the degree, i.e. the number of interpolation conditions, is not constant in the family. This is important for the treatment of rational functions and exponential sums.

In this section $X = [\alpha, \beta]$, $\alpha < \beta$, is a real interval and $C[\alpha, \beta]$ is understood to be the space of real valued continuous functions on $[\alpha, \beta]$ endowed with the uniform norm.

A. Varisolvent Families

3.1 Definition. Let M be a non-empty set in $C[\alpha, \beta]$.

(i) M is *locally solvent* of degree $m = m(u_0) \geq 1$ at u_0 if, given $\varepsilon > 0$ and m distinct points $x_i \in [\alpha, \beta]$, $i = 1, 2, \ldots, m$, there is a $\delta = \delta(\varepsilon, u_0, x_1, \ldots, x_m) > 0$ such that $|u_0(x_i) - y_i| \leq \delta$, $i = 1, 2, \ldots, m$, implies the existence of a function $u \in M$ satisfying $\|u - u_0\| < \varepsilon$ and

$$u(x_i) - y_i = 0, \qquad i = 1, 2, \ldots, m. \tag{3.1}$$

(ii) M has *Property Z* of degree $n = n(u_0)$ at $u_0 \in M$ if, for any $u \in M$, the difference $u - u_0$ either has at most $n - 1$ zeros or vanishes identically.

(iii) M is a *varisolvent* family if, at each point $u \in M$, both the local solvency condition and Property Z are defined and have the same degree $m = m(u)$.

(iv) M satisfies the *density property* if, given $u_0 \in M$ and $\varepsilon > 0$, there are $u_1, u_2 \in M$ such that $\|u_i - u_0\| < \varepsilon$, $i = 1, 2$, and

$$u_1(x) < u_0(x) < u_2(x) \quad \text{for } x \in [\alpha, \beta].$$

Obviously, each n-dimensional linear Haar space in $C[\alpha, \beta]$ is a varisolvent family of constant degree n. The example below shows the simplest non-trivial family.

3.2 Example. The family $E_1 := \{u: u(x) = ae^{tx}, a, t \in \mathbb{R}\}$ is varisolvent and the degree is

$$m(u) = \begin{cases} 2, & \text{if } u \neq 0, \\ 1, & \text{if } u = 0. \end{cases} \tag{3.2}$$

To verify this, consider first an element $u_0 \neq 0$. For each $u \in M$ the difference

$$u - u_0 = ae^{tx} - a_0 e^{t_0 x} = e^{t_0 x}[ae^{(t-t_0)x} - a_0]$$

has at most one zero or vanishes identically. Hence, Property Z of degree 2 holds.

Let y_1 and y_2 be non-zero real numbers with the same sign, and let $x_1 \neq x_2$. The interpolation problem

$$ae^{tx_i} = y_i, \qquad i = 1, 2, \tag{3.3}$$

can be solved explicitly

$$t = \frac{1}{x_2 - x_1} \log \frac{y_i}{y_1}, \qquad a = y_1 e^{-tx_1}.$$

Moreover the parameters a and t are continuous functions of y_1 and y_2. Therefore E_1 is locally solvent of degree 2 at $u_0 \neq 0$.

Next consider $u_0 = 0$. Then $u - u_0$ has no zero whenever $u \in E_1, u \neq u_0$. Since E_1 contains the one-dimensional linear space of constant functions, we obtain local solvability of degree 1 at $u_0 = 0$. □

The family E_1 contains the simplest exponential sums. It elucidates how essential it is to postulate the interpolation feature only locally. If $y_1 = -y_2 \neq 0$, then (3.3) has no solution.

Generally, it is easier to establish Property Z of the correct degree than to establish the solvability of the nonlinear interpolation problem (3.1). Appropriate tools for differentiable families will be given in the next section. If we have only continuous parametrizations, a topological argument may be used (see e.g. J. Schwartz, 1969, p. 77).

Brouwer's Theorem on the Invariance of Domain. *Let A be an open subset of $\mathbb{R}^n$. Assume that $f: A \to \mathbb{R}^n$ is continuous and one-one. Then $f(A)$ is open in $\mathbb{R}^n$.*

3.3 Lemma. *Let $M \subset C[\alpha, \beta]$, A be an open set in m-space, and $F: A \to F(A) \subset M$ be a continuous one-one mapping. If the difference $u_1 - u_2$ of each pair u_1, $u_2 \in F(A)$, $u_1 \neq u_2$, has at most $m - 1$ zeros, then M is locally solvent of degree m at each $u \in F(A)$.*

Proof. Given m distinct points $x_1, x_2, \ldots, x_m \in [\alpha, \beta]$ let $R: C[\alpha, \beta] \to \mathbb{R}^n$ be the restriction

$$Rf = (f(x_1), f(x_2), \ldots, f(x_m)).$$

The product map $R \circ F: A \to R \circ F(A) \subset \mathbb{R}^m$ is continuous and one-one. Then $R \circ F$ is a homeomorphism, and by Brouwer's theorem on the invariance of domain, $R \circ F(A)$ is an open set in $\mathbb{R}^m$. From this, the solvability follows immediately. □

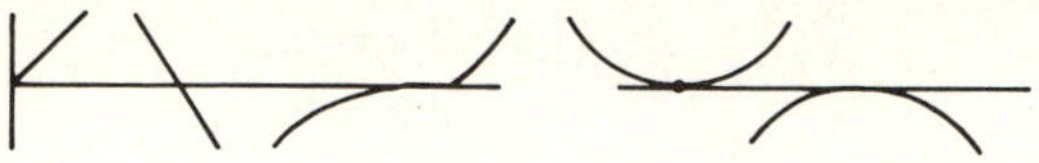

Fig. 16. Nodal and nonnodal zeros

B. Characterization and Uniqueness of Best Approximations

Varisolvent families may contain functions which are only continuous and not differentiable. Therefore, only simple and double zeros are distinguished (see Fig. 16).

3.4 Definition. Let x_0 be an interior point of the interval $[\alpha, \beta]$. The function $f \in C[\alpha, \beta]$ has a *nonnodal* zero at x_0, if either $f(x) \geq 0$ or $f(x) \leq 0$ holds in some neighborhood U of x_0 and if $f(x) \neq 0$ on the boundary of U. Nonnodal zeros are counted with multiplicity 2. All other zeros have multiplicity 1.

The following lemma is needed for the proof that Property Z remains true, if zeros are counted with their multiplicities.

3.5 Lemma. *Let u, u_1 be two elements of the varisolvent family M. Assume that there are $m + 1 = m(u) + 1$ points $x_0 < x_1 < \cdots < x_m$, such that*

$$(-1)^i s[u_1(x_i) - u(x_i)] \geq 0, \qquad i = 0, 1, \ldots, m, \tag{3.4}$$

where $s = +1$ or $s = -1$. Then $u_1 = u$.

Proof. For reasons of symmetry, we may assume that $m(u_1) \geq m(u)$. The proof is complete if we have equality in all $m + 1$ relations (3.4). Therefore, assume $(-1)^j s[u_1(x_j) - u(x_j)] > 0$ for some $j \leq m$.

Because of the local solvability, there is a $u_2 \in M$ satisfying

$$u_2(x_i) = u_1(x_i) + (-1)^i s \cdot \delta, \qquad 0 \leq i \leq m, \qquad i \neq j.$$

Here, $\delta > 0$ is assumed to be chosen such that $\|u_2 - u_1\| < \varepsilon := |(u_1 - u)(x_j)|$. Hence,

$$(-1)^i s[u_2(x_i) - u(x_i)] > 0, \qquad i = 0, 1, \ldots, m$$

and $u_2 - u$ has at least $m(u)$ zeros. This contradicts Property Z. □

3.6 Lemma. *Let u, u_1 be elements of a varisolvent family, $u \neq u_1$. Then $u - u_1$ has at most $m(u) - 1$ zeros counting nonnodal zeros twice.*

Proof. Assume that $u - u_1$ has at least m zeros, each zero being counted with its multiplicity. Let $\{x_i\}_{i=1}^k$ be the ordered set of points at which $u - u_1$ vanishes. The set is augmented: If x_i is a nonnodal zero, we add a point from the interval (x_i, x_{i+1}) or from (x_i, β), resp. Finally, we supplement the set by another point, which is located to the left of the first nonnodal zero. The augmented set contains at least $m+1$ points, and after reordering them, we have

$$(-1)^i s[u(x_i) - u_1(x_i)] \geq 0, \qquad i = 1, 2, \ldots, m+1,$$

where $s = +1$ or $s = -1$. Now Lemma 3.5 shows that $u = u_1$. □

3.7 Remark. Often the bound for the number of zeros can be reduced by one via a parity argument. Assume that

$$(-1)^m (u_1 - u)(\alpha) \cdot (u_1 - u)(\beta) > 0$$

where $m = m(u)$. Then the number of zeros in $[\alpha, \beta]$ must be of parity m. Thus the number of zeros can be at most $m - 2$ counting multiplicities.

The first step of the characterization of best approximations is the (nonlinear version of the)

3.8 Theorem of de la Vallée-Poussin. *Let $M \subset C[\alpha, \beta]$ have Property Z of degree m at u_0, and let $\varepsilon_0 := f - u_0$. Assume that there are $m+1$ ordered points $x_0 < x_1 < \cdots < x_m$ in $[\alpha, \beta]$ such that*

$$\operatorname{sgn} \varepsilon_0(x_i) = -\operatorname{sgn} \varepsilon_0(x_{i-1}), \qquad i = 1, 2, \ldots, m.$$

Then, for each $u \in M$ we have the lower bound

$$\|f - u\| \geq \min_{0 \leq i \leq m} |\varepsilon_0(x_i)|. \tag{3.5}$$

Proof. Denote the right hand side of (3.5) by the symbol c. Put $s = \operatorname{sign} \varepsilon_0(x_0)$. By assumption, $s(-1)^i \varepsilon_0(x_i) \geq c$. Suppose that $\|f - u\| < c$ for some $u \in M$. Then

$$s(-1)^i [f(x_i) - u(x_i)] \leq \|f - u\| < c$$
$$\leq s(-1)^i [f(x_i) - u_0(x_i)], \qquad i = 0, 1, \ldots, m.$$

Hence, $s(-1)^i [u_0(x_i) - u(x_i)] < 0$, $i = 0, 1, \ldots, m$, which contradicts Property Z. □

3.9 Alternation Theorem (Rice (1961). *Let u be an element of a varisolvent family $M \subset C[\alpha, \beta]$, and let $f \in C[\alpha, \beta]$. Assume that $f - u$ is not constant. Then u is a best approximation to f from M, if and only if $f - u$ has an alternant of length $m(u) + 1$.*

The assumption that $f - u$ is not a constant function may be abandoned, if M satisfies the density condition.

Proof. If $f - u$ has an alternant of length $m+1$, it follows from the de la Vallée-Poussin theorem that $\|f - u_1\| \geq \|f - u\|$ for any $u_1 \in M$. Therefore u is a best approximation.

Assume that $\varepsilon = f - u$ is a non-constant function and has an alternant of length exactly $j \leq m$. Then $[\alpha, \beta]$ can be split into j subintervals $I_1 = [\alpha, x_1]$, $I_2 = [x_1, x_2], \ldots, I_j = [x_{j-1}, \beta]$ such that

$$s\cdot(-1)^i\varepsilon(x) > -\|\varepsilon\| \quad \text{for } x\in I_i,\ i=1,2,\ldots,j,$$

where $s=+1$ or $s=-1$. Since the intervals are compact, we even have $s(-1)^i\varepsilon(x) \geq -\|\varepsilon\| + 2c$, $x\in I_i$, $i=1,2,\ldots,j$, with $0 < c < \frac{1}{2}\|\varepsilon\|$. Three cases are distinguished.

Case 1. $j = m$. Since M is locally solvent at u, there is a $u_1\in M$, satisfying

$$\begin{aligned} u_1(x_i) &= u(x_i), \qquad i=1,2,\ldots,m-1,\\ u_1(\alpha) &= u(\alpha) - s\delta(c,u,\alpha,x_1,\ldots,x_{m-1}). \end{aligned} \tag{3.6}$$

From Lemma 3.6, it follows that $u_1 - u$ changes its sign at the $m-1$ prescribed zeros and that there are no more zeros. Hence,

$$s(-1)^i(u_1-u)(x) > 0 \quad \text{for } x\in \operatorname{int} I_i,\ i=1,2,\ldots,j.$$

This and $\|u_1 - u\| < c$ implies that

$$\|\varepsilon\| \geq s(-1)^i\varepsilon(x) > s(-1)^i(f-u_1)(x) > -\|\varepsilon\| + c. \tag{3.7}$$

Since $x_1, x_2, \ldots, x_{m-1}$ are not extremal points, we get $\|f-u_1\| < \|f-u\|$ and u cannot be a best approximation.

Case 2. $j = m-1$. Since M is locally solvent at u, there is a $u_1\in M$ satisfying

$$\begin{aligned} u_1(\alpha) &= u(\alpha) - s\delta,\\ u_1(x_i) &= u(x_i), \qquad i=1,2,\ldots,m-2,\\ u_1(\beta) &= u(\beta) - s(-1)^m\delta. \end{aligned} \tag{3.8}$$

where $\delta = \delta(c,u,\alpha,x_1,\ldots,x_{m-2},\beta)$. Because $u_1 - u$ is given at the endpoints of the interval, we conclude from (3.8) and Remark 3.7 that the number of zeros must be precisely $m-2$. Therefore $u_1 - u$ changes its sign at the $m-2$ prescribed zeros and there are no more zeros. The rest of the proof proceeds as in Case 1.

Case 3. $j \leq m-2$. First put $i=j$. Since ε is not a constant function, we may choose a pair of points $x_i < x_{i+1}$, such that $[x_i, x_{i+1}]$ contains no point from $\{x_\nu\}_{\nu=1}^{i-1}$ and no point where $|\varepsilon(x)| \geq \|\varepsilon\| - 2c$. Replace i by $i+2$.

This procedure is repeated until $m-1$ or $m-2$ points are constructed. Then define u_1 by (3.6) or (3.8), respectively. Now (3.7) holds with the exception of the intervals $[x_j, x_{j+1}]$, $[x_{j+2}, x_{j+3}]$, ... Since $\|u_1 - u\| < c$, it follows, that $|f - u_1| < \|\varepsilon\|$ also holds in the exceptional intervals. Hence, u_1 is a better approximation than u.

Thus, in each of the three cases, we may find a better approximation in any neighborhood of u. □

Since best approximations are characterized in terms of alternants, varisolvent families are strict suns. Moreover, from the alternation theorem we immediately get the

3.10 Uniqueness Theorem for Varisolvent Families (Rice 1961). *Let $M \subset C[\alpha, \beta]$ be a varisolvent family with the density property. Then there is at most one (local) best approximation to each $f \in C[\alpha, \beta]$ from M.*

Proof. Let u be an *LBA* to f from M. Due to the density property, $f - u$ is not constant. By the alternation theorem, $f - u$ has an alternant $x_0 < x_1 < \cdots < x_{m(u)}$ and u is a best approximation. Suppose that u_1 is another solution. With the same arguments as in the proof of the de la Vallée-Poussin theorem, we obtain

$$s(-1)^i(u_1 - u)(x_i) \leq 0 \quad \text{for } i = 0, 1, \ldots, m(u),$$

so that Lemma 3.5 yields $u_1 = u$. □

C. Regular and Singular Points

As was already mentioned, it is characteristic of the varisolvent families of actual interest that the degree varies in the family. In this context we note that it was observed very early and proved later by Werner (1964) that the metric projection onto $R_{m,n}$ is not continuous at those points for which the best approximation does not have the maximal degree. We will obtain more insight into the structure of (the abstract) varisolvent families by investigating this phenomenon.

3.11 Definition. An element u is a *regular point* in a varisolvent family $M \subset C[\alpha, \beta]$ if one of the following equivalent properties holds:
1°. The degree is constant in a neighborhood of u.
2°. There is a neighborhood of u in M whose closure is compact.
3°. A neighborhood of u in M is a manifold, i.e., is homeomorphic to an open set in a Euclidean space.
An element $u \in M$ is *singular* if it is not a regular point.

A function $f \in C[\alpha, \beta]$ is said to be *normal* for M (or *M-normal*), if the best approximation to f from M exists, and it is a regular point. □

There are slightly different definitions and terms in the literature. In particular, distinct proofs for the same fact but using conditions 1°, 2°, or 3° resp. can be found (see e.g. Braess 1974d, Ling and Tornga 1974).

To be specific, we will start by defining regularity by Condition 1°, and later we will verify equivalence via the continuity of the metric projection at normal points.

Let M be varisolvent of degree m_0 at u_0. Then at each u in some neighborhood of u_0 the degree is at least m_0. Indeed choose m_0 distinct points x_i, $i = 1, 2, \ldots, m_0$ in and put $\delta = \delta(1, u_0, x_1, \ldots, x_{m_0})$. By 3.1(1) no $u \in M$ with $\|u - u_0\| < \delta$ is completely determined by its values at the $m_0 - 1$ points $x_2, x_3, \ldots, x_{m_0}$. In order to satisfy Property Z the degree must at least be m_0.

Therefore all elements whose degree equals the maximal degree $m_{\max} := \max\{m(u); u \in M\}$, are regular. In the families of rational functions, these are the only regular points. However, this is not always true.

3.12 Example. Let

$$M_0 = \{0\},\ M_1 = \{u(x) = a\colon a < 0\},\ M_2 = \{u(x) = ae^{tx}\colon a > 0, t \in \mathbb{R}\}.$$

It follows with the same arguments as in the discussion of Example 3.2 that $M := M_0 \cup M_1 \cup M_2$ is varisolvent and that

$$m(u) = \begin{cases} 2, & \text{if } u \in M_2, \\ 1, & \text{otherwise.} \end{cases}$$

The subsets M_1 and M_2 contain the regular points of degree 1 and 2, resp. and $u_0 = 0$ is singular. □

More generally, given M let M_n, $n \geq 1$, denote the subset of regular points of M at which the degree equals n. Choose $x_1 < x_2 < \cdots < x_n$ in $[\alpha, \beta]$. The restriction

$$R\colon M_n \to \mathbb{R}^n,$$

$$R(u) = (u(x_1), u(x_2), \ldots, u(x_n))$$

maps M_n one-one onto an open set in n-space. Moreover, from Definition 3.1(i) it follows that R^{-1} is continuous. Hence $(R(M_n), R^{-1})$ is an atlas for M_n which consists of only one chart.

Therefore, Condition 3.11(1°) implies 3.11(3°), which in turn implies 3.11(2°). The converse will be verified via the continuity of the metric projection.

3.13 Continuity Theorem for Normal Points. *Let M be a varisolvent family in $C[\alpha, \beta]$.*

(1) *The metric projection onto M is continuous at M-normal points.*

(2) *If a singular point u is a best approximation to $f \in C[\alpha, \beta]$ from M and if $f - u$ has an alternant of exact length $m(u) + 1$, then the metric projection onto M is not continuous at f.*

Proof. The first part follows immediately from Theorem 1.13.

To prove the second part, assume that in each neighborhood of u in M there is a $u_1 \in M$ with $m(u_1) \geq m + 1 := m(u) + 1$. Set $f_1 := f + u_1 - u$. Since $f_1 - u_1$ has only an alternant of length $m + 1 = m(u_1)$, the element u_1 is not a best approximation to f_1 from M. Moreover, by the theorem of de la Vallée-Poussin, $\|f_1 - u_2\| \geq \|f_1 - u_1\|$ holds for any u_2 from the subset $\{v \in M\colon m(v) \leq m\}$. Therefore, each better approximation to f_1 must have a degree $\geq m + 1$ and must be characterized by an alternant of length $\geq m + 2$. It cannot be close to u_1. □

If u is a singular point of an arbitrary varisolvent family M, then there cannot be a compact neighborhood of u in M. Otherwise, Lemma 1.13 and Theorem 3.13(2°) would lead to a contradiction. Hence, 3.11(2°) implies 3.11(1°) and all conditions in Definition 3.11 are indeed equivalent. □

Another property which is connected with regularity will be discussed in Section D. Also the sensitivity to a discretization of the interval (and the consideration of the approximation problem on a discrete point set), depends heavily on whether the solution is regular or singular (Werner 1967). Finally, the following shows that usually regularity and normality are generic properties.

3.14 Proposition. *Let M be a varisolvent family in $C[\alpha, \beta]$ and assume that* $\sup\{m(u)\colon u \in M\} < \infty$. *Then the set of regular points is open and dense in M. If moreover, M is an existence set, then the set of M-normal functions is open and dense in $C[\alpha, \beta]$.*

D. The Density Property

Originally, in 1961, Rice introduced the concept of varisolvency without the density property. In 1968, Dunham observed that a constant error curve cannot be excluded without tools as the density property. In 1978, Bartke actually constructed an example of a best approximation with a constant error curve.

On the other hand, the density property is not completely independent of the other conditions.

In order to verify the density at u, it is sufficient to prove that u is not a best approximation to $f(x) = u(x) + \varepsilon/2$ (and to $g(x) = u(x) - \varepsilon/2$, resp.). Because of the alternation theorem, it is sufficient to construct a $u_1 \in M$, $u_1 \neq u$ with

$$u(x) \leq u_1(x) < u(x) + \varepsilon. \tag{3.9}$$

If the degree is 1, 2 or 3, such elements are obtained immediately from the following interpolation problems (Barrar and Loeb 1968):

Case 1: $m = 1$, $\quad u_1(\alpha) = u(\alpha) + \delta, \quad \delta = \delta(\varepsilon, u, \alpha).$

Case 2: $m = 2$, $\quad u_1(\alpha) = u(\alpha) + \delta,\ u_1(\beta) = u(\beta), \quad \delta = \delta(\varepsilon, u, \alpha, \beta).$

Case 3: $m = 3$: $\quad u_1(\alpha) = u(\alpha),\ u_1(x_2) = u(x_2) + \delta,\ u_1(\beta) = u(\beta), \quad \delta = \delta(\varepsilon, u, \alpha, x_2, \beta)$, $x_2 \in (\alpha, \beta)$.

Since in each case $m - 1$ zeros are fixed, $u_1 - u$ does not change its sign and u_1 is a function with the desired properties.

3.15 Theorem (Braess 1974d, Ling and Tonga 1974). *The density property holds at each regular point in a varisolvent family M.*

Proof. If there is no $u_1 \in M$, $u_1 \neq u$ which satisfies (3.9), then u is a strict LBA to $f_0(x) = u(x) + 3\varepsilon$ from M. By the continuity lemma (Lemma 1.13) to each $f_1 \in C[\alpha, \beta]$ with $\|f_1 - f_0\| < \delta$ there is an LBA u_1 from M satisfying $\|u_1 - u\| < \varepsilon$, provided that $\delta < \varepsilon$ is sufficiently small. Then $(f_1 - u_1)(x) > \varepsilon$ and $f_1 - u_1$ must be constant. Now we fix f_1 such that $f_1 - f$ alternates $m + 2$

times. Then either $u_1 - u$ has m zeros or (3.9) holds. In either case we have a contradiction. □

The assumption that $m \le 3$ or that u is regular cannot be substantially relaxed. This is elucidated by a counterexample.

3.16 Example for a Varisolvent Family without the Density Property (Bartke 1978). Let $M = \{0\} \cup \{u \in \Pi_5 : u$ has at most 3 zeros and $u(x) > 0$ for an $x \in [-1, +1]\}$, counting zeros of polynomials with their algebraic multiplicity. Then M is a varisolvent family in $C[-1, +1]$ without the density property at 0 with

$$m(u) = \begin{cases} 4 & \text{if } u = 0, \\ 6 & \text{otherwise.} \end{cases}$$

The only property which is not obvious is the local solvability at $u = 0$. Consider the interpolation problem

$$u(x_i) = y_i, \qquad i = 1, 2, 3, 4, \qquad x_1 < x_2 < x_3 < x_4. \tag{3.10}$$

Let p be the interpolating polynomial in Π_3. If $\max p(x) \le 0$, we will construct a feasible u by modifying p. The constructed u cannot depend continuously on the data because otherwise the density property would already follow from a fixed point argument (Braess 1974d). So, it is natural that different cases have to be discussed. Let $w(x) = \prod_{i=1}^{4} (x - x_i) \in \Pi_4$.

Case 1. $p(x) < 0$ in $[-1, +1]$. Choose $\alpha > 0$ such that $\max\{p(x) + \alpha w(x)\} = 0$. Then $q := p + \alpha w$ has at most four zeros. Moreover, let $\tilde{q} \in \Pi_4$ satisfy the interpolation conditions (3.10) and $\tilde{q}(x_5) = 0$ for $x_5 = (x_2 + x_3)/2$. Then $\|\tilde{q}\|$ is small for small data and q being a convex combination of p and $\tilde{q}$ is also small.

Case 1a. q has at most 3 zeros counting multiplicities. Then $u = q + \varepsilon w$, ε small, is a solution.

Case 1b. q has at least two distinct zeros $z_1 < z_2$. Let $z_1 < \xi < z_2$. Then $u = p + [\alpha + \varepsilon(x - \xi)]w$ is a solution. It has no zeros for $x < \xi$ and for $x > \xi$ it has at most as many as q. Moreover $u(z_2) > 0$.

Case 1c. q has one zero z of multiplicity 4. We note that $q + \varepsilon h$ is strictly concave in a neighborhood of z if this is true for h. Then $q + \varepsilon h$ has at most two zeros. If $\pm w'(z) > 0$, then $h = \pm(z - x)w$ is an appropriate choice. If $w'(z) = 0$, then $w''(z) < 0$ and $h = w$ yields a solution.

Case 2. $\max p(x) = p(z) = 0$ for some $z \ne x_i$, $i = 1, 2, 3, 4$. Since p has at most 3 zeros, $u = p + \varepsilon[\operatorname{sgn} w(z)]w(x)$ is a solution.

Case 3. $\max p(x) = p(x_i) = 0$ and x_i is an interior point. Then $p'(x_i) = 0$ and $u = p + \varepsilon w$ is a solution.

Case 4. $\max p(x) = 0$ is attained at an x_i which is a boundary point. First, let $x_1 = \alpha$, $y_1 = 0$. Consider the solution of

$$v(x_i) = \frac{y_i}{x_i - \alpha}, \; i = 2, 3, 4,$$

with $v \in \Pi_4$, v having at most two zeros and $v(x) > 0$ for some $x \in [-1, +1]$. This may be done analogously to the cases 1, 2 and 3. Set $u = (x - \alpha)v$. Similarly, if β is the only crucial point, the problem is solved by a symmetry argument. If both α and β are involved, then two analogous reductions are performed. (The reader may check that the last case would break down, if the construction with $m = 3$ is tried). □

Exercises

3.17. Which of the following sets in $C[0, 1]$ is varisolvent?

$$M_1 = \left\{u(x) = \sum_{k=0}^{n} a_k x^k \colon a_k \geq 0\right\},$$

$$M_2 = \left\{u(x) = \sum_{k=0}^{n} a_k x^k \colon a_k > 0\right\}.$$

3.18. Show that in $C[0, 1]$ the family

$$M = \{u(x) = a_0 + a_1 \log(1 + tx) \colon a_0, a_1 \in \mathbb{R}, t > 0\}$$

is varisolvent of degree

$$m(u) = \begin{cases} 2 & \text{if } u \text{ is constant,} \\ 3 & \text{otherwise.} \end{cases}$$

Which are the regular points of M? Is M an existence set?

3.19. Let H be a Haar subspace of $C[\alpha, \beta]$ and let $\varphi \in C(\mathbb{R})$ be strictly monotonous. Show that

$$M = \{u \colon u(x) = \varphi(h(x)), h \in H\}$$

is varisolvent.

3.20. Let $M \subset C[\alpha, \beta] \subset L_2[\alpha, \beta]$ be a varisolvent family. Show that $f - u$ has at least $m(u)$ zeros, whenever u is a best L_2-approximation to f.

§4. Nonlinear Chebyshev Approximation: The Differentiable Case

The varisolvent families are one extension of Haar spaces. Another nonlinear theory was developed by Meinardus and Schwedt (1964). Instead of local solvency they postulated a local Haar condition, which means that either the tangent space satisfies the Haar condition, or the tangent cones contain linear Haar spaces. At first glance, this theory looks more restrictive, since the sets with the local and global Haar condition are varisolvent families with C^1-parametrizations. On the other hand, the local solvability is usually verified via the local Haar condition.

More important is the fact that the theory of varisolvent families does not apply to Haar cones and that the theory of Meinardus and Schwedt allows an extension to manifolds with boundaries. This point is crucial for the use of methods from global analysis in approximation theory and for the treatment of exponential sums and γ-polynomials.

A. The Local Kolmogorov Criterion

We start with a lemma, which often guarantees the differentiability of a mapping onto $C(X)$. Throughout this section X will be a compact set and $C(X)$ will be endowed with the uniform norm.

4.1 Lemma. *Let $A \subset \mathbb{R}^n$ be an open set, and let*

$$F: A \to C(X),$$
$$a \mapsto F(a)(x) := F(a, x).$$

Assume that the partial derivatives $(\partial/\partial a_j)F(a, x)$, $j = 1, 2, \ldots, n$, exist and are continuous in $A \times X$. Then the Fréchet-derivative of F exists for $a \in A$ and is a continuous function of a. Moreover

$$(d_a F \cdot b)(x) = \sum_{j=1}^{n} b_j \frac{\partial}{\partial a_j} F(a, x). \tag{4.1}$$

The lemma is easily verified by applying the mean-value theorem and noting that the derivatives are uniformly continuous in the compact set $B_\delta(a) \times X$ for some $\delta > 0$, cf. Krabs (1967).

Now, from linear Chebyshev approximation and the lemma above we get the

4.2 Local Kolmogorov Criterion. *Assume that u is a local best approximation to f from $M \subset C(X)$. Let $P[f - u] := \{x \in X: |(f - u)(x)| = \|f - u\|\}$ be the set of extremal points of the error function. Then for each $h \in C_u M$:*

$$\min_{x \in P[f-u]} \{(f(x) - u(x))h(x)\} \le 0. \tag{4.2}$$

If moreover, A is open in $\mathbb{R}^n$ and $F: A \to M$ is Fréchet-differentiable at $a = F^{-1}(u)$, then for each $b \in \mathbb{R}^n$:

$$\min_{x \in P[f-u]} \{[(f - u)\, d_a F \cdot b](x)\} \le 0.$$

Proof. Let $h \in C_u M$. By the lemma of first variation, zero is a best approximation to $f - u$ from the segment $[0, h]$ in $C_u M$. The Kolmogorov criterion for convex sets (Criterion I.2.8) yields the first statement. By recalling that $d_a F(\mathbb{R}^n) \subset C_{F(a)} M$, we obtain the second statement. □

B. The Local Haar Condition

The rest of this section will be concerned with the approximation of functions on a real interval $X = [\alpha, \beta]$.

The starting point for the theory of Meinardus and Schwedt is the following observation, which is immediate from the lemma of first variation and the characterization theorem for Haar subspaces.

4.3 Remark. Assume that the tangent cone $C_u M$ to $M \subset C[\alpha, \beta]$ contains an m-dimensional Haar subspace. If u is a local best approximation to f from M, then $f - u$ has an alternant of length $m + 1$.

4.4 Definition. Let $M \subset C[\alpha, \beta]$.

(i) M satisfies the *local Haar condition* of degree $m = m(u)$ at u, if there is an open set $A \subset \mathbb{R}^m$ and a C^1-mapping $F: A \to M$ such that $d_a F(\mathbb{R}^m)$ is an m-dimensional Haar subspace of $C[\alpha, \beta]$, where $a = F^{-1}(u)$.

(ii) M satisfies the *local* and the *global Haar condition*, if M satisfies both the local Haar condition and has Property Z at each element u with the same degree $m(u) \geq 1$.

4.5 Remark. For the treatment of actual examples, we often use the fact that the local Haar condition of degree m at u implies the local solvability at u of the same degree. Indeed, given $x_1 < x_2 < \cdots < x_m$, this condition implies that the Jacobian matrix

$$\left(\frac{\partial F(b, x_i)}{\partial b_k}\right)_{i,k=1}^{m}$$

is non-singular. It follows from the inverse function theorem that the system

$$F(b, x_i) = y_i, \qquad i = 1, 2, \ldots, m,$$

is, as required, solvable.

Thus, the condition on the local structure (in terms on conditions on the tangent spaces) and the global conditions on the zeros are once again consistent.

4.6 Characterization and Uniqueness Theorem (Meinardus and Schwedt 1964). *Assume that $M \subset C[\alpha, \beta]$ satisfies the local and global Haar condition. Then*

(1) *u is a best approximation to f from M, if and only if $f - u$ has an alternant of length $m(u) + 1$.*

(2) *To each $f \in C[\alpha, \beta]$ there is at most one best approximation from M.*

The characterization by alternants follows from Remark 4.3 and the de la Vallée-Poussin Theorem. The problem of a constant error curve does not arise. As for the uniqueness of the best approximation, we use the results from the theory of varisolvent families. Notice that the local and global Haar condition imply that M is varisolvent. □

C. Haar Manifolds

In sets with the local and the global Haar condition regularity may be defined as for varisolvent families. The regular points form manifolds which were called Haar embedded manifolds by Wulbert (1971a). The extension of this concept to manifolds with boundaries is possible here, while it seems to be impossible in the framework of varisolvency. (The reason is the fact that $\mathbb{R}_+ \times \mathbb{R}$ and $\mathbb{R}_+ \times \mathbb{R}_+$

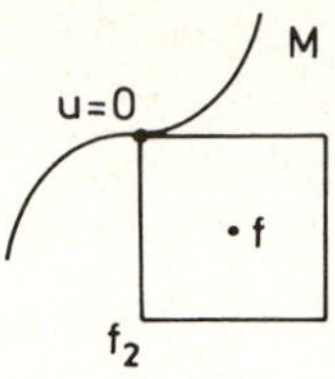

Fig. 17. Example for non strong uniqueness

are homeomorphic but not diffeomorphic and we need a framework in which the two sets are distinguished.)

4.7 Definition (Wulbert 1971, Braess 1973a). A C^1-manifold M of $C[\alpha, \beta]$ with boundary is called a *Haar manifold* if every tangent cone is a Haar cone.

Since any best approximation in a Haar cone is a strongly unique best approximation, the assumptions of Corollary 1.12 are satisfied (see also Barrar and Loeb 1970, Malozemov and Pevnyi 1973).

4.8 Theorem. *An element u is a strongly unique local best approximation to f from a Haar manifold M, if and only if it is a critical point to f in M.*

At this time, we mention that not each best approximation in a varisolvent family is a strongly unique one. Let $M = \{(a, a^3) \in \mathbb{R}^2; a \in \mathbb{R}\}$ be the smooth curve shown in Fig. 17. Then $u = 0$ is the best approximation to $f = (1, -1)$, but strong uniqueness fails.

A simple non-trivial Haar manifold is encountered when exponential sums with restrictions on the spectrum or rational functions with restrictions on the poles are considered.

4.9 Example. Let

$$M = \{u(x) = ae^{tx} \colon a \in \mathbb{R}, t \geq 0\} \subset E_1.$$

Then $M \backslash \{0\}$ is a Haar manifold. For $u = ae^{tx}$ in this subset, we gain by recalling (1.9)

$$C_u M = \begin{cases} \{(\delta_0 + \delta_1 x)e^{tx} \colon \delta_0, \delta_1 \in \mathbb{R}\}, & \text{if } t > 0, \\ \{(\delta_0 + \delta_1 x) \colon \delta_0 \in \mathbb{R}, \operatorname{sgn} \delta_1 = \operatorname{sgn} a\}, & \text{if } t = 0. \end{cases}$$

As a consequence, local best approximations can be characterized in terms of alternants with signs. Specifically, u is a (local) best approximation if $(f - u)$ has an alternant

of length 3	in case $a \neq 0$, $t > 0$,
of length 2 with sign s	in case $a \neq 0$, $t = 0$, $s = -\operatorname{sgn} a$,
of length 2	in case $a = 0$.

The first and the last case are clear from Example 3.2. For convenience, when analyzing the second case, we shall henceforth assume that $a > 0$.

Assume that there is a negative alternant $x_1 < x_2$ of length 2. If $u_1 = a_1 e^{t_1 x}$ is an equally good approximation, then it follows from the de la Vallée-Poussin type argument that

$$(u_1 - u)(x_1) \geq 0, \qquad (u_1 - u)(x_2) \leq 0. \tag{4.3}$$

Therefore $u_1 - u$ has one zero and $a_1 > 0$. If $t_1 > 0$, then $u_1 - u$ is a monotone increasing function, which contradicts (4.3). Hence $t = 0$, and noting (4.3) once more, we have $a_1 = a$.

Assume that there is at most a positive alternant of length 2. Then there is a polynomial $p \in \Pi_1$ with $\|f - u - p\| < \|f - u - 0\|$. By the de la Vallée-Poussin argument, we get

$$p(x_1) < 0, \; p(x_2) > 0, \; x_1 < x_2,$$

and $p(x) = \delta_0 + \delta_1 x$, $\delta_1 > 0$. Hence, $p \in C_u M$ and u is not a critical point. Consequently u is not a local best approximation. □

Note that there is at most one best approximation in M, because there is at most one on the boundary and each *LBA* in E_1 is a global solution.

D. The Local Uniqueness Theorem for C^1-Manifolds

The best uniform approximation in a linear subspace need not be unique if the Haar condition is violated. The analogous situation for nonlinear approximation theory is elucidated in Fig. 18. The tangent line at 0 is parallel to a coordinate axis, i.e., $T_0 M$ is not a Haar subspace. Then 0 is not the unique best approximation to f (and to $-f$) from the tangent line. For the nonlinear problem, 0 is a critical point to f (and to $-f$) in M, but 0 cannot be a strict *LBA* to both f and $-f$.

Moreover, Fig. 17 elucidates that even in Chebyshev sets not every critical point is an *LBA* if the Haar condition is violated (since 0 is not a best approximation to f_2).

The situation is similar in higher dimensional spaces.

4.10 Theorem on Local Uniqueness (Wulbert 1971b). *For u in an n-dimensional C^1-manifold $M \subset C(X)$ (without boundary) the following are equivalent:*

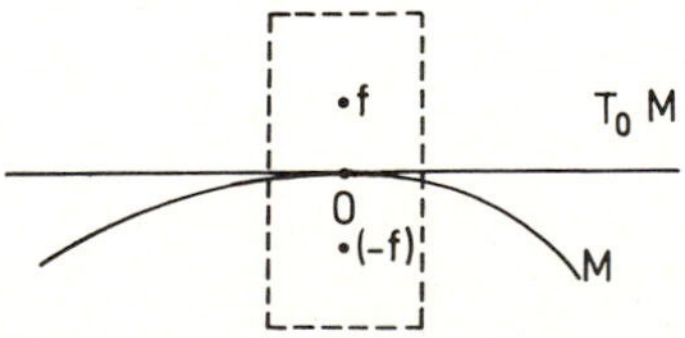

Fig. 18. Curve in 2-space, for which $T_0 M$ is not a Haar subspace

1°. *T_uM is a Haar subspace of $C(X)$.*
2°. Given any $f \in C(X)$, *u is a strict local best approximation to f from M if u is a critical point to f in M.*

Exercises

4.11. Is $M = \{u(x) = (a + bx)^3: a, b \in \mathbb{R}\} \subset C[0, 1]$ varisolvent and does it satisfy the local and the global Haar condition?

4.12. Show that $R_{0,1}$ and E_1 are families with the local and the global Haar condition.

4.13. Prove that critical points in a Haar manifold whose boundaries are consistent with the structure of a ∂-manifold, are isolated.
Hint. Show that strong uniqueness in the tangent space $\|f - F(0) - d_0Fa\| \geq \|f - F(0)\| + c_1\|a\|$ implies that $\|f - F(a) + d_aFa\| \leq \|f - F(a)\| - c_1\|a\| + o(\|a\|)$.

4.14. Let M be an n-dimensional C^1-manifold in $C[\alpha, \beta]$. Assume that T_uM satisfies the Haar condition. Prove that given $x_1 < x_2 < \cdots < x_n \in [\alpha, \beta]$, there is a neighborhood U of u, such that for $u_1, u_2 \in U$ the difference $u_1 - u_2$ vanishes at $\{x_i\}_{i=1}^n$ only if $u_1 = u_2$.

4.15. Assume that $M \in C[\alpha, \beta]$ satisfies the local and the global Haar condition. Show that a best approximation u to f from M is a strongly unique local best approximation whenever u is a regular point of M. If moreover each minimizing sequence for the approximation of f contains a convergent subsequence in M, then u is a strongly unique best approximation (cf. the comment after Theorem 4.8 and Theorem V.2.6).

§5. The Gauss-Newton Method

The proof of the lemma of first variation is, in principle, constructive. This means, that given a non-critical point for a nonlinear approximation problem, an improved approximant may be constructed. In this way, one gets an iterative procedure for the numerical determination of best approximations which generalizes Gauss' method for solving nonlinear least-squares problems (Gauss 1809). During the last decades many variants of the method have been considered.

As in many iterative methods for the numerical solution of nonlinear (optimization) problems, each cycle consists of two steps. 1. The determination of a *direction* for the improvement via the solution of an auxiliary linear (or convex) problem. 2. The determination of a *stepsize*.—As in Newton's method, when applied to a linear problem, only one step would yield the exact solution. But in nonlinear problems, a stepsize smaller than one may be necessary with the Gauss-Newton method, even in the neighborhood of a solution.

5.1 Example. Let $M = S^1$ be the unit sphere in Euclidean 2-space, and let $d(f, M) = 2$. Let v_0 be close to the nearest point u^*.

From Fig. 19 it is evident that $\|v_1 - u^*\| \approx 2\|v_0 - u^*\|$, if v_1 is obtained from the nearest point $v_0 + h_0$ on the tangent line by a projection onto S^1 (without damping).

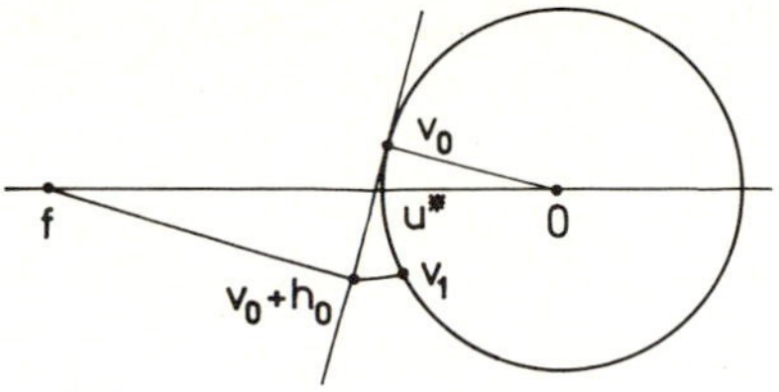

Fig. 19. Iteration step of the Gauss-Newton method with stepsize 1

Obviously, in the above example, the situation is the more favorable the shorter the distance of f from M. This is supported by general experience of the Gauss-Newton method. Furthermore, even better behaviour is encountered in the case where one has local strong uniqueness. Then one has quadratic convergence in a neighborhood of the solution.

Algorithms for special families, as the Remes algorithm or the differential correction algorithm for rational approximation will be discussed in the study of the corresponding families.

A. General Convergence Theory

There are many variants of the Gauss-Newton method (cf. the literature cited below). In Algorithm 5.2 below, only suboptimal solutions of the linearized problems are required. Moreover, the stepsize is determined by trial and no estimates of the second derivatives are used.

5.2 Gauss-Newton Algorithm with Damping. Let $v_0 \in M$. For $k = 0, 1, 2, \ldots$ perform the following:

Step 1 (*Determination of a direction of descent*). Determine a tangent vector $h_k \in C_{v_k} M$ which is a good approximation to $f - v_k$. Specifically, let

$$\|f - v_k - h_k\| \leq \|f - v_k\| - \tfrac{1}{2}\eta, \tag{5.1}$$

where $\eta := \|f - v_k\| - d(f - v_k, C_{v_k} M)$.

Step 2 (*Line search*). Evaluate $\varepsilon_k := \|f - v_k\| - \|f - v_k - h_k\|$. Let $(u_t)_{0 \leq t \leq 1}$ be the curve in M associated to the tangent vector h_k at v_k. For $t = 1, 2^{-1}, 2^{-2}, 2^{-3}, \ldots$ determine u_t until an element has been found for which

$$\|f - u_t\| \leq \|f - v_k\| - \varepsilon_k t/4. \tag{5.2}$$

When such an element has been found, set $v_{k+1} = u_t$ (and proceed to the next cycle). □

From the proof of the lemma of first variation, it follows that the line search terminates after a finite number of steps.

In the sequel, we will assume that there exists a C^1-parametrization

$F: A \to M \subset E$, with $A \subset \mathbb{R}^n$. Moreover, we will set $a_k = F^{-1}v_k$ and in view of (1.10), assume that

$$a_{k+1} = a_k + t_k b_k,$$

where $h_k = d_{a_k}Fb_k$ is the tangent vector determined in Step 1. The numbers ε_k are the *gains in the linearized problems*.

The analysis of the convergence properties will be based on the following:

5.3 Lemma. *Assume that $u_0 = F(a_0) \in M$ is an accumulation point of the Gauss-Newton iteration for the approximation of f from M. Then u_0 is a critical point, or $d_{a_0}F$ is not one-one.*

Proof. Suppose that u_0 is not a critical point and that $d_{a_0}F$ is one-one. Then

$$\|f - u_0\| - \|f - u_0 - d_{a_0}Fb\| = c_1 > 0$$

holds for some $b \in \mathbb{R}^n$. Let $(v_k)_{k \in J}$ be a subsequence which converges to u_0. For sufficiently large k, we have $\|f - v_k\| - \|f - v_k - d_{a_k}Fb\| \geq c_1/2$. The choice of $h_k = d_{a_k}Fb_k$ implies that

$$\varepsilon_k := \|f - v_k\| - \|f - v_k - h_k\| \geq c_1/4. \tag{5.3}$$

Moreover, from $\|h_k\| \leq 2\|f - v_k\| \leq 2\|f - v_1\|$ and $d_{a_k}F \to d_{a_0}F$, it follows that the sequence $(b_k)_{k \in J}$ is bounded. After passing to a subsequence (if necessary), we have $\lim_{k \in J} b_k = b_0$ for some $b_0 \in \mathbb{R}^n$, and (5.3) implies that

$$\|f - u_0\| - \|f - u_0 - d_{a_0}Fb_0\| = c_2 \geq c_1/4 > 0. \tag{5.4}$$

Since F is differentiable, one has

$$\|f - u_0\| - \|f - F(a_0 + tb_0)\| \geq \tfrac{3}{4}c_2 t \text{ for } 0 \leq t \leq \tau,$$

with some $\tau > 0$. Let $\tau/2 \leq t \leq \tau$. From the continuity of F, we get, for sufficiently large $k \in J$:

$$\begin{aligned}
&\|f - v_k\| - \|f - F(a_k + tb_k)\| \\
&\quad \geq \|f - u_0\| - \|f - F(a_0 + tb_0)\| - \|v_k - u_0\| - \|F(a_k + tb_k) - F(a_0 + tb_0)\| \\
&\quad \geq \frac{3}{4}c_2 t - \frac{1}{6}c_2\frac{\tau}{2} - \frac{1}{6}c_2\frac{\tau}{2} \geq \frac{5}{12}c_2 t \geq \frac{1}{4}\varepsilon_k \cdot t.
\end{aligned}$$

Therefore, Condition (5.2) holds for some $t \geq \tau/2$. For infinitely many $k \in J$, we have $\|f - u_k\| - \|f - u_{k+1}\| \geq \dfrac{1}{4}\varepsilon_k\dfrac{\tau}{2} \geq \dfrac{1}{12}c_2\tau > 0$. This is impossible. □

The following example shows that the convergence to critical points indeed cannot be guaranteed if d_aF is not always one-one.

5.4 Example (Schaback 1985). Let $\mathbb{R}^2$ be endowed with the Euclidean norm and let $M = F(\mathbb{R}^2)$ be given by

$$F\begin{pmatrix} x \\ y \end{pmatrix} = \begin{pmatrix} x^2 \\ y^4 \end{pmatrix}.$$

Consider the approximation of $f = (-\beta, 0)'$, where $\beta > 0$. Obviously, $v^* = 0$ is the only critical point. The solution of the linearized problem yields an increment

$$b_k = \begin{pmatrix} -\beta/2x_k(1 + O(x_k^2)) \\ -y_k/4 \end{pmatrix}.$$

The stepsize becomes small as $x_k \to 0$, a simple calculation yields $t_k \leq 3x_k^2/\beta(1 + O(x_k^2))$.

Let $x_0 = 1$ and $y_0 \neq 0$. At least, if $|y_k| \leq \min\{1, \beta\}$, one has $|x_{k+1}| \leq \frac{2}{3}|x_k|$ and t_k tends to zero so quickly that $\prod_{k=1}^{\infty} (1 - t_k/4) > 0$. Hence, (y_k) does not converge to zero. The Gauss-Newton algorithm does not yield the solution. □

In view of the example above, it is natural to assume nonsingularity of dF when sharper results are established. In particular, if both the solutions in M and in the tangent spaces are isolated, then the convergence of the whole sequence and not only of a subsequence can be guaranteed.

5.5 Lemma. *Let the assumptions of Lemma 5.3 prevail. Moreover assume that*
i) u_0 *is an isolated critical point,*
ii) 0 *is the unique best approximation to* $f - u_0$ *from* $T_{u_0}M$,
iii) $d_{a_0}F$ *is one-one.*
Then the sequence from the Gauss-Newton iteration converges to u_0.

Proof. Choose $\varepsilon > 0$ such that u_0 is the only critical point in $B_{2\varepsilon}(u_0)$ and that d_aF is one-one whenever $F(a) \in B_{2\varepsilon}(u_0)$. From ii) and arguments as in the proof of Theorem 1.11, it follows that $h_k \to 0$ for $k \to \infty$, $k \in J$, with J as in the proof of Lemma 5.3. Therefore, we have $\|u_{k+1} - u_k\| < \delta_0 \leq \varepsilon$ for all sufficiently large $k \in J$.

Suppose that $u_0 \neq \lim_{k\to\infty} v_k$. Then there is a neighborhood $B_\delta(u_0)$ with some $\delta < \delta_0$ such that infinitely many elements from (v_k) are in $B_\delta(u_0)$ but also infinitely many ones are outside $B_\delta(u_0)$. From $\|u_{k+1} - u_k\| < \varepsilon$ for $u_k \in B_\delta(u_0)$, we conclude that $u_{k+1} \in B_{2\varepsilon}(u_0) \backslash \mathring{B}_\delta(u_0)$ holds for infinitely many k's. The compact set $B_{2\varepsilon}(u_0) \backslash \mathring{B}_\delta(u_0)$ contains an accumulation point. Since u_0 is the only critical point in $B_{2\varepsilon}(u_0)$, this contradicts Lemma 5.3. □

From the preceding lemmas one gets a convergence result under the usual compactness conditions.

5.6 Theorem. *Let A be an open set of $\mathbb{R}^n$, and $F: A \to F(A) = M \subset E$ be a C^1-parametrization for M and let $f \in E$. Assume that the subset $\{u \in M: \|f - u\| \leq \|f - v_0\|\}$ is compact. Then a subsequence of the Gauss-Newton sequence starting at v_0 converges to some $u^* \in M$, such that u^* is a critical point, or $d_{a^*}F$ is not one-one. Further, if d_aF is one-one for $a \in A$, all critical points are isolated and the tangent spaces at critical points are uniqueness sets, and so the whole sequence converges.*

As was already mentioned, quadratic convergence is obtained in the case of strong uniqueness. We will say that the Gauss-Newton-iteration is performed with *optimal increments* if the best approximations in the tangent spaces and not only suboptimal solutions are computed.

5.7 Theorem (Cromme 1969). *Let A be an open set of $\mathbb{R}^n$ and $F\colon A \to F(A) = M \subset E$ be a C^2-parametrization for M. Let the Gauss-Newton iteration be performed with optimal increments. Assume that an accumulation point $v_0 = F(a_0)$ of the Gauss-Newton sequence is a strongly unique local best approximation to f and that $d_{a_0}F$ is one-one. Then the sequence (v_k) converges quadratically to v_0, i.e., for all sufficiently large k the stepsize is $t_k = 1$ and*

$$\|v_{k+1} - v_0\| \le c\|v_k - v_0\|^2$$

holds with some $c > 0$.

Proof. Without loss of generality, we may assume that $a_0 = 0$. Since d_0F is one-one, it is sufficient to prove that

$$\|a_{k+1}\| \le c\|a_k\|^2$$

for all sufficiently large k, where $a_k = F^{-1}(v_k)$.

From local strong uniqueness and Theorem 1.11, it follows that

$$\|f - v_0 - d_0Fb\| \ge \|f - v_0\| + c_1\|b\| \quad \text{for } b \in \mathbb{R}^n \tag{5.5}$$

with some $c_1 > 0$. Since F is of class C^2, there is a c_2 such that

$$\|d_aF - d_0F\| \le c_2\|a\| \quad \text{and} \quad \|F(a) - F(0) - d_aFa\| \le c_2\|a\|^2. \tag{5.6}$$

Given a, let $b = b(a)$ specify the solution of the linearized problem, i.e.,

$$\|f - F(a) - d_aF(-a + b)\| = \inf_{\tilde{b} \in \mathbb{R}^n} \{\|f - F(a) - d_aF\tilde{b}\|\}.$$

Note that $a_{k+1} = b(a_k)$ if the stepsize t_k equals 1. From (5.5) and (5.6), we obtain

$$\begin{aligned} &\|f - F(a) - d_aF(a - b)\| \\ &\quad \ge \|f - v_0 - d_0Fb\| - \|F(a) - F(0) - d_aFa\| - \|(d_0F - d_aF)b\| \\ &\quad \ge \|f - v_0\| + c_1\|b\| - c_2\|a\|^2 - c_2\|a\| \cdot \|b\| \\ &\quad \ge \|f - v_0\| + c_1\|b\|/2 - c_2\|a\|^2, \end{aligned} \tag{5.7}$$

whenever $\|a\| \le c_1/(2c_2)$. Furthermore, one has

$$\|f - F(a) + d_aFa\| \le \|f - v_0\| + \|F(0) - F(a) + d_aFa\| \le \|f - v_0\| + c_2\|a\|^2.$$

The last two inequalities and the optimality of $b = b(a)$ imply that $\|f - v_0\| + c_1\|b\|/2 - c_2\|a\|^2 \le \|f - v_0\| + c_2\|a\|^2$. Hence,

$$\|b(a)\| \le 4(c_2/c_1)\|a\|^2.$$

From this inequality, it follows that the stepsize $t = 1$ is accepted for large k and that we have quadratic convergence. □

We note that local convergence covers both the Chebyshev approximation from Haar submanifolds of $C[\alpha, \beta]$ and the zero residual case, i.e., $f - v_0 = 0$ (for any norm).

B. Numerical Stabilization

The above convergence results conceal that some extra devices are often necessary in numerical mathematics to obtain an efficient algorithm from a basic method. In particular, this is true for the Gauss-Newton method. Different ways for numerical stabilizations and accelerations are found in the available computer programs. Of course, which one yields the most effective algorithm, depends on the family of approximating functions and on the metric of the function space. Nevertheless, there are some common viewpoints.

1. *Recognition of singularities of dF.* In most actual approximation problems, there are parameters for which dF is not one-one. In many cases, it is not possible to eliminate such points à priori. For example, in the Chebyshev approximation by sums of exponentials, the tangent space at each critical point (but possibly one) suffers from a loss of dimension (see VII. 5B). Experience shows that (as in Example 5.4) the Gauss-Newton sequence has the tendency to run only to singular points and not to critical points as intended. If small stepsizes are observed during the iteration, this is often a hint to a singularity in the neighborhood. The algorithm of Deuflhard and Apostolescu (1980) takes steps to detect a loss of the rank of dF in order to overcome this bad behaviour.

2. *Bad conditioning of approximation problems.* Given an approximation problem, there is one natural metric; this is the metric induced by the norm of the underlying function space. Since the family M of approximating functions is usually given by a parametrization, the metric of the parameter set may also have an influence on the computation.

From the theoretical viewpoint, the metric of the parameter set does not seem to have any relevance to the problem. Indeed, we do not change a problem, if we replace, say the coordinate a_1 by $100a_1$ and adjust the parameter map to the change of the scale. Since the metric of the parameter space is changed by the transformation, it should not enter into the analysis.

This is true for the abstract problem, but it is not for the numerical problem, due to truncation and rounding errors. Usually, these errors can be roughly split into two parts; the first portion can be estimated in terms of the metric of the function space, while the other portion may be measured with the metric of the parameter space.

This phenomenon is encountered already in linear approximation problems (and is not typical for nonlinear problems). It has been long known that in least-squares problems with polynomials, the representation $u(x) = \sum_{k=0}^{n-1} a_k x^k$ leads to very ill-conditioned linear systems. If, on the other hand, basis functions $v_1, v_2, \dots v_n$ are chosen which are close to orthogonal polynomials, the numerical stability is substantially improved. A measure for the condition of a basis $\{v_i{}^i\}_{i=1}^n$ is the *condition number*:

$$\kappa = \frac{\max_{\|y\|=1} \|\sum_k y_k v_k\|}{\min_{\|y\|=1} \|\sum_k y_k v_k\|}.$$

Algorithms for large linear least-squares problems $\|Ay - b\|_{l_2} \to \min!$ do not compute the solution via the normal equation $A^T Ay = A^T b$, but perform a QR-decomposition of A or preconditioned conjugate gradient methods, see Björck (1981), Dennis, Gay and Welsch (1981), Watson (1980).

3. *The Marquardt-Levenberg-regularization.* Consider the nonlinear L_2-approximation. Set $A = d_{a_k} F$ where $v_k = F(a_k)$. One has

$$\begin{aligned}\varphi(a_k + b) &:= \|f - F(a_k + b)\|^2 \\ &= \|f - v_k - Ab - R_0(b)\|^2 \\ &= \varphi(a_k) + d_{a_k}\varphi \cdot b + b^T A^T A b + R(b),\end{aligned}$$

where the remainder term $R(b) = O(\|b\|^2)$ is small of second order. Given any positive definite matrix B, for all sufficiently large μ:

$$R(b) \leq \mu b^T B b.$$

Therefore, when minimizing the expression

$$d_{a_k}\varphi b + b^T(A^T A + \mu B)b$$

by setting $b = \frac{1}{2}(A^T A + \mu B)^{-1} d_{a_k}\varphi$, one has $\varphi(a_k + b) < \varphi(a_k)$.

Originally, the unit matrix I was chosen for the regularization by Levenberg (1944). The choice of the parameter μ played the same role as the choice of the damping factor t in Algorithm 5.2.

Now consider again the effect of linear transformations. The regularization with $B = I$ is only invariant under the group of orthogonal transformations of the parameter space. If in a generally applicable algorithm, invariance under all affine transformations is required, B must be a multiple of $A^T A$. In this case, the Marquardt-Levenberg regularization is equivalent to the damping of the Gauss-Newton iteration (see Marquardt (1963)).

Recall that rounding errors destroy affine invariance, but not invariance under scale transformations. The regularization may also employ non-invariant terms which are as large as rounding errors. In particular, the scale invariant choice $B = D$ is reasonable, where D is the diagonal matrix whose entries coincide with the diagonal elements of $A^T A$. In computations with single-precision numbers, the regularization

$$A^T A \to A^T A + \mu D \quad \text{where } \mu \approx 10^{-4}$$

makes the condition number of the matrix smaller and improves the stability of the Gauss-Newton iteration. On the other hand, values of μ which are larger than 10^{-3} may already deteriorate the convergence rate. Specifically, the regularization cannot replace the line search in the iteration.

Further affine invariant techniques were considered by Deuflhard and Heindl (1979).

4. *Readjustment of linear parameters.* In many problems, an acceleration of the Gauss-Newton sequence is possible due to the structure of the nonlinear problem. Assume that the approximating functions depend linearly on some of the parameters. For example, the parameters for the specification of the numerator of rational expressions or the factors α_ν in exponential sums

$$\sum_{\nu=1}^{n} \alpha_\nu e^{\beta_\nu x}$$

are of this type.

Algorithm 5.2 is modified as follows: Immediately before the test (5.2) is performed, all "linear" parameters of u_t are recalculated such that $\|f-.\|$ becomes minimal. This is done by solving a linear approximation problem with a small number of parameters.

It has turned out (Braess 1970) that in many cases the computation of best exponential sums for uniform approximation with the Gauss-Newton method was only successful for $n > 3$, when the acceleration was employed.

Chapter IV. Methods of Global Analysis

In the cases where the best approximation is not unique, information about the number of nearest points is desired. Statements on that number (even on the number of local solutions) are global results, in contrast with the characterization of local best approximations, which by definition refers to the behaviour in a neighborhood of the element under consideration.

In this chapter we will provide a brief introduction into the application of methods of global analysis in nonlinear approximation. Typical of the global methods is the following: Detailed knowledges of the approximating family are only required for local features, while assumptions on global properties are made only for general topological properties, such as connectedness and bounded compactness. This has been different up to now where we have assumed for example the linearity, convexity, solarity or a global property on zeros. For a brief discussion of the global point of view in nonlinear analysis, the reader is referred to Browder (1983).

The main result in this chapter is the uniqueness theorem for Haar manifolds. This will make it possible to estimate the number of critical points for the approximation by sums of exponentials in Chapter VII. A different method will be presented in § 3, where an example of a family with a computable nontrivial bound for the number of critical points will be treated. The example also shows the role of the generic viewpoint for obtaining optimal bounds.

In Section 1C some results on local homeomorphisms are derived. Other general tools from topology, functional analysis and nonlinear analysis are given without proofs when they are used, e.g., the Schauder Fixed Point Theorem in II, § 3B, the Lipschitz Inverse Function Theorem in III, § 2C, Brouwer's theorem on the invariance of domain in III, § 3A, the Borsuk Antipodality Theorem in I, § 4A, and degree theory in VIII, § 6B.

§ 1. Preliminaries. Basic Ideas

A. Concepts for the Classification of Critical Points

The definition of a critical point refers to local properties of the approximating family. Nevertheless, the critical points in a set are not independent of each other. There are relations between them which refer to topological properties of the family. This is obvious for one-parameter sets.

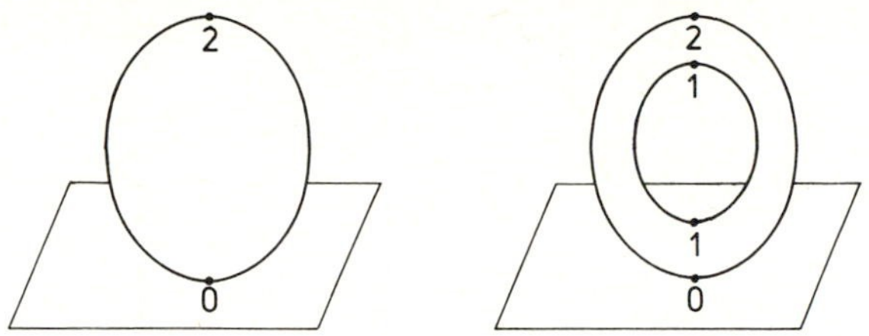

Fig. 20. Critical points with index of height function on a sphere and on a torus

1.1 Observation. Let T be an open interval in $\mathbb{R}$. Assume that $\varphi \in C(T)$ is not constant on any finite subinterval of T. Then the number of local minima differs from the number of local maxima by at most 1.

The treatment of multidimensional situations is the aim of several theories.

In the framework of Morse theory, an index is associated to each critical point of a C^2-function, (see Definition III.2.4). The theory provides relations between the number of critical points of all indices and the Betti numbers of the sets (Milnor 1963). Typically, each function on a 2-dimensional sphere has at least a (local) minimum and a (local) maximum, while each nondegenerate function on a 2-dimensional torus has two additional saddle points.

In nonlinear approximation problems the metric function is often not a C^2-function, and the Morse index of a critical point cannot be defined. In nonlinear Chebyshev approximation, the introduction of an index can be done on C^2-manifolds, see Jongen, Jonker and Twilt (1983). However, degeneracies, as in Example 1.6 below, mean that the Morse inequalities often do not provide as reasonable estimates as they do for *good* cases.

We will therefore describe a theory which has broader applications at the present time. Besides (strict) local best approximations, there is another kind of critical point which we distinguish: the critical points of mountain pass type.

The underlying idea is easily understood. Let x_0 be a strict local minimum of a C^1-function and let $x_1 \neq x_0$ be a point such that $h(x_1) \leq h(x_0)$. Consider any continuous curve γ from x_0 to x_1 and set $H(\gamma) = \sup\{h(x)\colon x \in \gamma\}$. The path for which $H(\gamma)$ is minimal (if it exists) contains a critical point which is a saddle point, or more precisely a critical point of mountain pass type, see Fig. 21.

For a rigorous treatment, the level sets for an approximation problem will be introduced. Consider the approximation from M in a normed linear space E. Given $f \in E$, and $\alpha \in \mathbb{R}$, the set

$$M^\alpha := \{u \in M\colon \|f - u\| \leq \alpha\} = M \cap B_\alpha(f) \tag{1.1}$$

is called the *level set* of the metric function $u \mapsto \|f - u\|$ for the level α.

1.2 Definition. Let u_0 be a critical point to f from M and let $\|f - u_0\| = \alpha$. Then u_0 is said to be a *critical point of mountain pass type* if there exists some ball $B_\varepsilon(u_0)$ with $\varepsilon > 0$ which contains no open neighborhood U of u_0 in M such

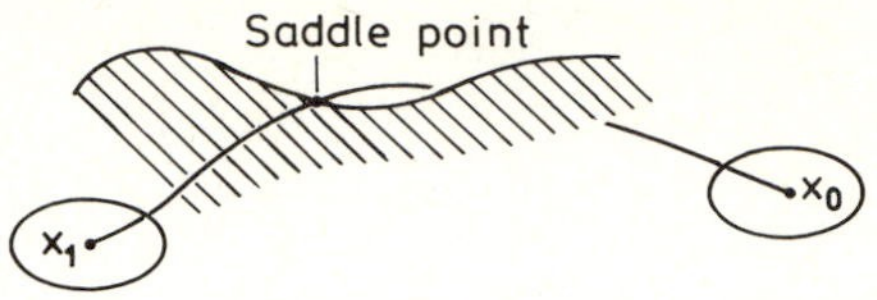

Fig. 21. Path with minimal maximum height for crossing the mountain

that the intersection

$$U \cap \mathring{B}_\alpha(f) \tag{1.2}$$

is connected.

These definitions of level sets and critical points of mountain pass type are specializations of obvious definitions for arbitrary real functions on a manifold. Usually, the existence of critical points of the mentioned type can be established under the following abstract compactness condition (Palais 1970) cf. Exercise 3.10.

1.3 Definition. Let $h \in C^1(M)$. Assume that any sequence (x_n) with

$$\sup_n |h(x_n)| < \infty \quad \text{and} \quad \lim_{n \to \infty} d_{x_n} h = 0$$

contains a convergent subsequence. Then h is said to satisfy the *Palais-Smale condition.*

1.4 Mountain Pass Theorem (Ambrosetti and Rabinowitz 1973). *Let $h \in C^1(M)$ satisfy the Palais-Smale condition. Assume that $x_0 \in M$ is a strict local minimum of h and that there is a continuous curve from x_0 to $x_1 \neq x_0$ with $h(x_1) \leq h(x_0)$. Then there is a critical point of mountain pass type.*

The commonly used tool for the proof of the mountain pass type is the construction of flows, (see Definition 2.3 below). Specifically, the flows (like gradient flows) generate orbits along which the functions of interest are decreasing.

Obviously, the Palais-Smale condition for the approximation problem may be considered to be somewhere between the approximative compactness and bounded compactness. Conversely, no compactness condition is required to show that in some Chebyshev sets there cannot exist critical points of mountain pass type.

1.5 Theorem (Vlasov 1973). *Any non-empty level set in a Chebyshev set in a uniformly convex Banach space is connected.*

B. An Example with Many Critical Points

Another application of level sets refers to the construction of functions with many *LBA*'s. If a level set is compact and consists of more than N components, then there are at least N local solutions.

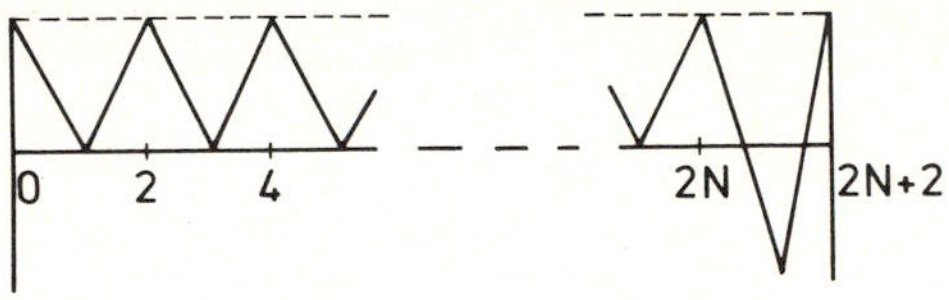

Fig. 22. Function with many local best Horner polynomials

1.6 Example. Let

$$M := \{u: u(x) = p^2(x) + c, p \in \Pi_2, c \in \mathbb{R}\}.$$

Given x, the evaluation of $u(x)$ is as expensive as the evaluation of a cubic polynomial as it requires the same number of arithmetic operations. M is the simplest nontrivial set of *Horner polynomials* introduced by Hettich and Wetterling (1973). Though $M \subset \Pi_4$ and M is a subset of a 5-dimensional linear set, functions with an arbitrarily large number of critical points for the uniform approximation in M may be constructed.

Given $N > 1$ define $f \in C[0, 2N + 2]$ by

$$f(x) = \begin{cases} |x - 2j + 1| & \text{for } 2j - 2 \leq x \leq 2j, j = 1, 2, \ldots, N, \\ 2|x - 2N - 1| - 1 & \text{for } 2N \leq x \leq 2N + 2. \end{cases}$$

The level set $M^{1-\varepsilon}$ contains only those elements for which the polynomial part p has no zero at $0, 2, 4, \ldots, 2N + 2$. Such a zero would imply that $c \geq \varepsilon$ and $\|f - u\| \geq u(2N + 1) - f(2N + 1) \geq \varepsilon + 1$, which is a contradiction. Therefore, the subsets

$$\{u \in M^{1-\varepsilon}: p \text{ has a zero in } (2j - 2, 2j) \text{ and a zero in } (2N, 2N + 2)\},$$

$$\text{with } j = 1, 2, \ldots, N,$$

are disjoint and moreover compact. If $0 < \varepsilon < (3N)^{-4}$, then the sets are nonempty, since the j-th set contains the function

$$u = \left[\frac{1}{4N^2}(x - 2j + 1)(x - 2N - 1)\right]^2 - \frac{1}{32N^4}.$$

Hence, f has at least N distinct local best approximations. □

C. Local Homeomorphisms

We conclude this section with some results on local homeomorphisms and covering spaces which are useful in the study of nonlinear interpolation and also, in a more indirect way, in nonlinear approximation.

1.7 Path Lifting Lemma. *Let $\Omega \subset \mathbb{R}^n$ be an open set whose closure is compact. Moreover let $f: \bar{\Omega} \to \mathbb{R}^n$ be a continuous map which is a local homeomorphism on Ω. Let $I = [0, 1]$. If $h: I \to \mathbb{R}^n$ is continuous, $h(0) = f(x_0) \in f(\Omega)$* and *$h(I)$ does not* *intersect* $f(\partial\Omega)$, *then there is a unique lifting* $g: I \to \Omega$ *such that* $h = f \circ g$ *and* $g(0) = x_0$.

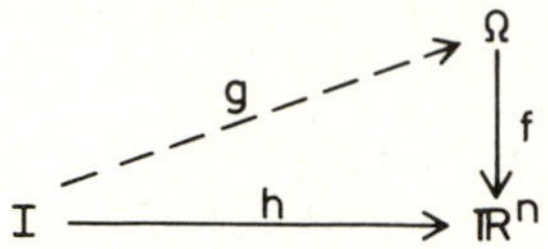

Proof. Assume that $0 \le t_0 < 1$ and g has been already constructed such that $h(t) = f \circ g(t)$ for $0 \le t \le t_0$. By the assumptions on f, there is an open neighborhood V of $g(t_0)$ in Ω which is mapped homeomorphically onto an open set $U \subset \mathbb{R}^n$ by f. By the continuity of h, we have $h[t_0, t_0 + \varepsilon) \subset U$ for some $\varepsilon > 0$. Hence, we can uniquely extend the domain of g to $[0, t_0 + \varepsilon)$ by setting $g(t) = f^{-1} \circ h(t)$ for $t_0 < t < t_0 + \varepsilon$.

Next assume that g has been defined for $0 \le t < t_0$. A subsequence of $\{g(t_0 - t_0/k)\}_{k \in \mathbb{N}}$ converges to some $x_1 \in \bar{\Omega}$. From the continuity of f, it follows that $f(x_1) = h(t_0) \in h(I)$. Hence $x_1 \in \Omega$. Set $g(t_0) = x_1$. Since f is a homeomorphism on some open neighborhood V of x_1 and $f(V)$ contains infinitely many points of $h([0, t_0))$, the construction of x_1 is independent of the chosen subsequence and g can be uniquely defined on $[0, t_0]$.

Hence the maximal domain of g is both open and closed in I. Therefore it equals I. □

The assumption that Ω is compact may be replaced by $\Omega = \mathbb{R}^n$ and $\lim_{\|x\| \to \infty} \|f(x)\| = \infty$. On the other hand, this condition cannot be omitted. This is obvious from the counter-example with $h: I \to \mathbb{R}$ defined by $h(x) = 1 - x$ and $f: [1, \infty) \to \mathbb{R}$ by $f(x) = 1/x$.

We note that the lifting theorem 1.7 is still true, if the domain I is replaced by I^m for some $m \ge 1$. To prove this, note that the boundary of the set $\{z \in I^m: \sum_i z_i \le \alpha\}$ is compact for $0 \le \alpha \le m$. Moreover, I^m is connected. A maximal extension argument yields the lifting map g. (For more details of the theory of covering spaces, see e.g. Croom 1978).

An immediate consequence is the following, cf. Ortega and Rheinboldt (1970), p. 136.

1.8 Norm-Coerciveness Theorem. *Let* $\Omega \subset \mathbb{R}^n$ *be an open, connected set. A local homeomorphism* $f: \Omega \to \mathbb{R}^n$ *with*

$$\lim_{x \to \partial\Omega} \|f(x)\| = \infty \tag{1.3}$$

is a homeomorphism of Ω *onto* $\mathbb{R}^n$.

Proof. The range of f is $\mathbb{R}^n$. Indeed, given $y \in \mathbb{R}^n$, it follows from (1.3) that there is an $x_0 \in \Omega$ for which the Euclidean distance $\|y - f(x_0)\|$ is minimal. Suppose that $f(x_0) \ne y$. Since f is a local homeomorphism, we have $f(x) = f(x_0) + \varepsilon(y - f(x_0))$ for some $x \in \Omega$ and some $\varepsilon > 0$. This contradicts the optimality of x_0 and proves that $f(x_0) = y$.

To show that f is globally one-one, assume that $f(x) = f(y)$ for some $x, y \in \Omega$. Since Ω is connected and locally path-connected, there is a continuous curve $(x_s)_{s\in I}$ with $x_0 = y$ and $x_1 = x$. The function $h: I^2 \to \mathbb{R}^n$, $(s,t) \mapsto (1-t)f(x_s) + tf(x)$ is then a continuous map for which

$$h(s,t) = f(x) \quad \text{for } (s,t) \in B := (\{0\} \times I) \cup (I \times \{1\}) \cup (\{1\} \times I). \tag{1.4}$$

The assumptions on f allow us to lift the map h. There is a continuous map $g: I^2 \to \Omega$ such that $h = f \circ g$ and $g(0,0) = y$. Since $g(B)$ is connected, it follows from (1.4) that g is constant on B. Hence, $g(B) = \{y\}$. Finally there is a unique lifting g_0 of $h_0: I \to \mathbb{R}^n$, $s \mapsto f(x_s)$ with $g_0(0) = y_0$. Therefore, $x_s = g_0(s) = g(s,0)$ for $s \in I$. Hence, $x = g(1,0) \in g(B)$ and $x = y$. □

From the special case $\Omega = \mathbb{R}^n$, one obtains

1.9 Corollary. *A local homeomorphism $f: \mathbb{R}^n \to \mathbb{R}^n$ with $\lim_{\|x\|\to\infty} \|f(x)\| = \infty$ is a homeomorphism of $\mathbb{R}^n$ onto $\mathbb{R}^n$.*

Exercises

1.10. Consider the family of Horner polynomials in Example 1.6. For which elements is the tangent cone a 4-dimensional tangent space, and for which points does the tangent space satisfy the Haar condition?

1.11. Let M be a connected n-dimensional Haar submanifold of $C[\alpha,\beta]$ without boundary for which the following (restrictive) compactness property holds: Each sequence of functions in M, which are bounded at n points $\{x_i\}_{i=1}^n \subset [\alpha,\beta]$ contains a subsequence which converges to an element from M. Prove that M is varisolvent and that given $\alpha \le x_1 < x_2 < \cdots < x_n \le \beta$ and $y \in \mathbb{R}^n$, the interpolation problem $u(x_i) = y_i$, $i = 1, 2, \ldots n$ always has a (unique?) solution u in M.

1.12. Let $f: \Omega \to X$ be a continuous map. An open subset U of X is said to be *evenly covered* by f if $f^{-1}(U)$ is the disjoint union of open subsets of Ω each of which is mapped homeomorphically onto U by f. Let the conditions of Lemma 1.7 prevail. Show that each $y \in h(I)$ has an open neighborhood in X evenly covered by f.

1.13. Let $\bar{\Omega}$ be a compact set in n-space which is homeomorphic to the unit ball, and let $f: \bar{\Omega} \to \mathbb{R}^n$ be continuous. Assume that the restriction of f to Ω is a local homeomorphism and that the restriction of f to $\partial\Omega$ is one-one. Then f is a homeomorphism.

§2. The Uniqueness Theorem for Haar Manifolds

The aim of this section is the

2.1 Uniqueness Theorem for Haar Manifolds (Wulbert 1971 and Braess 1973a). *Each connected boundedly compact Haar submanifold M of $C[\alpha,\beta]$ is a Chebyshev set. Moreover, to each $f \in C[\alpha,\beta]$ there is no other critical point in M than the unique nearest point.*

This theorem was first given by Wulbert (1971) for manifolds without boundaries and under the more restrictive compactness assumption mentioned in

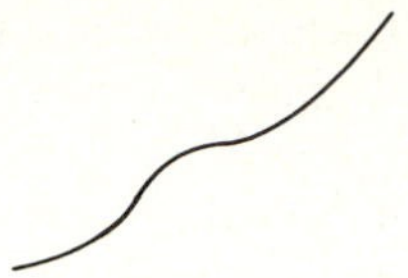

Fig. 23. Haar submanifold of $\mathbb{R}^2$

Exercise 1.11. The extension to manifolds with boundaries which are only boundedly compact and the proof with critical point theory is due to Braess (1973a). The difference in the compactness assumptions is elucidated by an example.

Example. The set

$$M = \left\{u(x) = \operatorname{tg}\left(\frac{\pi}{4}ax\right) : -1 < a < 1\right\} \tag{2.1}$$

is a 1-dimensional connected, boundedly compact Haar submanifold of $C[1,2]$. The sequence of elements in M which correspond to $a_n = 1 - n^{-1}$, $n = 1, 2, \ldots$ is bounded at $x = 1$, but does not contain a convergent subsequence.

In the particular case of a 1-dimensional manifold in $\mathbb{R}^2$ endowed with the supremum norm, Theorem 2.1 is obvious (see Fig. 23). If the tangent spaces satisfy the Haar condition, i.e., if the tangent is never parallel to either the x_1 or the x_2-axis, then each coordinate is a strictly monotone function on the curve. We must exclude the possibility that the manifold consists of two or more separated curves. This is done by the connectedness assumption.

Moreover, one cannot do without a compactness assumption. Consider the two-dimensional sphere in 3-space endowed with the uniform norm. After eliminating the 6 points $(\pm 1, 0, 0)$, $(0, \pm 1, 0)$, $(0, 0, \pm 1)$, one gets a path-connected locally compact Haar submanifold of $\mathbb{R}^3$ which is obviously not a uniqueness set.

A. The Deformation Theorem

The essential tools of the theory are deformations between level sets (1.1). They will be constructed by glueing together maps which are defined locally.

2.2 Lemma. *Let M be a C^1-manifold in a normed linear space E. Assume that $v \in M$ is not a critical point to $g \in E$. Then there exist a neighborhood U of v in M, two numbers $c, \delta > 0$ and a homotopy $\psi: I \times U \to M$ such that*

$$\begin{gathered} \psi(0, u) = u, \\ \|f - \psi(t, u)\| \leqq \|f - \psi(s, u)\| - c(t - s), \quad 0 \le s < t \le 1, \quad u \in U, \end{gathered} \tag{2.2}$$

whenever $\|f - g\| < \delta$.

Proof. Since v is not a critical point, there is a tangent vector $h \in C_v M$ such that

$$\|g - v - h\| < \|g - v\|. \tag{2.3}$$

Let $F: C \to M$ be a parametrization with $v = F(0)$. The set of elements in $C_v M$ for which (2.3) holds is open. Therefore, we may assume that $h = d_0 F b$ for some $b \in C$.

Set $c = \frac{1}{6}[\|g - v\| - \|g - v - h\|]$. By continuity, there is a neighborhood V_1 of 0 in C such that

$$\left.\begin{aligned} \|g - F(\tilde{a})\| - \|g - F(\tilde{a}) - d_0 F(b - a)\| &> 4c \\ \|d_{\tilde{a}} F(b - a) - d_0 F(b - a)\| &< c \end{aligned}\right\} \quad a, \tilde{a} \in V_1. \qquad \begin{aligned}(2.4)\\(2.5)\end{aligned}$$

From (2.4) and since $\|f - g\| < \delta$, where $\delta < c$, it follows that

$$\|f - F(a)\| - \|f - F(a) - d_a F(b - a)\| > 2c, \qquad a \in V_1. \tag{2.6}$$

Recall that C is convex. Choose $\delta_1 > 0$ and a neighborhood V_2 of 0 in C such that

$$a + \tau(b - a) \in V_1, \text{ for } a \in V_2, 0 \leq \tau \leq \delta_1.$$

With this and $0 \leq s \leq t \leq \delta_1$, we compute

$$\begin{aligned} \|f - F(a + t(b - a))\| &= \left\| f - F(a + s(b - a)) - \int_s^t d_{a+\tau(b-a)} F(b - a)\, d\tau \right\| \\ &\leq \|f - F(a + s(b - a)) - (t - s) d_0 F(b - a)\| \\ &\quad + \left\| \int_s^t (d_{a+\tau(b-a)} F - d_0 F)(b - a)\, d\tau \right\|. \end{aligned}$$

When decomposing the first term in the standard way and using (2.5) and (2.6) after inserting $a + s(b - a)$ for a, we get

$$\begin{aligned} \|f - F(a + t(b - a))\| &\leq \|f - F(a + s(b - a))\| - (t - s)\cdot 2c + (t - s)c \\ &\leq \|f - F(a - s(b - a))\| - (t - s)c. \end{aligned} \tag{2.7}$$

Now we can define a homotopy ψ on $I \times U$ with $U = F(V_2)$ by transforming the shift on the parameter set back to the manifold:

$$u \overset{F^{-1}}{\longmapsto} a \longmapsto a + t\delta_1(b - a) \overset{F}{\longmapsto} \psi(t, u)$$

which in closed form becomes $\psi(t, u) = F[F^{-1}(u) + t\delta_1(b - a)]$. After replacing $c \cdot \delta_1$ by c, we get (2.2) from (2.7). □

Note that ψ has the semigroup property, i.e.

$$\psi(s, \psi(t, u)) = \psi(s + t, u)$$

whenever $u, \psi(t, u) \in U$ and $0 \leq s \leq s + t \leq 1$. This property will be lost when the homotopy will be modified to become a flow on M (cf. Exercise 2.9).

2.3 Definition. A homotopy $\phi: I \times M \to M$ is said to be a (*descending*) *flow* if

$$\phi(0,u) = u \quad \text{for } u \in M$$

and

$$\|f - \phi(t,u)\| \leq \|f - \phi(s,u)\|, \qquad 0 \leq s < t \leq 1, \qquad u \in M, \tag{2.8}$$

with equality only if $\phi(t,u) = \phi(s,u)$. The smallest closed subset $N \subset M$ such that $\phi(1,u) = u$ holds for $u \in M \backslash N$ is called the support of ϕ. Moreover, $\phi_t\colon M \to M$, $0 \leq t \leq 1$ is defined by $\phi_t(u) := \phi(t,u)$.

Now we are in a position to construct local flows.

2.4 Lemma. *Let M be a C^1-manifold in E. Assume that v is not a critical point to f and that $\alpha = \|f - v\|$. Then, given a neighborhood U of v, there is a flow ϕ with support in $\bar{U}$ such that $\|f - \phi(1,v)\| < \|f - v\|$. Moreover, there is a truncated flow $\tilde{\phi}$ with support in $\overline{U \backslash M^\alpha}$ such that for each $u \in M$ and $0 \leq t \leq 1$*

$$\textit{either} \quad \tilde{\phi}(t,u) = \phi(t,u) \quad \textit{or} \quad \tilde{\phi}(t,u) \in M^\alpha. \tag{2.9}$$

Proof. Let $\psi\colon I \times U_1 \to M$ be the homotopy given in Lemma 2.2. Since $\|f - \psi(1,v)\| < \|f - v\|$, there is a neighborhood $U_3 \subset U \cap U_1$ of v such that

$$\|f - \psi(1,u)\| < \|f - v\|, \qquad u \in U_3. \tag{2.10}$$

Then we have $\psi(t,u) \in U$ for $u \in U_3$ and $0 \leq t \leq t_0$ with some $t_0 > 0$. Choose a cutting-off function $\chi \in C(M)$ satisfying

$$\chi(v) = t_0,$$

$$0 \leq \chi(u) \leq t_0 \quad \text{for } u \in U_3,$$

$$\chi(u) = 0 \quad \text{for } u \in M \backslash U_3.$$

Obviously,

$$\phi(t,u) = \begin{cases} \psi(\chi(u) \cdot t, u), & \text{if } u \in U_3, \\ u & \text{otherwise} \end{cases}$$

defines a flow as stated in the lemma. Moreover, it follows from the compactness of I, (2.2), and (2.10) that $s(u) = \min\{t\colon t \in I, \psi(t,u) \in M^\alpha\}$ is continuous in U_3. Therefore

$$\tilde{\phi}(t,u) = \begin{cases} \psi(\min\{s(u), \chi(u)\} \cdot t, u) & \text{for } u \in U_3, \\ u & \text{otherwise} \end{cases}$$

is a suitable truncation of ϕ. □

A set B is said to be a *deformation retract* of A if there is a continuous mapping from A into $B \subset A$ which is homotopic to the identity map $A \to A$. In this case, A and B have the same homotopy type. The following theorem means that the homotopy type of the level sets does not change unless one passes a level which contains a critical point.

2.5 Deformation Theorem (Noncritical Interval Theorem). *Let M be a C^1-manifold in E and let $\alpha < \beta$. If $N := \{u \in M: \alpha \leq \|f - u\| \leq \beta\}$ is compact and contains no critical points, then M^α is a (strong) deformation retract of M^β.*

Proof. Given $v \in N$, we may choose a neighborhood U_v and a flow ϕ_v such that

$$\|f - \phi_v(1, u)\| < \|f - u\| \quad \text{for } u \in U_v. \tag{2.11}$$

In particular, if $v \notin M^\alpha$ we may choose ϕ_v with a support which does not intersect M^α. Let $\{U_{v_i}, i = 1, 2, \ldots, m\}$ be a finite cover of N. We define retractions r_i: $M \to M$, $i = 0, 1, 2, \ldots, m$, recursively.

$$\begin{aligned} r_0(u) &= u, \\ r_i(u) &= \phi_{v_i}(1, r_{i-1}(u)), \qquad i = 1, 2, \ldots, m. \end{aligned} \tag{2.12}$$

Since the retractions r_i are generated by flows, we have

$$\|f - r_m(u)\| \leq \cdots \leq \|f - r_i(u)\| \leq \|f - r_{i-1}(u)\| \leq \cdots \leq \|f - u\|. \tag{2.13}$$

Let $u \in U_{v_j}$. Then either $u \neq r_{j-1}(u)$ or $u = r_{j-1}(u) \neq r_j(u)$. Thus, given $u \in N$, there is an index $k \leq m$ such that $r_{k-1}(u) \neq r_k(u)$. From the definition of flows, it follows that the strict inequality holds in (2.13) for some $i \leq j$. Since N is compact,

$$\varepsilon := \inf\{\|f - u\| - \|f - r_m(u)\|: u \in N\} > 0.$$

Note that $r_m(M^\beta) \subset M^{\max\{\beta - \varepsilon, \alpha\}}$. Choose k such that $k\varepsilon \geq \beta - \alpha$. Then

$$(r_m)^k(M^\beta) \subset M^\alpha.$$

Hence, M^α is a retract of M^β. By using the second statement of Lemma 2.3, we see that M is a strong retract, i.e., the elements of M^α are not moved by an appropriate deformation. To see this, we modify (2.12):

$$\tilde{r}_0(u) = u,$$

$$\tilde{r}_i(u) = \begin{cases} \varphi_{v_i}(1, \tilde{r}_{i-1}(u)), & \text{if } \|f - v_i\| > \alpha, \\ \tilde{\varphi}_{v_i}(1, \tilde{r}_{i-1}(u)), & \text{if } \|f - v_i\| = \alpha, \end{cases} \qquad i = 1, 2, \ldots, m.$$

The elements of M^α are invariant under the retraction $(\tilde{r}_m)^k$. On the other hand, by construction either $r_m(u) \in M^\alpha$ or $\tilde{r}_m(u) = r_m(u)$. Thus, in this case we also have $(\tilde{r}_m)^k(M^\beta) \subset M^\alpha$. □

B. The Mountain Pass Theorem

A local best approximation is said to be a *strict LBA* if it is the unique nearest point to f from some neighborhood U in M.

2.6 Mountain Pass Theorem for Nonlinear Approximation (Braess 1973a). *Let M be a C^1-manifold in a normed linear space E and let $\alpha < \beta$. Assume that M^β is connected and that $N := \{u \in M: \alpha \leq \|f - u\| \leq \beta\}$ is compact. If N contains two* *strict local best approximations to f, then there is a critical point in N which is not a strict local best approximation. If moreover, all critical points in N are isolated, then N contains a critical point of mountain pass type.*

Remark. If a Morse index can be associated to a critical point and if the point is not degenerate, then the theorem may be stated in terms of indices. In that case, local minima are characterized by an index 0 and critical points of mountain pass type by an index 1 (see Problem 2.12). Since the notation of mountain pass was not commonly used in 1973, Theorem 2.6 was originally called the Non-zero-index theorem.

Proof of Theorem 2.6. (1) We will prove the existence of a critical point as stated in the first part under the weaker assumption that N contains a strict LBA v_1 and an element v_2 with $\|f - v_2\| \leq \|f - v_1\|$.

The set of all critical points in a C^1-manifold is closed. Therefore, the metric function $\|f - \cdot\|$ attains its maximum on the set of critical points in N, say at $v \in N$. We may assume that v is a strict LBA, because otherwise the proof is complete. There is a neighborhood U of v such that

$$\|f - u\| > \|f - v\| \text{ for } u \in U \backslash \{v\}.$$

Choose $\delta > 0$ such that the intersection of M and the sphere $S_\delta(v) := \{u \in U: \|u - v\| = \delta\}$ is contained in U. Then

$$\varepsilon := \frac{1}{2} \min_{u \in M \cap S_\delta(v)} \{\|f - u\| - \|f - v\|\} > 0.$$

Since N is connected and locally path-connected, there is a curve with endpoints v and v_2. Since this curve passes $S_\delta(v)$, we have $\alpha + 2\varepsilon \leq \beta$. By construction, the set $\{u \in M: \alpha + \varepsilon \leq \|f - u\| \leq \beta\}$ contains no critical points and is compact. By the deformation theorem, $M^{\alpha+\varepsilon}$ is a retract of M^β. In particular, the retraction sends the curve from v to v_2 in N to a curve in $M^{\alpha+\varepsilon}$. Since the curve must cross $S_\delta(v)$, this is a contradiction.

(2) If all critical points in N are isolated, there are only finitely many of them. Let γ be the smallest number such that given $\varepsilon > 0$ there is a curve in $M^{\gamma+\varepsilon}$ connecting the strict LBA v_1 with v_2. By the deformation theorem, there exists a critical point v_3 with $\|f - v_3\| = \gamma$. For convenience, we assume that v_3 is the only critical point at that level. With the arguments as in the proof of Theorem 2.5, we may construct a retraction from $M^{\gamma+\varepsilon}$ into $M^{\gamma-\varepsilon} \cup U$, with U being a given neighborhood of v_3. Note that the curve enters U at a point below the level γ and the same holds for the point where it leaves U. By (1.2), the part of the curve in U could be replaced by a curve below the critical level, if v_3 were not of mountain pass type. □

Since all critical points in a Haar submanifold of $C(X)$ are strongly unique LBA's, the uniqueness theorem 2.1 is a corollary of the mountain pass theorem. □

C. Perturbation Theory

There can be more than one critical point in a (non-compact) Haar submanifold M of $C(X)$. Nevertheless, the critical points are isolated. Indeed a little more is true. Namely:

2.7 Local Continuity Theorem. *Let v be a local best approximation to f from a Haar submanifold $M \subset C(X)$, X being a compact interval. Then there is an open neighborhood U of v in M and $r > 0$ such that to each $g \in B_r(f) \subset C(X)$ there is exactly one critical point in U.*

Proof. Let v be a strongly unique best approximation to f from $U_1 := M \cap B_\delta(v)$, and let $c > 0$ be the constant of strong uniqueness. Since C^1-manifolds are locally compact, we may assume that U_1 is compact. Set $\alpha = \|f - v\|$. Since C^1-manifolds are locally path-connected, there is a path-connected neighborhood U of v in $M \cap B_{c\delta/2}(v)$. By definition, U contains $M \cap B_{2r/c}(v)$ for some $0 < r \leq c^2\delta/4$.

Let $\|f - g\| < r$. Then $d(g, U_1) \leq \|g - v\| \leq \|f - v\| + \|f - g\| \leq \alpha + r$. On the other hand, from the strong uniqueness of v, we conclude that $\|g - u\| \geq \|f - v\| + c\,\|v - u\| - \|g - f\| > \alpha + r$ for each $u \in U_1 \setminus \mathring{B}_{2r/c}(v)$. Hence the best approximation to f from the compact set U_1 belongs to U.

In the same way we conclude that a level set for the metric function $u \mapsto \|g - u\|$ is contained in U_1. Set $\beta = \alpha + 3c\delta/4$. For any $u \in U$, we have $\|g - u\| \leq \|g - f\| + \|f - v\| + \|v - u\| < \beta$. On the other hand, for any $u \in U_1 \setminus \mathring{B}_\delta(v)$, we have $\|g - u\| \geq \|f - v\| + c\,\|v - u\| - \|g - f\| \geq \alpha + c\delta - r > \beta$. Hence, the connected component of $M \cap B_\beta(g)$ which contains U belongs to U_1 and is compact. From the mountain pass theorem, we conclude that there is only one critical point to g in U. □

By setting $r = \|f - g\| + \varepsilon$, we get $\|v - P_{U_1}(g)\| \leq 2c^{-1}\|f - g\|$ as in Freud's theorem for the linear case. Moreover, if M has no boundary, a simpler proof is possible (see Exercise III.4.10).

An extension of the preceding theorem refers to the situation where not only the given function $f \in C(X)$ but also the manifold is perturbed.

2.8 Perturbation Lemma. *Assume that M and N are Haar manifolds in $C(X)$, X compact. Assume that v is a local best approximation to f from M. Let U and r be as in Theorem 2.7. Also assume that there is a homeomorphism Ψ from U to an open set V in N such that*

$$\|\Psi(u) - u\| < \frac{r}{2} \quad \text{for } u \in U.$$

Then, to each $g \in B_{r/2}(f) \subset C(X)$ there is exactly one critical point in $V \subset N$.

For the proof of the perturbation lemma observe that $d(g, \Psi(U)) \leq d(f, U) + r/2 + r/2$. Therefore the proof can proceed with the same arguments as the proof of the preceding theorem. □

Exercises

2.9. Assume that v is not a critical point to f from the C^1-manifold M. Referring to the notation as in Lemma 2.4 consider the differential equation in the parameter set C;

$$\frac{d}{dt}a_t = \chi(F(a_t))(b - a_t), \qquad t \geq 0$$

$$a_0 = a.$$

Show that the orbits do not leave C and that $\phi(t, u) = (F^{-1}u)_t$ defines a descending flow which has the semigroup property as stated in Lemma 2.4.

2.10. Verify that the flow constructed in the preceding problem satisfies

$$\|f - \phi(t,u)\| \leq \|f - \phi(s,u)\| - c\|\phi(t,u) - \phi(s,u)\|, \qquad 0 \leq s \leq t \leq 1$$

with some $c > 0$.

2.11. If $F(0)$ is a nondegenerate critical point of a C^2-function on an n-dimensional manifold, then by the Morse lemma (Milnor 1963), there are local coordinates such that

$$f(x) = \alpha - x_1^2 - x_2^2 - \cdots - x_\lambda^2 + x_{\lambda+1}^2 + \cdots + x_n^2. \tag{2.13}$$

Here λ is the index. Prove that there is a retraction from $\{x \in B_1 : f(x) < \alpha\}$ onto S^λ for $\lambda \geq 1$.

2.12. Consider the function (2.13). Show that $x = 0$ is a critical point of mountain pass type if and only if $\lambda = 1$.

2.13. Endow $\mathbb{R}^n$ with the uniform norm. What is the maximal number of critical points which exist when one considers the approximation from $M = \{x = (x_1, x_2, \ldots, x_m): \sum_i x_i^2 = 1\}$?

2.14. Show that each approximatively compact Chebyshev set M in a normed linear space E is a strong retract of E.
Hint: The metric projection P_M is continuous.

§3. An Example with One Nonlinear Parameter

The estimation of the number of critical points is often performed in an inductive way. Typical difficulties in that procedure will be elucidated by a transparent example. We consider the families of exponential sums with confluent characteristic numbers

$$E_n^c := \left\{u\colon u(x) = \sum_{j=0}^{n-1} a_j x^j e^{tx}, a_j, t \in \mathbb{R}\right\}. \tag{3.1}$$

This approximation problem may be reduced to a problem with one (nonlinear) parameter. Then, by making use of Observation 1.1, the following result is deduced:

3.1 Theorem (Braess 1975a). *To each $f \in C[\alpha, \beta]$, there are at most n local best approximations from E_n^c.*

The analogous family of rational functions with confluent poles is considered in Exercise 3.9.

The elements in the family E_n^c are characterized by one nonlinear parameter. The linear subsets $E_n^c(t)$ that contain those elements for which the parameter t in the natural parametrization (3.1) is fixed, are Chebyshev sets. The family E_n^c is a one-parameter collection of (linear) Chebyshev sets.

Given $t \in \mathbb{R}$, set $u_t = P_{E_n^c(t)} f$ and

$$r(t) := \|f - u_t\| = d(f, E_n^c(t)). \tag{3.2}$$

The investigation will be done in four steps. 1. Study of the manifold $E_n^c \backslash E_{n-1}^c$. 2. Reduction to one parameter. 3. Evaluation of a simple and crude bound. 4. Improvement of the bound by the elimination of exceptional cases.

A. The Manifold $E_n^c \backslash E_{n-1}^c$

Let $n \geq 1$. The natural parametrization

$$F\colon (a, t) \mapsto \sum_{j=0}^{n-1} a_j x^j e^{tx}$$

is obviously one-one, if one excludes the points in $\mathbb{R}^{n+1}$ which are mapped to the zero function. Then F^{-1} is also continuous. Therefore, $E_n^c \backslash \{0\}$ is a topological manifold (without boundary).

The derivatives of F are

$$\begin{aligned} \frac{\partial F}{\partial a_j} &= x^j e^{tx}, \qquad j = 0, 1, \ldots, n-1, \\ \frac{\partial F}{\partial t} &= \sum_{j=0}^{n-1} a_j x^j \cdot x e^{tx}. \end{aligned} \tag{3.3}$$

If $a_{n-1} \neq 0$, the span of the derivatives contains all functions of the form $\sum_{j=0}^{n} \delta_j x^j e^{tx}$, $\delta_j \in \mathbb{R}$, $j = 0, 1, \ldots, n$. Thus dF is injective at $F^{-1}(u)$ whenever $u \in E_n^c \backslash E_{n-1}^c$, and the tangent space is an $(n+1)$-dimensional Haar subspace of $C[\alpha, \beta]$. Consequently, $E_n^c \backslash E_{n-1}^c$ is an $(n+1)$-dimensional Haar submanifold of $C[\alpha, \beta]$. Here, the convention $E_0^c = \{0\}$ has been adopted.

It is crucial to note that on the other hand E_n^c itself is not a C^1-submanifold of $C[\alpha, \beta]$. The *order* of a confluent exponential sum u is defined to be the smallest number k such that $u \in E_k^c$.

3.2 Lemma.
(1) *For $u \in E_n^c \backslash E_{n-1}^c$ the following are equivalent:*
  1°. *u is a strongly unique local best approximation to f from E_n^c.*
  2°. *There is an alternant of length $n + 2$ to $f - u$.*
(2) *If $u(t) := P_{E_n^c(t)} f$ has the order $k < n$, then u is a strongly unique local best approximation from E_k^c.*

Proof. Since $E_n^c \backslash E_{n-1}^c$ is a Haar submanifold of $C[\alpha, \beta]$, the first part is immediate from Theorem III.4.8. To prove the second part, we note that u_t is the nearest point from an n-dimensional Haar subspace. There is an alternant of length $n + 1 \geq k + 2$ to $f - u$. We now get local optimality from the characterization in the first part of the lemma. □

Consider a solution u_t of a linear subproblem and assume that $u_t \notin E_{n-1}^c$. The essential step in the analysis is the study of the subsets $E_n(t+) := \bigcup_{\tau \geq t} E_n^c(\tau)$ and $E_n^c(t-) := \bigcup_{\tau \leq t} E_n^c(\tau)$ which contain the elements with a spectral number $\tau \geq t$ and $\tau \leq t$, respectively. The corresponding tangent cones have the form

$$C_{u_t} E_n^c(t+) := C = \left\{ h(x) = \sum_{j=0}^{n-1} \tilde{\delta}_j \frac{\partial F}{\partial a_j} + \tilde{\delta}_n \frac{\partial F}{\partial t} : \tilde{\delta}_n \geq 0 \right\}$$

$$= \left\{ h(x) = \sum_{j=0}^{n} \delta_j x^j e^{tx} : s\delta_n \geq 0 \right\}, \tag{3.4}$$

$$C_{u_t} E_n^c(t-) = -C,$$

where $s = \operatorname{sgn} a_{n-1}$. The leading coefficient δ_n is sign-restricted. In the following, the symbol δ_n will always refer to the representation of the tangent rays in (3.4). The tangent cones are Haar cones.

3.3 Proposition. *Assume that $u_t = P_{E_n^c(t)} f$ has order exactly n. Then u_t is a strongly unique local best approximation in $E_n^c(t+)$ or in $E_n^c(t-)$.*

Proof. First we note that 0 is the best approximation to $f - u$ either from C or from $(-C)$. Indeed, suppose to the contrary that there are $h_1 \in C$ and $h_2 \in (-C)$ such that

$$\|f - u_t - h_i\| < \|f - u_t\|, \qquad i = 1, 2. \tag{3.5}$$

The segment $[h_1, h_2]$ intersects $E_n^c(t) = C \cap (-C)$, say in h_0. From (3.5) we conclude that $\|f - u_t - h_0\| < \|f - u_t\|$, which contradicts the optimality of u_t.

Therefore 0 is the nearest point to $f - u_t$ from the tangent cone to one of the restricted sets. From Theorem I.3.10, one obtains strong uniqueness and by Corollary III.1.12, u_t is a strongly unique *LBA* from the corresponding subset. □

B. Reduction to One Parameter

First, some properties of the one-parameter mappings defined in (3.2) will be summarized.

3.4 Remarks.

(1) The mappings $t \mapsto u_t$ and $t \mapsto r(t)$ are continuous.

(2) If u_t is a strict local best approximation to f from E_n^c, then r has a strict local minimum at t.

(3) If $u_t \notin E_{n-1}^c$, then r does not have a local maximum at t.

Proof. (1) Though the continuity of the mapping $t \mapsto u_t$ may be deduced from Theorem 2.8, we prefer a direct proof.

Set $u_t = p_t e^{tx}$ with $p_t \in \Pi_{n-1}$. Moreover, note that $\|u_t\| \le 2\|f\|$, since it is optimal in a linear space.

Given $\varepsilon > 0$ and $t \in \mathbb{R}$, we have

$$\|1 - e^{(t-\tau)x}\| < \varepsilon, \qquad \|1 - e^{(\tau-t)x}\| < \varepsilon.$$

for all τ in some neighborhood of t. By making use of the triangle inequality and the optimality of p_t and p_τ for some linear problems, we obtain

$$\begin{aligned}\|f - p_\tau e^{tx}\| &\le \|f - p_\tau e^{\tau x}\| + \|p_\tau e^{\tau x}(1 - e^{(t-\tau)x})\| \\ &\le \|f - p_t e^{\tau x}\| + 2\|f\| \cdot \varepsilon \\ &\le \|f - p_t e^{tx}\| + \|p_t e^{tx}(1 - e^{(\tau-t)x}\| + 2\|f\|\varepsilon \\ &\le \|f - p_t e^{tx}\| + 4\|f\|\varepsilon.\end{aligned}$$

From strong uniqueness for the approximation in the linear space $\{q \cdot e^{tx}: q \in \Pi_{n-1}\}$, we conclude that

$$\|p_\tau - p_t\| \le \text{const} \cdot 4\|f\|\varepsilon.$$

This implies the continuity of the mapping $t \mapsto u_t = p_t e^{tx}$. As a byproduct it follows that $r(t) = \|f - u_t\|$ is also continuous.

(2) Assume that u_t is a strict *LBA* to f from E_n^c. Then by definition there is no equally good element in some neighborhood of u_t. From (1), we know that this neighborhood contains u_τ whenever τ is close to t. Hence, $r(\tau) > r(t)$ if $\tau \ne t$ is close to t, and r has a strict local minimum at t.

(3) Let u_t have the exact order n. With the same arguments as in Part (2), we get from Proposition 3.3

$$r(\tau) > r(t)$$

for $\tau > t$ or for $\tau < t$ in some neighborhood of t. Hence, t is not a local maximum of r. □

Now we define

$$c_n = \sup_{f \in C[\alpha,\beta]} \{\text{number of } LBA\text{'s to } f \text{ from } E_n^c\}.$$

From the preceding investigation we conclude that

$$c_1 = 1, \qquad c_n \le 1 + \sum_{k<n} c_k, \qquad n = 2, 3, \ldots \tag{3.6}$$

Indeed, $E_1^c = E_1$ is a varisolvent family and a uniqueness set, (see Example III.3.2). Next, let $n \ge 2$ and count the local maxima of r. By Remark 3.4(3) and Lemma 3.2(2), r has a local maximum at $t = \tau$ only if u_τ is a local solution for some E_k^c where $k < n$. Hence the number of local maxima is at most $\sum_{k<n} c_k$. Now Observation 1.1 implies (3.6). Moreover, it follows that r is not constant on any finite interval and that

$$c_n \le 2^{n-1}, \qquad n = 1, 2, \ldots. \tag{3.7}$$

C. Improvement of the Bounds

The bound (3.7) is not optimal because some exceptional cases are treated in a too conservative manner. The number of local solutions is not the same for all $f \in C[\alpha, \beta]$, and therefore the number cannot be a continuous function of f. So it is natural that exceptional cases occur.

The remedy for this is the consideration from the generic viewpoint (see Definition II.1.5). First, the problem is treated for a residual set of functions in $C[\alpha, \beta]$. Then the exceptional cases are excluded by definition, which thus cannot spoil the bound. Afterwards we verify via perturbation arguments that the improved bound is even correct for each f in $C[\alpha, \beta]$, and not just for a residual set.

Note that a local maximum of $r(t)$ may only arise from a solution $\hat{u}$ in E_k^c for $k \leq n - 2$, if $f - \hat{u}$ has an alternant with more than the $k + 2$ points which are necessary for a solution in E_k^c. We will eliminate such cases with an exceptional length without changing the maximal number of solutions. To this end, consider the set

$$\mathrm{Alt}_n(f) := \{u \in E_n^c\text{: there is an alternant of length} \geq n + 2 \text{ to } f - u\}.$$

3.5 Assertion. *Assume that* $c_n < \infty$. *Then the set of functions* f *for which* $\mathrm{Alt}_n(f)$ *does not intersect* E_{n-1}^c *is dense in* $C[\alpha, \beta]$.

Proof. Given f, let N be the number of elements of $\mathrm{Alt}_n(f) \backslash E_{n-1}^c$. Furthermore, assume that $u_t = p_t e^{tx} \in E_{n-1}^c \cap \mathrm{Alt}_n(f)$.

Let $g(x) = f(x) + \varepsilon x^{n-1} e^{tx}$ and $v^{(N+1)} = (p_t + \varepsilon x^{n-1}) e^{tx}$ for $\varepsilon > 0$. By construction $v^{(N+1)} \in \mathrm{Alt}_n(g)$. Moreover, it follows from Lemma III.1.13, i.e. the continuity lemma, that we have N distinct LBA's to g from $E_n^c \backslash E_{n-1}^c$ which are close to the given N LBA's to f. Therefore we have increased the number of critical points at least by one.

This procedure may be repeated if there is an $u \in E_{n-1}^c \cap \mathrm{Alt}_n(g)$. Each repetition produces an additional LBA. Since the number is bounded by c_n, the process must stop after at most c_n steps.

Since each perturbation may be arbitrarily small, the density property is established. □

Proof of Theorem 3.1. By (3.6), the theorem is true for $n = 1$. Assume that it has already been proved for $n - 1$. Given $f \in C[\alpha, \beta]$, let g be a function which will be fixed later. Set

$$r(t) = d(f, E_n^c(t)), \qquad \tilde{r}(t) = d(g, E_n^c(t)).$$

From the triangle inequality, one gets $|r(t) - \tilde{r}(t)| \leq \|f - g\|$, $t \in \mathbb{R}$. Therefore $\tilde{r}$ has at least as many local minima as r, whenever $\|f - g\|$ is sufficiently small. Since $c_{n-1} < \infty$, by Assertion 3.5 there is some $g \in C[\alpha, \beta]$ which is arbitrarily close to f, such that $g - u$ has not an alternant of length $n + 1$ whenever $u \in E_{n-2}^c$.

Table 1. Approximation of $f(x) = \cos(\pi/2)\,x$ on $[-1, +1]$

n	Characteristic numbers (upper numbers) and degree of approx. (lower numbers) of LBA's					
1	0					
	0.5					
2	−1.786	1.786				
	0.2762	0.2762				
3	−3.040	0	3.040			
	0.2073	0.02801	0.2073			
4	−4.198	−0.6589	0.6589	4.198		
	0.1706	0.00612	0.00612	0.1706		
5	−5.316	−1.152	0	1.152	5.316	
	0.1471	0.00169	0.000597	0.00169	0.1471	
6	−6.412	−1.583	−0.4226	0.4226	1.583	6.412
	0.1306	0.000521	0.000083	0.000083	0.000521	0.1306

This implies that $u_t = P_{E_n^c(t)} g \notin E_{n-2}^c$ for each $t \in \mathbb{R}$, and that $\tilde{r}$ has at most c_{n-1} local maxima. Hence, r and $\tilde{r}$ have at most $c_{n-1} + 1$ local minima, i.e.,

$$c_n \leq 1 + c_{n-1},$$

and Theorem 3.1 is proved. □

The bound for the number of local solutions given in Theorem 3.1, is sharp. This is elucidated by a numerical example.

3.6 Example. Let $f(x) = \cos(\pi/2)x$ on the interval $[-1, +1]$. The numerical results in Table 1 show that there are exactly n LBA's to f from E_n^c for $n = 1, 2, \ldots, 6$. We conjecture that it is also true for $n > 6$.

The numbers given in the table are rounded. The numerical results were computed by the Gauss-Newton method. Some precautions against rounding errors are required when solving the auxiliary linear problems. The acceleration mentioned in part 3 of III.5.B turned out to be useful for the improvement of numerical stability.

Exercises

3.7. Prove that the functions for which each local best approximation from E_n^c has order at least $n - 1$, form a residual set in $C[\alpha, \beta]$.

3.8. Given t, the set of exponential sums of the form

$$\left.\begin{array}{l} \alpha_1 e^{tx} + \alpha_2 e^{\tau x},\ \tau \neq t \\ \text{or} \\ (\alpha_1 + \alpha_2 x)e^{tx},\ \tau = t \end{array}\right\} \quad \alpha_1, \alpha_2 \in \mathbb{R},\ \tau \in \mathbb{R}$$

is a varisolvent Chebyshev set. By taking the union over $t \in \mathbb{R}$, define a one-parameter family and prove that there are at most 4 local minima.

3.9. Consider the family of rational functions with confluent poles:

$$\{u = p(x)\cdot(x - t)^{-n}\colon p \in \Pi_{n-1}, t \in \mathbb{R}\backslash[\alpha, \beta]\}.$$

Note that each function in the family may be written in the form

$$u(x) = \sum_{\nu=0}^{n-1} \alpha_\nu (\partial/\partial t) \frac{1}{x - t}.$$

Kaufmann and Taylor (1978) considered the approximation problem (with a different notation) and provided examples with more than one local best approximation. They posed the question: how many local best approximations can exist? Answer their question.

3.10. Let $n \geq 2$. The origin in n-space is the unique critical point and a strict local minimum of

$$f(x) = x_1^2(1 + x_2)^3 + \sum_{k=2}^{n} x_k^2.$$

Why does there exist no critical point of mountain pass type although $x = 0$ is not a global minimum?

Chapter V. Rational Approximation

The approximation of functions by rational expressions is important in different disciplines of analysis. First one thinks of the representation of special functions by rational approximations for the use in computers, but the applications go far beyond this point. First, rational approximations arise quite naturally in the numerical solution of ordinary and parabolic differential equations and in the study of other numerical methods. Furthermore, the Stieltjes and the Hamburger moment problem can be well understood via methods from rational approximation. The latter shows that nonlinear approximation theory may be helpful even for problems which originally are convex problems and not nonlinear in a strict sense.

Uniform approximation on an interval $X = [\alpha, \beta]$ by rational functions from*

$$R_{m,n} = \{u = p/q\colon p \in \Pi_m,\ q \in \Pi_n,\ q(x) > 0 \text{ in } X\} \tag{0.1}$$

was considered first by Chebyshev, although other approximation processes and rational interpolation were considered in different theories earlier. Since, for each $m, n \geq 0$, the family R_{mn} is a Chebyshev set in $C[\alpha, \beta]$, all theories from nonlinear approximation have been tested with rational approximation. There is one essential difference from uniform approximation in linear Chebyshev sets: the possibility of degeneracy. The rational fractions p_1/q_1 and p_2/q_2 are deemed *equivalent* if $p_1 q_2 = p_2 q_1$. A rational function of degree (m, n), where $n \geq 1$, is called *degenerate* if $u = 0$ or $u \in R_{m-1,n-1}$. Though degeneracy does not affect uniqueness of the best approximation, it does spoil the continuity of the metric projection, which in turn implies an anomalous behaviour in many directions.

The algebraic properties may be studied in the framework of generalized rational approximation, i.e., given two linear families M_1, M_2 consider

$$M = \{u = p/q\colon p \in M_1,\ q \in M_2,\ q(x) \geq 0 \text{ in } X\}.$$

Such generalized rational families are always suns in $C(X)$.

Recalling from III.2 that there may be many local best approximations in the L_2-case, the qualitative difference between uniform and L_p-approximation is obviously significant.

---

* For convenience, we will write R_{mn} instead of $R_{m,n}$ when no confusion is possible.

§ 1. Existence of Best Rational Approximations

A. The Existence Problem in $C(X)$

The families of rational functions R_{mn} for $n \geq 1$ are not boundedly compact. For $C[0,1]$, this is obvious from the sequence:

$$u_k(x) = \frac{1}{1+kx}, \qquad k = 1, 2, \ldots, \tag{1.1}$$

since

$$\lim_{k\to\infty} u_k(x) = \begin{cases} 1, & x = 0, \\ 0, & x > 0, \end{cases}$$

no subsequence converges to a continuous function. The situation for L_p-approximation is similar. The sequence (v_k), $v_k := u_k/\|u_k\|_p$, where u_k is defined in (1.1), $k = 1, 2, \ldots$, is bounded, but does not converge in L_p to a rational function.

The existence of best approximations in $C(X)$ can be obtained by applying Lemma II.1.4. For a unified approach we note that the supremum norm and the L_p-norms are monotone norms. This means that

$$|f(x)| \leq |g(x)| \quad \text{for } x \in X \text{ implies } \|f\| \leq \|g\|. \tag{1.2}$$

1.1 Lemma. *Let X be a compact real interval. Assume that $(u_k) \subset R_{mn}$ is bounded for the supremum-norm in $C(X)$ or an L_r-norm, $1 \leq r < \infty$. Then there are $l \leq n/2 + 1$ points $z_1, z_2, \ldots, z_l$ and a subsequence (u_{k_j}) which converges to some $u^* \in R_{mn}$ uniformly on each compact interval $[\alpha, \beta] \subset X \setminus \{z_i\}_{i=1}^l$. If u^* is not degenerate, then the subsequence converges uniformly in X.*

Proof. Assume that $\|u_k\| \leq c_1, k = 1, 2, \ldots$ One may normalize the denominators q_k, $u_k = p_k/q_k$ such that

$$\|q_k\|_\infty = \max_{x\in X} |q_k(x)| = 1 \quad \text{and} \quad q_k(x) > 0, x \in X.$$

It follows that $\|p_k\| \leq \|u_k\| \leq c_1$. Since all norms on Π_m are equivalent, we have $\|p_k\|_\infty \leq c_2$, $k = 1, 2, \ldots$ Therefore, we may choose a subsequence such that $p_{k_j} \to p^* \in \Pi_m$ and $q_{k_j} \to q^* \in \Pi_n$ uniformly in X. Hence $\|q^*\|_\infty = \lim \|q_{k_j}\|_\infty = 1$ and $q^*(x) \geq 0$ for $x \in X$. Therefore q^* has at most $l \leq n/2 + 1$ zeros in X, say $z_1, z_2, \ldots, z_l$.

Let $[\alpha, \beta]$ be a subinterval of X without a zero of q^*. Then (p_{k_j}/q_{k_j}) converges to p^*/q^* uniformly in $[\alpha, \beta]$.

If (u_k) is $\|.\|_\infty$-bounded, then $\max\{|p^*(x)/q^*(x)|: x \in [\alpha, \beta]\} \leq c_1$. If (u_k) is $\|.\|_r$-bounded, then

$$\int_\alpha^\beta |p^*/q^*|^r \, d\mu \leq \overline{\lim_{k\to\infty}} \int_\alpha^\beta |p_k/q_k|^r \, d\mu \leq \overline{\lim_{k\to\infty}} \|u_k\|^r \leq c_1^r.$$

The bounds are independent of α and β. Since α and β may be chosen arbitrarily close to the zeros of q^*, it follows that p^*/q^* has no poles in X but only removable singularities. We obtain

$$\frac{p^*(x)}{q^*(x)} = \frac{p_1(x)}{q_1(x)}, \quad \text{whenever } q^*(x) \neq 0,$$

where $q_1(x) > 0$ in X and p_1/q_1 is irreducible, i.e., p_1 and q_1 have no common factors other than constants. By setting $u^* = p_1/q_1 \in R_{mn}$ we get the first part of the theorem.

If u^* is not degenerate, then no division has been performed, i.e., $p_1 = p^*$ and $q_1 = q^*$. The second statement of the lemma is also correct. □

From the preceding lemma and Lemma II.1.4, the existence of best approximations is immediate.

1.2 Existence Theorem for Rational Approximation (Walsh 1931). *Let $m, n \geq 0$ and let X be a compact real interval. Then the set of rational functions R_{mn} is an existence set in $C(X)$.*

On the negative side, R_{mn}, $n \geq 1$, is not approximatively compact. In the next section we will see that there are functions $f \in C[0, 1]$ to which $u_0 = 0$ is the unique nearest point from R_{mn} and for which $f(0) > 0$. Here we have a minimizing sequence by choosing $u_k(x) = f(0) \cdot (1 + kx)^{-1}$, which does not contain a convergent subsequence.

The next negative result refers to discrete sets. For $n \geq 1$, R_{mn} is not an existence set in $C(X)$, if X is a discrete set. Let $N > m + n$ and

$$X = \{j/N : j = 0, 1, \ldots, N\}.$$

Put $f(x) = 1$ for $x = 0$ and $f(x) = 0$ otherwise. Obviously, $f \notin R_{mn}$ but from the (minimizing) sequence (1.1) it follows that $d(f, R_{mn}) = 0$. No nearest point exists.

In R_{mn}, we have only to consider functions without zeros common to the numerator and denominator. This is no longer possible in generalized rational approximation (see Meinardus 1967).

1.3 Example. Let $B = B_1(0) = \{x, y \in \mathbb{R}^2 : x^2 + y^2 \leq 1\}$. The function

$$f(x, y) = \frac{x^4 + y^4 + (x^2 + y^2) T_n(x)}{x^2 + y^2}$$

where T_n is the n-th Chebyshev polynomial, is to be approximated by rational expressions of the form p/q, where

$$p(x, y) = \sum_{\substack{i,k \geq 0 \\ i+k \leq 4}} a_{ik} x^i y^k,$$

$$q(x, y) = \sum_{\substack{i,k \geq 0 \\ i+k \leq 2}} b_{ik} x^i y^k.$$

Moreover $q(x, y) \geq 0$ for $(x, y) \in B$ is required. If $n \geq 7$, then a best approximation is

$$u^* = \frac{p^*}{q^*} := \frac{x^4 + y^4}{x^2 + y^2}.$$

To verify this, notice that $\varepsilon^* := f - u^* = (x^2 + y^2)T_n(x)$ and $|\varepsilon^*|$ attains its maximum on ∂B at $2n$ points: $x_k = \cos k\pi/n$, $y_k = \sin k\pi/n$, $k = 0, 1, \ldots, 2n - 1$. Suppose that $p_1/q_1 \neq p^*/q^*$ is a better or an equally good approximation. Then $\varepsilon^*[p^*/q^* - p_1/q_1] \leq 0$ holds on the $2n$ points and since $q^*, q \geq 0$, we have

$$[\varepsilon^*(p^*q_1 - p_1q^*)](x_k, y_k) \leq 0. \tag{1.3}$$

If polar coordinates $x = r\cos\varphi$, $y = r\sin\varphi$ are introduced, the restriction $p^*q_1 - p_1q^*$ to ∂B is a trigonometrical polynomial of degree ≤ 6 in the φ-variable and cannot have more than 12 zeros. This contradicts (1.3), because ε^* has $2n \geq 14$ points between which the error curve alternates. □

B. Rational L_p-Approximation. Degeneracy

With similar arguments, the existence of best approximations in L_p can be established. Here even approximative compactness is verified (cf. Blatter 1968). We emphasize that the proof for the latter makes use of the fact (cf. Example III.2.2) that no degenerate rational function can be a best approximation.

1.4 Existence Theorem for Rational L_p-approximation. *Let $m, n \geq 0$ and let X be a compact real interval. The set of rational functions R_{mn} is an existence set in $L_p(X)$ for $1 \leq p < \infty$. If moreover, $1 < p < \infty$, then R_{mn} is approximatively compact.*

Proof. Let (u_k) be a minimizing sequence for the approximation of f. By Lemma 1.1, there are $l \leq n/2 + 1$ points $z_1, z_2, \ldots, z_l$, such that, after passing to a subsequence if necessary, (u_k) converges to $u^* \in R_{mn}$ uniformly on $X_\delta = \{x \in X: |x - z_i| \geqq \delta \text{ for } i = 1, 2, \ldots, l\}$ for each $\delta > 0$. Hence,

$$\int_{X_\delta} |f - u^*|^p \, d\mu \leq \lim_{k\to\infty} \int_{X_\delta} |f - u_k|^p \, d\mu \leq \lim_{k\to\infty} \|f - u_k\|^p = d(f, R_{mn})^p.$$

Since this holds for each $\delta > 0$ and $f - u^*$ is integrable, it follows that

$$\int_X |f - u^*|^p \, d\mu \leq d(f, R_{mn})^p.$$

Therefore u^* is a nearest point to f.

Finally, let $1 < p < \infty$. From Example III.2.2 it is known that a best approximation u^* to $f \in L_p$ cannot be a degenerate rational function (Cheney and Goldstein 1967).

By Lemma 1.1, a uniformly convergent subsequence may be obtained from a minimizing sequence. It also converges in $L_p(X)$ and thus R_{mn} is approximatively compact. □

Since R_{mn} is approximatively compact but not convex for $n \geq 1$, it cannot be a Chebyshev set in the uniformly convex space L_p for $1 < p < \infty$. This argument from the functional analytic approach in II.3 was first mentioned by Efimov and Stechkin. On the other hand, functions with at least two nearest points in R_{mn} for odd n, are easily obtained with the invariance principle:

1.5 Example (Braess 1967, Lamprecht 1967, 1970). Consider the approximation in $L_r[-1, +1]$ for $1 < r < \infty$ and recall that the symmetry of the Lebesgue measure implies that $\|f\| = \|g\|$ if $g(x) = f(-x)$. Choose $f \notin R_{mn}$ such that

$$f(-x) = (-1)^{m+1} f(x). \tag{1.4}$$

Each $u = p/q \in R_{mn}$, n odd, with the symmetry $u(-x) = (-1)^{m+1} u(x)$ has representing polynomials such that $p(-x) = (-1)^{m+1} p(x)$, $q(-x) = q(x)$. It follows that $p/q \in R_{m-1,n-1}$. Since no best approximation is degenerate, no best approximation has the mentioned symmetry property. By the invariance principle I.1.3, f has at least two nearest points in R_{mn}. □

The following theorem shows that there are many functions with more than one nearest point in R_{mn} whenever $n \geq 1$. The restriction to the symmetric functions as in the example above can be abondoned due to an idea by Diener (1986) who applied a fixed point theorem in the investigation of families with one nonlinear parameter.

1.6 Theorem (Braess 1986). *Let $1 < r < \infty$, $m \geq 0$, and $n \geq 1$. Each $(m + 2)$-dimensional subspace E_1 of $L_r[\alpha, \beta]$ such that $E_1 \cap R_{mn} = \{0\}$, contains a function with at least two best approximations from R_{mn}.*

Proof. Set $S^{m+1} := \{f \in E_1 \colon \|f\| = 1\}$. Suppose that to each $f \in E_1$ there is a unique nearest point from R_{mn}. Since R_{mn} is approximatively compact, it follows as in the proof of Theorem II.3.1 that the metric projection $P \colon S^{m+1} \to R_{mn}$ is continuous. Moreover, we recall that the range of P contains no degenerate functions from R_{mn}. Let the denominators be normalized by $\max_{x \in [\alpha, \beta]}\{q(x)\} = 1$. Then the mapping $\Phi \colon R_{mn} \backslash R_{m-1,n-1} \to \mathbb{R}^{m+1}$ which sends $u = p/q$ with $p = \sum_{k=0}^{m} a_k x^k$ to the vector of coefficients $(a_0, a_1, \ldots, a_m)$, is continuous. Obviously, $\Phi \circ P$ is an odd mapping, i.e. $\Phi \circ P(-f) = -\Phi \circ P(f)$. By the Borsuk Antipodality Theorem there is an $f_0 \in S^{m+1}$ such that $\phi \circ P(f_0) = 0$. But then Pf_0 is degenerate and we get a contradiction.

Therefore, P cannot be a continuous mapping and not each $f \in E_1$ has a unique nearest point in R_{mn}. □

By combining the arguments of Example 1.5 and Theorem 1.6, a function with (at least) 4 best rational L_r-approximations may be constructed, see Exercise 1.14.

The possibility of degenerate best approximations in the L_1-case has been settled by Dunham (1971). Here, the solution may belong to $R_{m-1,n-1}$. Consider $f \in L_1[-1, +1]$ defined by

$$f(x) = \begin{cases} -4x^2 + 1, & \text{if } |x| \leqq 1/2, \\ 0, & \text{otherwise.} \end{cases}$$

The best L_1-approximation to f from $R_{0,1}$ is $u_0 = 0$. Indeed, each positive function $u \in R_{01}$ is convex, and therefore

$$\int_{|x| \leq 1/2} |u|\, dx \leq \int_{1/2 \leq |x| \leq 1} |u|\, dx.$$

Hence,

$$\begin{aligned} \|f - u\| &\geq \int_{|x| \leq 1/2} |f|\, dx - \int_{|x| \leq 1/2} |u|\, dx + \int_{1/2 \leq |x| \leq 1} |u|\, dx \\ &\geq \int_{|x| \leq 1/2} |f|\, dx = \|f - u_0\|. \end{aligned}$$

On the other hand, if f is continuous and $f \notin R_{mn}$, no best L_1-approximation from R_{mn} will be contained in $R_{m-2,n-2}$. Otherwise

$$\frac{p(x)}{q(x)} + \frac{\varepsilon}{q(x)[(x - x_0)^2 + \delta]}$$

with some $\varepsilon, \delta \in R$, $x_0 \in X$ would be a better candidate in R_{mn} than p/q. □

Exercises

1.7. Are the subsets of rational functions having only real poles (no real poles, resp.) existence sets in $C[\alpha, \beta]$ and $L_p[\alpha, \beta]$?

1.8. When $C[0, 1]$ is equipped with the translation invariant metric

$$d(f, g) = \sup\left\{\frac{|f(x) - g(x)|}{1 + |f(x) - g(x)|} : x \in [0, 1]\right\}$$

it becomes a Fréchet space. Is R_{mn} still an existence set for the approximation problem?

1.9. Prove that the set of functions of the form $ax/(b|x| + c)$ where $a, b, c \in \mathbb{R}$, is not an existence set in $C[-1, +1]$.
Hint: Consider a piecewise linear function f such that $f(-1) = f(1) = 0$, $f(1/2) = -f(-1/2) = 1$ (Cheney 1966).

1.10. Prove that the set of logarithmic antisymmetric rational functions $\{u(x) = p(x)/p(-x): p \in \Pi_n, p(x) \geq 0\}$ is an existence set.

1.11. Fix $m \in \mathbb{N}$. Prove that $\bigcup_{n>0} R_{mn}$ is dense in the cone $C_+[\alpha, \beta] = \{f \in C[\alpha, \beta]: f(x) > 0 \text{ in } [\alpha, \beta]\}$, but not in $C[\alpha, \beta]$.

1.12. Is the set of rational functions p/q with p and q being trigonometric polynomials of degree $\leq m$ and n resp., where $q(x) > 0$, an existence set in $C_{2\Pi}$?
Hint: Trigonometric polynomials can be written as rational expressions of e^{ix}.

1.13. Let u be a non-degenerate function in R_{mn}. Does $(-u)$ belong to the same component of $R_{mn} \backslash R_{m-1,n-1}$?
Hint: Three cases are distinguished by Bartke (1984) for the enumeration of components of $R_{mn} \backslash R_{m-1,n-1}$. The number is

$$\begin{cases} n+1, & \text{if } n > m, \\ m+1, & \text{if } n-m \text{ is a non-negative, even number,} \\ m+2, & \text{if } n-m \text{ is a positive, odd number.} \end{cases}$$

1.14. Let $1 < r < \infty$ and $m \geq 0$. Let E_1 be an $(m+2)$-dimensional subspace of $L_r[\alpha, \beta]$ such that $E_1 \cap R_{m1} = \{0\}$. Moreover. assume that each $f \in E_1$ satisfies (1.4). Show that E_1 contains a function f with at least 4 best L_r-approximations from R_{m1}.
Hint: Show that some $f \in E_1$ has two best rational functions with negative poles by making use of the Borsuk Antipodality Theorem.

1.15. Let $1 < r < \infty$, $m \geq 0$, and $n \geq 1$. Assume that E_1 is an $(m+n+2)$-dimensional subspace of $L_r[\alpha, \beta]$. Then the number of best L_r-approximations from R_{mn} is not constant for $E_1 \backslash \{0\}$.
Another problem with rational functions is found in Exercise VI.2.14.

§2. Chebyshev Approximation by Rational Functions

A. Uniqueness and Characterization of Best Approximations

The sets of rational functions (0.1) are the most important non-convex Chebyshev sets in $C(X)$. Since these are suns and are asymptotically convex (see Exercises III.1.15 and III.1.16), the characterization of nearest points may be successfully treated in almost every theory of nonlinear approximation. Here it will be done by showing that R_{mn} for $m, n \geq 0$, is varisolvent.

In order to have a measure for the degeneracy, the definition of *defect* of rational functions has been introduced by Achieser (1930):

$$d(u) = \begin{cases} \min(m - \partial p, n - \partial q), & \text{if } u = p/q \neq 0, \\ n, & \text{if } u = 0. \end{cases} \tag{2.1}$$

Obviously, u is degenerate in R_{mn} if $d(u) > 0$. Moreover $d = d(u)$ is the greatest number such that $u \in R_{m-d,n-d}$. This is clear for $u \neq 0$ and is also true for $u = 0$, with the usual convention that $\Pi_m = R_{mn} = \{0\}$ whenever $m < 0$, $n \geq 0$.

2.1 Lemma. *Let $m, n \geq 0$. Then R_{mn} is a varisolvent family in $C[\alpha, \beta]$ with the density property, the degree of varisolvency is*

$$m + n + 1 - d(u) \tag{2.2}$$

at $u \in R_{mn}$. Moreover, R_{mn} satisfies the local and the global Haar condition.

Proof. (1) Let $u_0 = 0$. Obviously, $u_0 - p/q = -p/q$ has at most $m = m + n + 1 - d(u) - 1$ zeros whenever $p/q \in R_{mn}$. This proves Property Z of the degree (2.2).

Since $u_0 \in \Pi_m \subset R_{mn}$ and Π_m is an $(m+1)$-dimensional Haar subspace of $C[\alpha, \beta]$, R_{mn} is locally solvent of degree $m+1$ at u_0.

(2) Let $u_0 = p_0/q_0 \neq 0$. Recall that $p_0 \in \Pi_{m-d}$, $q_0 \in \Pi_{n-d}$, where $d = d(u_0)$. For $u = p/q \in R_{mn}$, we write $(u - u_0) = (pq_0 - p_0 q)/q \cdot q_0$. Since $pq_0 - p_0 q \in \Pi_{m+n-d}$, the difference $u - u_0$ has at most $m + n + 1 - d(u_0) - 1$ zeros. This yields Property Z.

To establish local solvability, we will prove that R_{mn} satisfies the local Haar condition (cf. Definition III.4.4). Consider the parametrization:

$$F(a, b, x) = \frac{\sum_{k=0}^{m} a_k x^k}{\sum_{k=0}^{n} b_k x^k}.$$

Obviously,

$$\frac{\partial F}{\partial a_k} = \frac{x^k}{q}, \qquad k = 0, 1, \ldots, m,$$

$$\frac{\partial F}{\partial b_k} = -\frac{p x^k}{q^2}, \qquad k = 0, 1, \ldots, n.$$

Setting $P = \sum_k \delta_k x^k$, $Q = \sum_k \theta_k x^k$, we get

$$\sum_{k=0}^{m} \delta_k \frac{\partial F}{\partial a_k} + \sum_{k=0}^{n} \theta_k \frac{\partial F}{\partial b_k} = \frac{1}{q_0^2}(q_0 P - p_0 Q). \tag{2.3}$$

The dimension of the space $H := [q_0^{-2}(q_0 \Pi_m - p_0 \Pi_n)]$ equals $\dim \Pi_m + \dim \Pi_n - \dim W$, where W is the subset of $\Pi_m \times \Pi_n$ which is mapped onto the zero function in H. Since p_0/q_0 is irreducible,

$$q_0 P - p_0 Q = 0$$

implies that there is a polynomial g such that

$$P = p_0 \cdot g, \quad \text{hence } \partial g \leq m - \partial p_0,$$

$$Q = q_0 \cdot g, \quad \text{hence } \partial g \leq n - \partial q_0.$$

Thus $\partial g \leq d(u_0)$ and the dimension of the kernel is $d(u_0) + 1$. This implies that $\dim H = m + n + 1 - d(u_0)$. Moreover, since $\partial(q_0 P - p_0 Q) \leq m + n - d(u_0)$, the space H is a Haar subspace of $C[\alpha, \beta]$. From Remark III.4.5 we know that R_{mn} is locally solvent of degree $m + n + 1 - d(u_0)$ at u_0.

(3) Finally, given $u = p/q \in R_{mn}$, the elements $(p + \varepsilon)q^{-1}$ and $(p - \varepsilon)q^{-1}$, $\varepsilon > 0$, also belong to R_{mn}. Therefore R_{mn} has the density property.

Thus, the proof of varisolvency and of the local and global Haar condition is complete. □

From Theorem 1.2, Lemma 2.1, and the results on varisolvent families we get the

2.2 Uniqueness and Alternation Theorem (Achieser 1930). *Let $m, n \geq 0$. Then R_{mn} is a Chebyshev set in $C[\alpha, \beta]$. A rational function $u \in R_{mn}$ is a best approximation to f from R_{mn} if and only if $f - u$ has an alternant of length $m + n + 2 - d(u)$.*

From this characterization of nearest points in terms of alternants, it is also clear that R_{mn} is a sun.

Given f, the best approximations $u_{mn} = P_{R_{mn}} f$, $m, n = 0, 1, \ldots$ may be arranged in a table with double entries, called the *Walsh Table*.

Walsh Table of Best Approximations

$n \backslash m$	0	1	2	3	
0	u_{00}	u_{10}	u_{20}	u_{30}	$\cdots$
1	u_{01}	u_{11}	u_{21}	u_{31}	$\cdots$
2	u_{02}	u_{12}	u_{22}	u_{32}	$\cdots$
	$\cdots$				

The top line contains the best polynomials. A rational function may enter into several places. Typically, identical (non-zero) elements are found in square blocks. Let $u_{mn} = p/q$ with $\partial p = m$, $\partial q = n$, and assume that $f - u_{mn}$ has an alternant of exact length $m + n + 2 + d$ where $d \geq 0$. Then

$$u_{n+i,m+j} = u_{mn}, \qquad 0 \leq i, j \leq d, \tag{2.4}$$

and all other elements in the table are different. Blocks with a length greater than one refer to degeneracies.

If f has a symmetry: $f(-x) = f(x)$ or $f(-x) = -f(x)$ then all blocks for the approximation on $X = [-1, +1]$ have an even length, with a possible exception at the left hand boundary of the table.

B. Normal Points

Since R_{mn} is a Chebyshev set in $C[\alpha, \beta]$ and the nearest point can always be characterized in terms of alternants, there seems, at first glance, to be no difference from polynomial approximation. But it will turn out that many properties known from polynomial approximation hold only for functions having a non-degenerate best rational approximation. The crucial point will be the discontinuity of the metric projection at non-normal points.

A function $f \in [\alpha, \beta]$ is said to be (m, n)-*normal* (or *normal* for short) if its best approximation is not degenerate, or if $f \in R_{mn}$. This means that a rational function is *singular* in the sense of the general theory of varisolvent functions, if and only if it is *degenerate*. [The difference in notation is for historical reasons. In the general theory, the denotation of degenerate elements has to be avoided, since there are other kinds of degeneracies in some other special varisolvent families.]

The set of nondegenerate elements in R_{mn} is dense. Indeed, if p/q is degenerate, then $(p + \varepsilon x^m)/(q + \varepsilon x^n)$ is not, for any small ε.

2.3 Continuity Theorem. *Let $m, n \geq 0$. The metric projection from $C[\alpha, \beta]$ onto R_{mn} is continuous at f if and only if f is (m, n)-normal.*

Proof. (1) From the general continuity theorem III.3.13(1), the continuity of P at a normal point $f \in R_{mn}$ is immediate. Moreover, if $f \in R_{mn}$, then $\|Pg - f\| \leq \|Pg - g\| + \|g - f\| \leq 2\|g - f\|$ implies continuity at f.

(2) Assume that f is not (m, n)-normal. Suppose that $f - u_0$, where $u_0 = Pf$, has an alternant of exact length $m + n + 2 - r$ with $0 < r \leq d(u_0)$. In each neighborhood of u_0, there is a non-degenerate function $u_1 \in R_{mn}$. Set $g = f + (u_1 - u_0)$. By the alternation theorem $u_2 := Pg \neq u_1$. From $\|g - u_2\| < \|g - u_1\|$ we conclude with a de la Valllée-Poussin type argument that $u_2 - u_1$ has at least $m + n + 1 - r$ zeros. This and Property Z imply $d(u_2) < r$, and $g - u_2$ has an alternant of length at least $m + n + 2 - (r - 1)$. Therefore $g - u_2$ is not close to $g - u_1 = f - u_0$ and P is not continuous at f.

(3) Finally assume that $f - u_0$ has an alternant $x_1 < x_2 < \cdots < x_{m+n+2}$ though $u_0 = Pf$ is degenerate. For ease of notation let $[\alpha, \beta] = [0, 1]$. First we consider the case where $x = 0$ does not belong to the alternant, i.e., $x_1 > 0$. Put $s = +1$ if $(f - u_0)(0) \geq 0$ and $s = -1$ otherwise. Next select a $\xi \in (0, x_1)$ such that

$$s \cdot (f - u_0)(x) \geq -\tfrac{1}{2}\|f - u_0\| \text{ for } 0 \leq x \leq \xi. \tag{2.5}$$

Moreover, set $a = s\|f - u_0\|/2\|q_0^{-1}\|$ and for $k = 1, 2, \ldots$

$$h_k(x) = \frac{a}{q_0(x)} \frac{1}{1 + kx},$$

$$\tilde{h}_k(x) = \frac{a}{q_0(x)} \min\left\{\frac{1}{1 + kx}, \frac{1}{1 + k\xi}\right\},$$

$$g_k = f - \tilde{h}_k,\ u_k = u_0 - h_k.$$

By construction, $g_k - u_k$ has the same alternant as $f - u_0$. Hence, $u_k = Pg_k$. Obviously, $\|u_k - u_0\| = \|h_k\| \geq |a| q_0^{-1}(0) > 0$, while $\|g_k - f\| = \|\tilde{h}_k\| \to 0$. Therefore P is not continuous at f.

In the case where $x_1 = 0$ belongs to the alternant, we may construct a function f_1 by an arbitrarily small perturbation such that $Pf_1 = Pf_0 = u_0$ but $x = 0$ does not belong to the alternant of $f_1 - u_0$ (cf. Fig. 24). From the discussion above we know that in each neighborhood of f_1 there is a function of the form $g_k = f_1 + \tilde{h}_k$ such that $\|Pg_k - u_0\| > |a| q_0^{-1}(0)$. Thus, P is discontinuous at f also in this case. □

When Maehly and Witzgall (1960) proved the continuity of the metric projection at normal points, they already presented an example of discontinuity. The non-normal case with an alternant of length $\leqq m + n + 1$ was treated by Cheney and Loeb (1964). The completion of the proof is due to Werner (1964).

From Lemma 1.1 and the discontinuity of P at non-normal points we get

2.4 Corollary. *For $f \in C[\alpha, \beta]$ the following are equivalent:*

1°. *f is (m, n)-normal.*

2°. *Each minimizing sequence for f in R_{mn} contains a convergent subsequence.*

Another consequence of the continuity theorem is

2.5 Corollary. *Let $m, n \geq 0$. Then the set of (m, n)-normal points is residual in $C[\alpha, \beta]$.*

Proof. Note that $R_{m-1,n-1}$ is closed in R_{mn}. Since the metric projection P is continuous at normal points, the set $P^{-1}(R_{mn} \backslash R_{m-1,n-1}) \backslash R_{mn}$ is open in $C[\alpha, \beta]$.

The proof of the continuity theorem shows that, in each neighborhood of a non-normal f, there is a g with $d(Pg) \geq d(Pf) - 1$. By induction, it follows that there is also a g with $d(Pg) = 0$. Therefore $P^{-1}(R_{mn} \backslash R_{m-1,n-1}) \backslash R_{mn}$ is dense in $C[\alpha, \beta] \backslash R_{mn}$ and also dense in $C[\alpha, \beta]$.

Since $P^{-1}(R_{mn} \backslash R_{m-1,n-1}) \backslash R_{mn}$ is open and dense in $C[\alpha, \beta]$ and is a subset of the set of (m, n)-normal points, the proof is complete. □

On the other hand, each $(m+2)$-dimensional subspace of $C[\alpha, \beta]$ contains an element which is not (m, n)-normal whenever $n \geq 1$. As becomes apparent from Exercise 2.15, there is an analogy between non-normality in Chebyshev approximation and non-uniqueness in L_p-approximation.

The subset of nondegenerate points $R_{mn} \backslash R_{m-1,n-1}$ is a Haar embedded manifold. Therefore, each non-degenerate best approximation is a strongly unique local best approximation. This and Corollary 2.4 yield strong uniqueness globally.

2.6 Theorem (Barrar and Loeb 1970). *The best approximation to $f \in C[\alpha, \beta]$ from R_{mn} is strongly unique if and only if f is (m, n)-normal.*

Proof. (1) If f is normal, there is an $r > 0$ such that u_0 is a strongly unique best approximation in $R_{mn} \cap B_r(u_0)$ with a constant $c_1 > 0$. Next we have

$$\|f - u\| \geq \|f - u_0\| + \delta \quad \text{for } u \in R_{mn}, r \leq \|u - u_0\| \leq 3d(f, R_{mn}) \tag{2.6}$$

with $\delta > 0$, because by Corollary 2.4 each minimizing sequence converges. We put $c_2 = \delta / 3d(f, R_{mn})$ and replace (2.6) by the weaker inequality:

$$\|f - u\| \geq \|f - u_0\| + c_2 \|u - u_0\| \quad \text{for } u \in R_{mn} \cap (B_{3d}(u_0) \backslash B_r(u_0)).$$

Finally, if $\|u - u_0\| \geq 3d(f, R_{mn})$, then

$$\begin{aligned} \|f - u\| &\geq \|u - u_0\| - \|u_0 - f\| \\ &\geq \tfrac{1}{3}\|u - u_0\| + 2d(f, R_{mn}) - \|u_0 - f\| = \|f - u_0\| + \tfrac{1}{3}\|u - u_0\|. \end{aligned}$$

This proves strong uniqueness with $c = \min\{c_1, c_2, 1/3\}$.

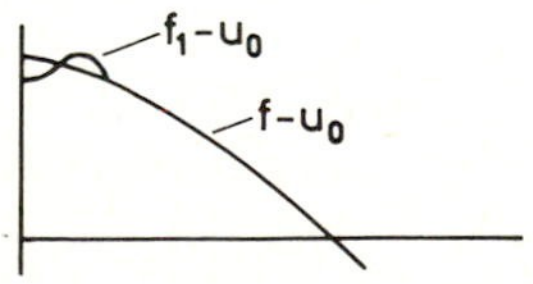

Fig. 24. Shift of the first alternation point

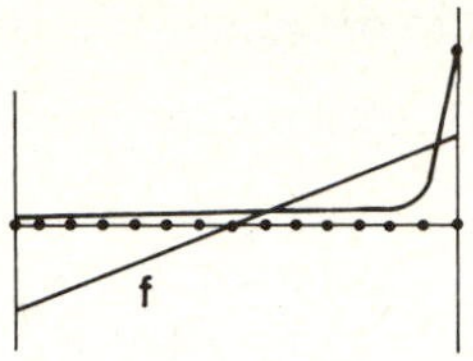

Fig. 25. Influence of discritization in the nonnormal case

(2) A degenerate best approximation to $f \notin R_{mn}$ cannot be strongly unique, since not every minimizing sequence converges. □

When a best rational approximation is to be determined numerically, the approximation problem is usually solved on a discrete point set and not on an interval. Whether or not the problem is drastically changed by discretization, is also a consequence of normality (cf. Werner 1967).

If f is not normal, then a best approximation if it exists need not be close to the solution. Consider, for example $f(x) = x$ on $X = [-1, +1]$. Because of the symmetry, 0 is the nearest point from $R_{0,1}$. Given $N > 1$, the best approximation on the discrete set

$$\left\{x_i = \frac{i - N}{N}; i = 1, 2, \ldots, 2N\right\}$$

is a positive rational function with $u(1) \geq 2 - N^{-1}$.

On the other hand, let f be normal, $x_1, x_2, \ldots$ be an infinite sequence which is dense in X, and let $\delta_k \to 0$ as $k \to \infty$. Put $X_k = \{x_i\}_{i=1}^k$ and choose $u_k \in R_{mn}$ such that

$$\max_{x \in X_k} |f(x) - u_k(x)| \leq d(f, R_{mn}) + \delta_k.$$

In particular, u_k may be the best approximation on X_k. Although it is not clear a priori that (u_k) is bounded in $C(X)$, it follows from the same arguments as in Lemma 1.1 and the existence theorem that u_k converges to $R_{mn}f$ on a dense set. Since $R_{mn}f$ is not degenerate, the convergence is uniform on X.

Another peculiarity connected with degeneracy was detected by Saff and Varga (1977): There are non-normal real functions on a real interval for which a rational function with complex coefficients is optimal. In that case, there exist at least two best approximations, since f and R_{mn} are invariant under the involution $(Tf)(x) = \overline{f}(x)$.

Example. Let $f \in C[-1, +1]$, $f(-1) = f(+1) = 1 + d$, $f(1/2) = 1 - d$, for $d > 0$ and f piecewise linear in between. Then $u_0 = 1$ is the best rational function in R_{01} with real coefficients. Consider approximations of the form

$$u(x) = \frac{1 + i\alpha}{1 + i\beta x}$$

with $\alpha, \beta \to 0$.

$$\begin{aligned}|f(x) - u(x)|^2 &= |f - u_0|^2 - 2\,\mathrm{Re}\overline{(f - u_0)}(u - u_0) + |u - u_0|^2 \\ &= |f - u_0|^2 - 2(f - u_0)\,\mathrm{Re}\frac{i\beta x - i\alpha}{1 + i\beta x} + |u - u_0|^2 \\ &= |f - u_0|^2 - 2(f - u_0)\frac{\beta x(\beta x - \alpha)}{1 + \beta^2 x^2} + |u - u_0|^2.\end{aligned}$$

The middle term is negative at $P[f - u_0] := \{-1, 1/2, 1\}$ when $\beta = 3\alpha/2$. At these points we have $|(f - u)(x)| < \|f - u_0\|$ provided that $d = d(f, R_0)$ is sufficiently large. Obviously, the inequality holds also for the points in a neighborhood U of $P[f - u_0]$. By standard arguments one has $|(f - u)(x)| < \|f - u_0\|$ for $x \in [-1, +1] \backslash U$ provided that α is sufficiently small. Thus u is a better approximation. □

There are more differences between real and complex rational approximation, see Trefethen and Gutknecht (1983a), Ruttan (1985).

We conclude this section with a criterion which is often useful for the proof of normality.

2.7 Normality Criterion (Werner 1966). *If f satisfies the (m,n)-normality condition, i.e., if*

$$\{1, x, \ldots, x^{m-1}, f, xf, \ldots, x^{n-1}f\}$$

spans an $(m + n)$-dimensional Haar subspace, then f is (m,n)-normal.

Proof. Assume that f satisfies the normality condition. Then $P - fQ$ has at most $m + n - 1$ zeros, whenever $P \in \Pi_{m-1}$, $Q \in \Pi_{n-1}$, and $Q \neq 0$. Given a degenerate rational function p/q, $q \neq 0$, we choose a polynomial $g \in \Pi_{d-1}$, $d = d(p/q)$, with $d - 1$ zeros distinct from the zeros of $p - fq$. Noting that $P := gp \in \Pi_{m-1}$, $Q := gq \in \Pi_{n-1}$, we conclude that $(gp) - f(gq)$ has at most $m + n - 1$ zeros. Consequently $p - fq$ has at most $m + n - d(p/q)$ zeros.

By the alternation theorem, the degenerate function p/q cannot be a best approximation. □

The criterion applies to the function $f(x) = e^x$. Given $p \in \Pi_{m-1}$ and $q \in \Pi_{n-1}$, we have

$$(p - qe^x)^{(m)} = g(x)e^x$$

with some $g \in \Pi_{n-1}$. Hence, $(p - qe^x)^{(m)}$ has at most $n - 1$ zeros. By Rolle's theorem, $p - qe^x$ has at most $m + n - 1$ real zeros. Thus e^x is (m,n)-normal for all $m, n \geq 0$. □

C. The Lethargy Theorem and the Lip 1 Conjecture

Due to Bernstein's Lethargy Theorem, there are functions in $C[-1, +1]$ for which the error of the polynomial approximation converges to zero arbitrarily slowly. Though the approximation by rational functions is better than that by polynomials in most cases, there is an analogous negative result (cf. Shapiro 1964).

2.8 Lethargy Theorem *If (ε_n) is any positive sequence converging monotonically to zero, then there is an $f \in C[-1, +1]$ such that $E_{nn}(f) \geq \varepsilon_n$ for $n = 1, 2, \ldots$.*

Proof. Put $\alpha_k = \varepsilon_{3^{k-2}} - \varepsilon_{3^{k-1}}$, for $k = 2, 3, \ldots$ and $f = \sum_{k=2}^{\infty} \alpha_k T_{3^k}$. By construction, $\alpha_k \geq 0$ and $\sum_k \alpha_k = \varepsilon_1 < \infty$. Since $\|T_n\| = 1$, the series for f converges to a continuous function. Let $n = 3^m$, $m \geq 1$. We will show that the best approximation in R_{nn} is $p = \sum_{k=2}^{m} \alpha_k T_{3^k}$. Consider the points

$$x_i = \cos \frac{i\pi}{3^{m+1}}, \qquad i = 0, 1, \ldots, 3^{m+1}.$$

Then $T_{3^k}(x_i) = (-1)^i$ for $i = 0, 1, \ldots, 3^{m+1}$, whenever $k \geq m + 1$. We conclude that $\|f - p\| = \sum_{k=m+1}^{\infty} \alpha_k = \varepsilon_{3^{m-1}}$, and that the points mentioned form an alternant of length $3^{m+1} + 1 = 3n + 1 \geq 2n + 2$. By the Alternation Theorem, p is a nearest point.

Now for any $n \geq 2$, with $3^{m-1} < n \leq 3^m$, we get

$$E_{nn}(f) \geq E_{3^m, 3^m}(f) = \varepsilon_{3^{m-1}} \geq \varepsilon_n.$$

Finally, we observe that $E_{11}(f) \geq E_{33}(f) = \varepsilon_1$ and the proof is complete. □

We note that infinitely many diagonal entries in the Walsh table for the functions constructed above contain certain polynomials.

While one may obtain, in this way, functions for which rational approximation is asymptotically not better than polynomial approximation, the situation is different in the class of Lip 1 functions. We recall that f is said to be a *Lip* α *function* where $0 < \alpha \leq 1$, if

$$|f(x) - f(y)| \leq M|x - y|^{\alpha}$$

for some $M > 0$.

2.9 Lip 1 Theorem (Popov 1977). *Let f be a Lip 1 function on a compact real interval. Then*

$$\lim_{n \to \infty} n E_{nn}(f) = 0.$$

An interesting proof of the theorem which for a long time has been a conjecture was given by Newman (1979).

On the other hand, $nE_{nn}(f)$ can converges to zero arbitrarily slowly for some Lip 1 function. Let (ε_n) be a positive sequence converging monotonically to zero. Then, as in the proof of the lethargy theorem, one can obtain a Lip 1 function f from a series of gap polynomials such that $nE_{nn}(f) \geq \varepsilon_n$ holds for infinitely many $n \in \mathbb{N}$. If moreover $\sum_n \varepsilon_n/n < \infty$, then the construction yields a Lip 1 function f with $nE_{nn}(f) \geq \varepsilon_n$ even for all $n \in \mathbb{N}$ (Exercise 2.12).

More recent results of Popov and Petrushev (1984, 1986) and of Peller (1983) establish connections between the asymptotics for rational L_p-approximation and for L_q-approximation by spline functions, where $p \neq q$.

Exercises

2.10. Consider the set of even functions in $C[-1, +1]$. Show that the metric projection from this subset into R_{mn}, $n \geq 1$, is continuous at f if and only if Pf belongs to a square in the Walsh table of length 2 or if $f \in R_{mn}$.

2.11. For which pairs (m, n) does $\sqrt{x}$ satisfy the (m, n)-normality condition for $x > 0$? Hint: Put $x = z^2$ and apply Descartes' Rule of signs for polynomials.

2.12. Let (ε_n) be a sequence converging downward to zero such that $\sum_n \varepsilon_n/n < \infty$. Show that

$$f = \sum_{k=1}^{\infty} \varepsilon_{3^k} \cdot 3^{-k} T_{3^k}$$

is a Lip 1 function and estimate $E_{nn}(f)$ from below.

2.13. Let $0 < \alpha < 1$ and let (ε_n) be a positive sequence converging monotonically to zero. Does there exist a Lip α function f such that $E_{nn}(f) \geq \varepsilon_n \cdot n^{-\alpha}$ for all n?

2.14. Show that the length of the blocks in the Walsh table may be unbounded even if f is not a rational function.

2.15. Show that each $(m + 2)$-dimensional subspace of $C[\alpha, \beta]$ contains a nonzero function which is not (m, n)-normal.
Hint: There is a simple proof which makes use of Exercise II.3.21. However, the analogy with Theorem 1.6 becomes more apparent if the Borsuk Antipodality Theorem is used directly.

§ 3. Rational Interpolation

A. The Cauchy Interpolation Problem

Given $f \in C[a, b]$ and $m + n + 1$ points

$$a \leq z_0 < z_1 < \cdots < z_{m+n} \leq b, \tag{3.1}$$

a rational fraction $u = p/q$, $p \in \Pi_m$, $q \in \Pi_n$ is to be determined such that

$$u(z_i) = f(z_i), \qquad i = 0, 1, \ldots, m + n. \tag{3.2}$$

A formula for a solution, (if it exists), was first described by Cauchy (1821). Though there is not always a solution $u \in R_{mn} \cap C[a, b]$, the linearized interpolation problem

$$p(z_i) - q(z_i) f(z_i) = 0, \qquad i = 0, 1, \ldots, m + n, \tag{3.3}$$

is always solvable with $p \in \Pi_m$, $q \in \Pi_n$, and $q \not\equiv 0$. Indeed, (3.3) may be considered as a homogeneous linear system of equations for $n + m + 2$ polynomial coefficients. For a nontrivial solution, we have $q \not\equiv 0$ because $q \equiv 0$ implies $p \equiv 0$. Moreover, all solutions of (3.3) are equivalent.

There are two possible cases in which a rational function p/q satisfying (3.3) will not be considered as a solution of the original problem:

1. There may arise quotients of the form $\frac{0}{0}$, e.g., for

$$z_0 = 0, \qquad z_1 = 1, \qquad z_2 = 2, \qquad y_0 = y_1 = 1, \qquad y_2 = -2. \tag{3.4}$$

The interpolation problem with $f(z_i) = y_i$ is solved formally in R_{11} by the degenerate rational expression $(x - 2)/(x - 2)$.

2. There may be poles between the points of interpolation. The rational fraction from R_{02} which interpolates the data (3.4) is $1/[1 - \frac{3}{4}x(x - 1)]$.

The first pathological case is strongly connected with the normality condition already met in Criterion 2.7.

3.1 Lemma (Werner and Braess 1969). *The following conditions on $f \in C[a, b]$, are equivalent*:

1°. *f satisfies the (m, n)-normality condition.*

2°. *For each set of knots satisfying* (3.1), *the solution of the interpolation problem* (3.3) *is not degenerate.*

3°. *For each set of knots $a \le z_0 \le z_1 < \cdots z_{m+n} \le b$, the denominator q of the solution of* (3.3) *satisfies*

$$s \cdot q(z_k) > 0, \qquad k = 0, 1, \ldots, m + n, \tag{3.5}$$

where s is either $+1$ or -1.

Proof. Assume that p/q is a degenerate solution of (3.3). After dividing p and q by a common factor $(x - z)$ we get polynomials $P \in \Pi_{m-1}$ and $Q \in \Pi_{n-1}$ such that $P - Qf$ has at least $n + m$ zeros, contradicting the normality condition. Conversely, let $P - Qf$ have $n + m$ zeros $\{z_i\}_{i=1}^{n+m}$, then obviously by $p = (z - z_0)P$ and $q = (z - z_0)Q$, a degenerate solution for some Cauchy problem is given. Consequently, 1° and 2° are equivalent.

For the same reason, 3° implies 2° and we have only to verify the converse.

Assume that f satisfies the normality condition. Since the best approximation $u_0 = P_{R_{mn}} f$ is not degenerate, $f - u_0$ has $m + n + 1$ zeros $x_0 < x_1 < \cdots < x_{m+n}$. Given $z_0 < z_1 < \cdots < z_{m+n}$, consider the one-parameter family $u_t = p_t/q_t$ of interpolants for the knots

$$z_k(t) = (1 - t)x_k + tz_k, \qquad 0 \le t \le 1, \qquad k = 0, 1, \ldots, m + n. \tag{3.6}$$

Since $q_t(z_k(t)) \neq 0$, we may fix the denominators by the normalization $q_t(z_0(t)) = 1$. From the nondegeneracy it follows that the mapping $t \mapsto (p_t, q_t)$ is continuous. Hence, the mapping $t \mapsto q_k(z_k(t))$, for $k = 0, 1, \ldots, m + n$, is also continuous. This proves (3.5). □

As a consequence, one obtains a sufficient condition for rational functions with linear denominators.

3.2 Corollary (Werner 1963). *If* $f \in C^m[z_0, z_{m+1}]$ *and* $f^{(m)}$ *has no zero in* (z_0, z_{m+1}), *then the interpolant in* R_{m1} *is non-degenerate and continuous between the knots.*

Proof. If $p \in \Pi_{m-1}$, then $(f-p)^{(m)} = f^{(m)} \neq 0$. By Rolle's theorem, $f-p$ has at most m zeros and $\{1, x, \ldots, x^{m-1}, f\}$ spans a Haar space. By Lemma 3.1, q does not change its sign between the knots if p/q interpolates f. Since q has at most one zero between the knots, it cannot have any. □

On the other hand, the normality condition is not sufficient for $n \geq 2$. The function e^x satisfies the condition for all m and n. But e^x is interpolated by

$$u(x) = \left(\frac{x+5}{x-5}\right)^2$$

at 5 points $z_2 = 0$, $-z_1 = z_3 \in [3,4]$ and $-z_0 = z_4 \in (5,6)$. The interpolant has a pole in (z_3, z_4).

B. Rational Functions with Real Poles

The study of the regularity of rational interpolants for $n \geq 2$ leads to the consideration of certain Hankel matrices and to a moment problem. Throughout this section we will assume that

$$l := m + 1 - n \geq 0 \tag{3.7}$$

and that $f \in C^{m+n-1}[a,b]$. Set

$$M_{mn} := M_{mn}(x,f) := \begin{pmatrix} D^{m-n+1}f & D^{m-n+2}f & \ldots & D^m f \\ D^{m-n+2} & D^{m-n+3}f & \ldots & D^{m+1}f \\ & \ldots & & \\ D^m f & D^{m+1}f & \ldots & D^{m+n-1}f \end{pmatrix}, \tag{3.8}$$

where $D^k f := f^{(k)}(x)/k!$ is the k-th coefficient of Taylor's series for an expansion at x (and $D^k f = 0$ whenever $k < 0$). In particular, if $m = n$ then $M_{nn}(x,f)$ coincides with the matrix $M_n(x,f)$ of Donoghue (1974).

3.3 Proposition. *If* $M_{mn}(x,f)$ *is definite in* $[a,b]$, *then* $Q^2 f - P$ *has at most* $m+n-1$ *zeros in* $[a,b]$ *counting multiplicities, whenever* $Q \in \Pi_{n-1}$, $P \in \Pi_{m+n-2}$, $Q \neq 0$.

Proof. We recall Leibniz' rule for the derivative of products:

$$D^k(g \cdot h) = \sum_{j=0}^{k} (D^{k-j}g) D^j h.$$

By applying it twice, we get

$$\begin{aligned} D^{m+n-1}(fQ)^2 &= \sum_{i=0}^{m-1} (D^{m+n-1-i}fQ)D^iQ \\ &= \sum_{i,j=0}^{m-1} (D^{m+n-1-i-j}f)(D^iQ)D^jQ. \end{aligned} \tag{3.9}$$

The last expression is understood to be the value of the quadratic form $M_{mn}(x, f)$ evaluated for the vector $(q(x), Dq(x), \ldots, D^{n-1}(x))$. The value is positive, since $(q(x), q'(x), \ldots, q^{(n-1)}(x)) \neq 0$ unless $q = 0$. Therefore $D^{m+n-1}(fQ^2 - P) = D^{m+n-1}(fQ^2) > 0$ for $x \in [a, b]$. By Rolle's theorem, $fQ^2 - P$ has at most $m + n - 1$ zeros. □

3.4 Theorem. *If $M_{mn}(x, f)$ is positive definite for $x \in [a, b]$, then the solution of the Cauchy problem* (3.2) *exists and is continuous in $[z_0, z_{m+n}]$. If moreover, $M_{m+1,n+1}$ is definite, then the solution is continuous in $[a, b]$.*

Proof. Let $p \in \Pi_{m-1}$, $q \in \Pi_{n-1}$. By Proposition 3.3, $q(qf - p)$ has at most $m + n + 1$ zeros. Therefore f satisfies the (m, n)-normality condition.

Let p/q be a solution of the linearized problem (3.3). First we observe that q has no zero of multiplicity ≥ 2. Suppose to the contrary that $q(x) = (x - \tilde{x})^2\tilde{q}(x)$. Then $\tilde{q}(qf - p) = Q^2f - P$, (where $Q = (x - \tilde{x})\tilde{q}$ and $P = p\tilde{q}$), has at most $n + m - 1$ zeros and p/q cannot be a solution.

Now given a Cauchy problem, we obtain a continuous solution by an adaptation of the path lifting lemma. Consider the one-parameter family p_t/q_t from the proof of Lemma 3.1. By construction, $q_0(x) > 0$ in $[z_0, z_{m+n}]$ and $q_t(z_0(t)) > 0$, $q_t(z_{m+n}(t)) > 0$. If $q_1(x) \leq 0$ for some x from $[z_0, z_k]$, there is a $t \in (0, 1)$ such that $q_t(x) \geq 0$, $x \in [z_0(t), z_{m+n}(t)]$, with equality in at least one point $\tilde{x}$. Therefore q_t has a double zero. But this possibility has been excluded in the first part of the proof.

The assumption that $M_{m+1,n+1}(x, f)$ is definite implies that the function $q(qf - p)$ has at most $m + n + 1$ zeros in $[a, b]$. Since the factor $(qf - p)$ already has that many zeros, q cannot have any. □

Two special cases may be described in an easier way. If $n = 1$, then only Corollary 3.2 is reproduced. For $n = 2$ positive definiteness of M_{m2} is equivalent to $[f^{(m-1)}]^{1/(m+1)}$ being positive and concave.

To get some more insight, we consider a subset of rational functions with real poles which are related to *Pick functions*. For convenience, choose $X = [-1, +1]$:

$$\begin{aligned} R^p_{mn} := \Big\{u(x) = p(x) + x^l \sum_{k=1}^{n} \frac{a_k}{1 - t_k x}: \\ p \in \Pi_{l-1}, a_k \geq 0, t_k \in (-1, +1), k = 1, 2, \ldots, n\Big\}. \end{aligned} \tag{3.10}$$

If $n = m$, then R^p_{mn} contains Pick functions, i.e., there is an analytic continuation into the upper half plane such that the imaginary part is positive (cf. Donoghue

1974). For $m > n$ the derived rational function $(u - p)/x^{m-n}$, p being as in (3.10) belongs to the Pick class.

At this point, we note that in this framework, the case $m \geq n$ may always be reduced to $m = n$. Assume that $M_{mn}(x, f)$ is definite and that $p = \sum_{j=0}^{m-n-1} D^j f(0) x^k$ is the Taylor polynomial for f. Then, for the reduced function

$$g(x) = [f(x) - p(x)]/x^{m-n} \tag{3.11}$$

the matrix $M_{nn}(x, g)$ is definite. On the other hand, definiteness of M_{mn} implies the same property for $M_{m+1,n-1}$.

Recalling that $(1 - tx)^{-1} = 1 + tx + \cdots + t^l x^l (1 - tx)^{-1}$, by elementary calculations one verifies

$$M_{mn}\left(x, \frac{x^l}{1 - tx}\right) = \left(\frac{t^{i+j}}{(1 - tx)^{l+i+j}}\right)_{i,j=0}^{n-1} = \frac{1}{(1 - tx)^l} b_t^T b_t, \tag{3.12}$$

where $b_t = (1, t(1 - t)^{-1}, \ldots, t^{n-1}(1 - tx)^{-n+1}) \in \mathbb{R}^n$. The matrix (3.12) has rank one and is positive semi-definite. The vectors $b_{t_1}, b_{t_2}, \ldots, b_{t_n}$ are linearly independent whenever $t_1, t_2, \ldots, t_n$ are disjoint. Therefore $M_{mn}(x, u)$ is positive definite if u is a non-degenerate function from R_{mn}^p.

The position of the poles is strongly connected with the Hankel matrix $M_{m+1,n}$. As in (3.12), we get

$$M_{m+1,n}\left(x, \frac{x}{1 - tx}\right) = \frac{t}{(1 - tx)^{l+1}} b_t^T b_t.$$

Consequently, given $u \in R_{mn}^p$, the number of positive eigenvalues of $M_{m+1,n}(x, u)$ equals the number of poles on the positive real axis.

3.5 Theorem (Braess 1974c). *If $M_{mn}(x, f)$ is positive definite in $[z_0, z_{m+n}]$ the solution of the Cauchy problem belongs to R_{mn}^p. If moreover $M_{m+1,n}(x, f)$ is positive or negative definite resp., then all poles of the interpolant are contained in (z_{m+n}, ∞) or in $(-\infty, z_0)$, resp.*

Proof. (1) Choose a non-degenerate rational function $u_0 \in R_{mn}^p$ and put $f_t = (1 - t)u_0 + tf, 0 \leq t \leq 1$. Since the set of functions with a positive definite Hankel matrix M_{mn} is a convex cone, it follows that $M_{mn}(x, f_t)$ is positive definite for $0 \leq t \leq 1$. Given $z_0 < z_1 < \cdots < z_{m+n}$, let p_t/q_t interpolate f_t. Since R_{mn}^p is closed in R_{mn} and $p_0/q_0 \in R_{mn}^p$, it follows from a continuity argument that $p_t/q_t \in R_{mn}^p$, $0 \leq t \leq 1$. In particular, the interpolant to $f = f_1$ belongs to R_{mn}^p.

(2) If moreover, $M_{m+1,n}$ is positive definite, then we may choose u_0 with all poles in (z_{m+n}, ∞). Then $M_{m+1,n}(x, u_0)$, and consequently $M_{m+1,n}(x, f_t)$ for $0 \leq t \leq 1$, are positive definite. It follows that f_t for $0 \leq t \leq 1$ satisfies the $(m + 1, n)$-normality condition. This implies that $\partial q_t = n$ whenever $p_t - f_t q_t$ has $n + m + 1$ zeros. During the deformation no pole can leave the positive real axis. This proves the second statement of the theorem. □

3.6 Representation Theorem. *Let $l \geq 0$. The following conditions on a function $f \in C[-1, +1]$ are equivalent:*

1°. *There is a nonnegative measure* μ *on* $(-1, +1)$ *which is not concentrated on a finite set and* $p \in \Pi_{l-1}$, *such that*

$$f(x) = p(x) + \int_{-1}^{+1} \frac{x^l}{1 - tx} d\mu(t). \tag{3.13}$$

Moreover $\int_{-1}^{+1} (1 - |t|)^{-1} d\mu(t) < \infty$.

2°. *For each pair* (m, n) *with* $m - n + 1 = l$, *the matrix* $M_{mn}(x, f)$ *is positive definite for* $x \in (-1, +1)$.

3°. *For each pair* (m, n) *with* $m - n + 1 = l$, *the best approximation to* f *from* R_{mn} *belongs to* R^p_{mn}.

Proof. From the representation (3.13), we conclude that the derivatives exist, and that

$$M_{mn}(x, f) = \int_{-1}^{+1} M_{mn}\left(x, \frac{x^l}{1 - tx}\right) d\mu(t)$$

is positive definite. Therefore 1° implies 2°.

If $M_{mn}(x, f)$ is definite, the nearest point u to f from R_{mn} is not degenerate. Hence $f - u$ has $m + n + 1$ zeros, and by Lemma 3.5 it follows that $u \in R^p_{mn}$. Consequently, 2° implies 3°.

Assume that condition 3° holds. By the Weierstraß theorem, the rational functions u_n which are best in $R_{n+l-1,n}$ converge to f. Since $u_n \in R^p_{n+l-1,n}$, we have a representation of the form

$$u_n(x) = p_n(x) + \int_{-1}^{+1} \frac{x^l}{1 - tx} d\mu_n \tag{3.14}$$

where $p_n \in \Pi_{l-1}$ and μ_n is a nonnegative measure concentrated onto n points.

First consider the case $l = 0$. Then

$$u_n(1) + u_n(-1) = \int_{-1}^{1} \frac{2}{1 - t^2} d\mu_n(t)$$

Hence, putting $d\tilde{\mu}_n = (1 - t^2)^{-1} d\mu_n$ we get a sequence of measures $\tilde{\mu}_n$ with bounded mass. By Helly's theorem (see e.g. Billingsley 1968) there is a non-decreasing function $\tilde{\mu}$ such that

$$\lim_{n_k \to \infty} \tilde{\mu}_{n_k}(t) = \tilde{\mu}(t)$$

holds for a subsequence (n_k) and all points at which $\tilde{\mu}$ is continuous. Moreover, by Helly's theorem, one has $\lim \int \varphi(t)\, d\tilde{\mu}_{n_k}(t) = \int \varphi(t)\, d\tilde{\mu}(t)$ as $n_k \to \infty$ for any $\varphi \in C[-1, +1]$. By putting $\varphi(t) = (1 - t^2)/(1 - tx)$, $|x| \leq 1$, we get the representation (3.13) for $l = 0$.

Let $l > 0$. If the polynomial part p_n from the sequence (3.14) is bounded, then by applying the arguments from the preceeding discussion to the reduced function (3.11), the representation is proved. If on the other hand, $\|p_n\| \to \infty$, then from $v_n := u_n/\|p_n\|$, $n = 1, 2, \ldots$, a limit with the representation like the right hand side of (3.13) is obtained, which has a polynomial part with $\|p\| = 1$. Since this is a representation for the zero function, a contradiction is obtained. The case $\|p_n\| \to \infty$ is thus excluded and the proof is complete. □

3.7 Examples.

(1) For $0 < a < 1$ one has

$$z^{-1}\log(1 + az) = \int_0^a \frac{dt}{1 + zt}, \quad z \in \mathbb{C}\backslash(-\infty, a^{-1}]$$

(2) Let $0 < \alpha < 1$, $a > 1$. Consider

$$\left.\begin{aligned} f(z) &= (z + a)^\alpha \\ \text{or} \quad f(z) &= -(z + a)^{-\alpha} \end{aligned}\right\} \quad \text{on } \mathbb{C}\backslash(-\infty, -a].$$

Obviously, the imaginary part of $f(z)$ is positive in the upper half plane. When the Cauchy integral for $f(z)$ and for $[f(z) - f(0)]z^{-1}$ is formed, after some standard deformation of the contour, only the integral along the cut on the negative real axis survives. Specifically, one gets

$$(1 + z)^{1/2} = 1 + \frac{1}{\pi}\int_0^1 \sqrt{t^{-1}(1 - t)}\,\frac{z}{1 + tz}\,dt,$$

$$(1 + z)^{-1/2} = \frac{1}{\pi}\int_0^1 [t(1 - t)]^{-1/2}\,\frac{1}{1 + tz}\,dt.$$

We conclude this section with a remark on the uniqueness of the measure in (3.13). Assume that f is given in the form (3.13). Let $|x| < 1$. Then the power series for $(1 - tx)^{-1}$ converges uniformly for $-1 \le t \le +1$, and it follows that

$$\begin{aligned} f(x) &= p(x) + \sum_{k=0}^{\infty} c_k x^{k+l}, \\ c_k &= \int_{-1}^{+1} t^k\, d\mu(t). \end{aligned} \tag{3.15}$$

Consequently, the derivatives of f at $x = 0$ are given by the moments.

It follows that the measure in (3.13) is unique. Suppose to the contrary that (3.13) holds with $\mu = \mu_1$ and $\mu = \mu_2$ where $\mu_1 \neq \mu_2$. Then there is a continuous function $\varphi \in C[-1, +1]$ such that $\int \varphi\, d(\mu_1 - \mu_2) \neq 0$. By the Weierstraß theorem, there is a polynomial q such that $\int q(t)\, d(\mu_1 - \mu_2)(t) \neq 0$. This contradicts the fact that μ_1 and μ_2 have the same moments. □

C. Comparison Theorems

By Bernstein's well known comparison theorem, a function $f \in C^{n+1}[a, b]$ can be approximated by polynomials from Π_n at least as well as the function g, if

$|f^{(n+1)}(x)| \le g^{(n+1)}(x)$, $x \in [a,b]$. A similar but weaker result will be established for rational approximation.

3.8 Comparison Lemma for Rational Interpolation. *Let $m \ge n-1$, and assume that*

$$0 \le M_{m+1,n+1}(x,f) \le M_{m+1,n+1}(x,g) \text{ for } x \in [a,b]. \tag{3.16}$$

If u_f and $u_g \in R_{mn}$ interpolate f and g, resp., at the points $a \le z_0 < z_1 < \cdots < z_{n+m} \le b$, then

$$|f(x) - u_f(x)| \le |g(x) - u_g(x)| \quad \text{for } x \in [a,b]. \tag{3.17}$$

Proof. The relations (3.16) mean that $M_{m+1,n+1}(x,f)$ and $M_{m+1,n+1}(x,g-f)$ are positive semi-definite. Firstly, we establish (3.17) under the stronger assumption that the matrices are positive definite.

Note that u_g also interpolates $h = g - (f - u_f)$ at the given points. From $u_f \in R^p_{mn}$ it follows that $M_{m+1,n+1}(x,h)$ is positive definite.

By Proposition 3.3, $f - u_f$ and $h - u_g$ have simple zeros at $z_0, z_1, \ldots, z_{n+m}$ and they change sign at these points. Moreover, both functions are positive for $x > z_{m+n}$. Hence,

$$f - u_f \quad \text{and} \quad (g - u_g) - (f - u_f)$$

have the same sign for $x \in [a,b]$. Thus $|(g-u_g)(x)| > |(f-u_f)(x)|$ for all $x \in [a,b] \backslash \{z_i\}_{i=0}^{n+m}$. This proves (3.17).

If the matrices are only semi-definite, choose a function φ with $M_{m+1,n+1}(x,\varphi)$ being positive definite. From the discussion above the inequalities apply to $f_\varepsilon := f + \varepsilon\varphi$ and $g_\varepsilon := g + 2\varepsilon\varphi$. By letting $\varepsilon \to 0$ we get the lemma. □

The preceding lemma will now be applied to the case where $z_0, z_1, \ldots, z_{m+n}$ are the zeros of the error function of the best approximation to g. Then we obtain

$$E_{mn}(f) \le \|f - u_f\| \le \|g - u_g\| = E_{mn}(g).$$

With this we have proved the

3.9 Comparison Theorem for Rational Approximation (Braess 1974c). *Let $m \ge n-1$, $f, g \in C^{m+n+1}[a,b]$ and assume that* (3.16) *holds. Then*

$$E_{mn}(f) \le E_{mn}(g).$$

In view of Theorem 3.6 the following is immediate

3.10 Corollary. *Let $m \ge n-1$. Assume that for f and g there are integral representations of the from* (3.13) *with non-negative measures μ_f and μ_g. If also $\mu_g - \mu_f$ is a non-negative measure, then*

$$E_{mn}(f) \le E_{mn}(g).$$

The asymptotic rate of rational approximation for Stieltjes functions can be given in terms of the *complete elliptic integrals of the first kind with modulus k*:

$$\mathsf{K}(k) = \int_0^1 \frac{dt}{\sqrt{(1-t^2)(1-k^2t^2)}} \quad \text{for } 0 < k < 1. \tag{3.18}$$

Also k' with $k^2 + k'^2 = 1$ is the *complementary modulus* and $\mathsf{K}'(k) := \mathsf{K}(k')$. In the following, $E = [a, b]$ and $F = [c, d]$ will be disjoint real intervals. The function

$$f(z) = \int_c^d \frac{d\mu(x)}{z - x},$$

where μ is a positive measure with $\mu(F) < \infty$, is analytic in $\mathbb{C}\backslash F$. Barrett's result from 1971 referred to symmetric domains, while Gončar (1978) determined the asymptotic rate without the condition on the symmetry. Moreover, Ganelius (1982) established slightly sharper estimates than (3.19) under slightly more restrictive conditions.

3.11 Theorem. *If $\mu' = d\mu/dx > 0$ almost everywhere on the interval F (with respect to the Lebesgue measure), then*

$$\lim_{n\to\infty} E_{n-1,n}(f, [a, b])^{1/n} = \frac{1}{\omega^2}, \tag{3.19}$$

where $\omega = \omega(E, F)$ is the Riemann modulus of the doubly-connected domain $G = \mathbb{C}\backslash(E \cup F)$, given by

$$\omega = \exp \frac{\pi \mathsf{K}(k)}{\mathsf{K}'(k)} \quad \text{and} \quad k^2 = \frac{(d-a)(c-b)}{(d-b)(c-a)}. \tag{3.20}$$

Gončar's interesting proof makes use of a different technique for the treatment of Stieltjes functions. He shows that rational interpolants are bounded on compact subsets of $\mathbb{C}\backslash F$. Moreover, some results of Zolotarov on rational functions, which are large on F and small on E, are used. The reader will find them in Section 5D, in Exercises 5.14–5.16, and in Theorem VI.3.4.

There is a simple proof for the special case where $\mu' = d\mu/dx$ is bounded from above and from below. Theorem 5.5 provides sharp bounds for $E_{n-1,n}(g_1)$ and $E_{n-1,n}(g_2)$ with $g_1(z) = [(z - c \pm \varepsilon)/(z - d \pm \varepsilon)]^{1/2}$ and $g_2(z) = 1/(1 + g_1(z))$. Then the comparison theorem yields (3.19).

Exercises

3.12. Let p/q be the best uniform approximation to $\sqrt{x}$ from $R_{m,m}$ or from $R_{m,m-1}$ (on an interval on the positive real line). Show that all coefficients of p and q are positive.
Hint: Use Descartes' rule of signs.

3.13. Verify the integral representations

$$\log z = \int_\infty^0 \left[\frac{1}{t-z} - \frac{1}{t^2+1}\right] dt,$$

$$\sqrt{z} = \frac{1}{\sqrt{2}} + \frac{1}{\pi} \int_{-\infty}^0 \left[\frac{1}{t-z} - \frac{1}{t^2+1}\right] \sqrt{t}\, dt.$$

3.14. Prove that the imaginary part of $\tan z$ is positive in the upper half plane.
Hint: Use the addition theorem for the tangent function.

3.15. Show that the rational function which interpolates e^x has no poles between the given knots, provided that the knots lie in a sufficiently small interval.

3.16. Verify the equivalence of the following conditions on $f \in C[a,b]$ (Haverkamp 1977):
1°. For each choice of $m + n + 1$ points in $[a,b]$, the solution of the interpolation problem (3.3) is continuous in $[a,b]$.
2°. For each $t \in [a,b]$ the induced function $f_t(x) = (x - t)f(x)$ satisfies the $(m + 1, n)$-normality condition.

3.17. Let $m \geq n - 1$ and $-1 \leq z_0 < z_1 < \cdots < z_{m+n} \leq 1$. Define $h\colon R^p_{mn} \to \mathbb{R}^{m+n+1}$ by $h(u) := (u(z_i))_{i=0}^{m+n}$. Show that the range of h is a convex set.

§4. Padé Approximation and Moment Problems

A. Padé Approximation

Let f be $(m + n)$ times differentiable at $x = 0$ and

$$f(x) = \sum_{k=0}^{m+n} c_k x^k + O(x^{m+n+1}). \tag{4.1}$$

We look for a rational function in R_{mn} whose power series expansion agrees with f as far as possible. Specifically, we look for a rational function p/q, $p \in \Pi_m$, $q \in \Pi_n$, such that

$$f(x) - p(x)/q(x) = O(x^{m+n+1}). \tag{4.2}$$

If $q(0) \neq 0$, then (4.2) is equivalent to

$$q(x)f(x) - p(x) = O(x^{m+n+1}). \tag{4.3}$$

Although, p and q are not uniquely determined by (4.3), the ratio p/q (in lowest terms) does, however, determine a unique rational fraction with $p \in \Pi_m$ and $q \in \Pi_n$. It is called the (m, n) *Padé approximant* of f.

Since the Padé approximant depends only on $c_0, c_1, \ldots, c_{n+m}$, often it is admitted that the given function $f(x) = \sum_{k=0}^{\infty} c_k x^k$ is only a formal power series.

The Padé approximation problem may be understood to be an interpolation problem with Hermite-type data, while the Cauchy problem refers to Lagrangean data. There is much resemblance with the Cauchy problem, e.g., (4.3) is considered as the linearized version of (4.2) just as (3.3) represented a linearized version of (3.2). When putting $p(x) = \sum a_i x^i$ and $q(x) = \sum b_j x^j$, one gets from (4.3):

$$\left(\sum_{k=0}^{n+m} c_k x^k\right)\left(\sum_{j=0}^{n} b_j x_j\right) = \sum_{i=0}^{m} a_i x_i + O(x^{m+n+1}). \tag{4.4}$$

This leads to the linear system

$$\sum_{j=0}^{n} c_{m-n+i+j} b_{n-j} = 0, \qquad i = 1, 2, \ldots, n, \tag{4.5}$$

$$\sum_{j=0}^{\min\{i,n\}} c_{i-j} b_j = a_i, \qquad i = 0, 1, \ldots, m, \tag{4.6}$$

with the convention that $c_k = 0$ for $k < 0$. This system of homogeneous equations for $n + m + 2$ coefficients always has a nontrivial solution. Specifically one has $q \not\equiv 0$.

Since the numerator and the denominator of a rational fraction are defined apart from a multiplying factor, we can put $b_0 = 1$. Then (4.5) becomes:

$$M_{mn}(0, f) \begin{pmatrix} b_n \\ b_{n-1} \\ \cdots \\ b_1 \end{pmatrix} = \begin{pmatrix} c_{m+1} \\ c_{m+2} \\ \cdots \\ c_{m+n} \end{pmatrix}. \tag{4.7}$$

Having the coefficients of the denominator, the a_i's can be calculated from (4.6).

The solution of (4.4) yields a Padé approximant. It is called *degenerate* if $q(0) = 0$ or if $\partial p \le m - 1$ and $\partial p \le n + 1$.

We mention that the notation in the literature is not always the same. Sometimes, only a solution of (4.2) is called a Padé approximant. Moreover, degeneracy is sometimes defined by the weaker condition that $\partial p \le m - 1$ *or* $\partial q \le n - 1$.

The rational approximations above were named after Henri Padé, who studied them in his thesis (1892), though he was not their discoverer. The foundations were already laid by Cauchy (1821), and older studies are cited in the historical remarks by Brezinski (1985). Cauchy's representation formula for rational interpolation was the starting point for Jacobi (1846) to discover Padé approximants. Frobenius (1881) and Padé (1892) arranged the approximants in a table with double entries, now known as the *Padé table.*

4.1 Padé's Block Theorem. *The Padé table can be completely dissected into $r \times r$ blocks with horizontal and vertical sides $r \ge 1$. The solutions of the linearized problem* (4.3) *within any block are equivalent. Specifically, let u_{mn} for $M \le m < M + r$, $N \le n < N + r$ form a block. Then we have, within the block:*

1°. *Solutions $u_{n,m}$ with $q(0) \ne 0$ exist for $m + n \le M + N + r - 1$ and equal u_{MN}. Moreover u_{mn} is non-degenerate if and only if $m = M$ or $n = N$.*

2°. *There is no solution of* (4.2) *in R_{mn} for $m + n \ge M + N + r$, and a formal solution of* (4.3) *is $(x^j/x^j)u_{MN}$ with $j = m + n - M - N - r$.*

Padé's block theorem follows from the fact that $u_{MN} = p/q$ with $q(0) \ne 0$ and $q(x)f(x) - p(x) = O(x^{M+N+r})$, with r being maximal.

The analogy to the Walsh table is obvious. The main point is that Padé as well as Chebyshev approximation are characterized by an equioscillation property, and from this the square block structure follows. Another table with the

M, N	$M+1, N$	$M+2, N$	$M+3, N$	$M+4, N$
$M, N+1$	$M+1, N+1$	$M+2, N+1$	$M+3, N+1$	
$M, N+2$	$M+1, N+2$	$M+2, N+2$		
$M, N+3$	$M+1, N+3$			
$M, N+4$				

Fig. 26. 5×5 block in Padé table. The labelled entries refer to existent Padé approximants with $q(0) \neq 0$.

same structure, i.e., the Carathéodory-Fejer table, was studied by Trefethen and Gutknecht (1985).

4.2 Proposition. *The* (m, n) *Padé approximant is non-degenerate if and only if*

$$\det M_{mn}(0, f) \neq 0. \tag{4.8}$$

If it is non-degenerate, it is unique up to a multiplying factor.

Proof. Condition (4.8) means that there is no nontrivial q_1 in Π_{n-1} such that at $x = 0$:

$$(q_1 f)^{(m)} = (q_1 f)^{(m+1)} = \cdots = (q_1 f)^{(m+n-1)} = 0.$$

This, in turn, is equivalent to the statement that $p_1 - q_1 f$ does not have a zero of order $m + n$, whenever $p_1 \in \Pi_{m-1}$, $q_1 \in \Pi_{n-1}$, and neither p_1/q_1 nor $(xp_1)/(xq_1)$ is a solution of (4.3). □

Examples for functions to which the proposition is easily applied, are e^x and $\sqrt{x + a}$ for $a > 0$.

In the *normal* case, the Padé approximant depends continuously on the given data (see Werner and Wuytack 1983, Trefethen and Gutknecht 1985). Moreover, Padé approximants with real poles may be treated with the same arguments as in the proof of Theorem 3.5:

4.3 Theorem. *Let* $m - n + 1 \geq 0$. *If*

$$(c_{m-n+i+j})_{i,j=0}^{n-1} \tag{4.9}$$

is positive definite, then the (m, n) *Padé approximant belongs to the family* R_{mn}^p. *If moreover*

$$(c_{m-n+1+i+j})_{i,j=0}^{n-1}$$

is positive definite (*or negative definite, resp.*), *then all poles of the approximant lie on the positive* (*on the negative, resp.*) *real axis.*

Proof. Let u_0 be a non-degenerate function from R_{mn}^p with positive poles. Let $c_k(t) = (1 - t)D^k u_0(0) + tc_k$ for $0 \leq t \leq 1$ and $k = 0, 1, \ldots, m + n$, and consider the matrix which results from (4.9) when c_k is replaced by $c_k(t)$. The matrix is positive definite for $0 \le t \le 1$. By Proposition 4.1, the Padé approximants to these data are non-degenerate. With this the statement of the theorem follows from a continuity argument as in Theorem 3.5. □

For other aspects of Padé approximation, see Gragg (1972), Brezinski (1980), Baker and Graves-Morris (1981), Cuyt and Wuytack (1986).

B. The Stieltjes and the Hamburger Moment Problem

Now we are in a position to solve the *Stieltjes moment problem.*

4.4 Theorem (Stieltjes 1894). *The following conditions on a sequence* $(c_0, c_1, c_2, \ldots)$ *are equivalent:*

1°. *For all* $n \ge 1$ *the Hankel determinants*

$$\det(c_{i+j})_{i,j=0}^{n-1} \quad \textit{and} \quad \det(c_{i+j+1})_{i,j=0}^{n-1}$$

are positive.

2°. *There is a nonnegative Borel measure which is not concentrated on a finite number of points such that*

$$c_k = \int_0^\infty t^k \, d\mu(t), \qquad k = 0, 1, 2, \ldots$$

3°. *There is a function analytic in the right hand plane of the form*

$$f(z) = \int_0^\infty \frac{d\mu(t)}{1 + tz}, \tag{4.10}$$

with a measure not concentrated on a finite number of points such that

$$c_k = \frac{(-1)^k}{k!} f^{(k)}(0), \qquad k = 0, 1, \ldots \tag{4.11}$$

Proof. Assume that (4.10) holds. For any z with $\operatorname{Re} z > 0$ we have

$$(-1)^k f^{(k)}(z) = k! \int_0^\infty \frac{t^k}{(1 + tz)^{k+1}} \, d\mu(t), \qquad k = 0, 1, \ldots$$

Note that the right hand side is positive for $z = x \in \mathbb{R}$ and that is is a monotone function in x. Hence,

$$f^{(k)}(0) = (-1)^k k! \int_0^\infty t^k \, d\mu(t), \qquad k = 0, 1, \ldots$$

and 2° is established.

Assume that c_k for $k = 0, 1, \ldots,$ is the k-th moment. Given $b \in \mathbb{R}^n$, $b \ne 0$, then $q(t) = \sum_{i=0}^{n-1} b_i t^i$ has at most $n - 1$ zeros and

$$0 < \int_0^\infty q^2(t)\,d\mu(t) = \sum_{i,j=0}^{n-1} b_i b_j \int_0^\infty t^{i+j}\,d\mu(t) = \sum_{i,j=0}^{n-1} c_{i+j} b_i b_j.$$

Hence, the matrix with the entries c_{i+j} is positive definite. Similarly, by integrating the non-negative polynomial $tq^2(t)$, the definiteness of (c_{i+j+1}) is established. Noting that the determinant of a positive definite matrix is positive, we have Condition 1°.

Assume that 1° holds. Then the matrices $(c_{i+j})_{i,j=0}^{n-1}$ and $(c_{i+j+1})_{i,j=0}^{n-1}$ are positive definite. By Theorem 4.3, the $(n-1, n)$ Padé approximant u_n has its poles on the positive real axis and it solves the truncated moment problem $D^k u_n(0) = c_k$ for $k = 0, 1, \ldots, 2n-1$. We may write

$$u_n(-x) = \int_0^\infty \frac{1}{1+tx}\,d\mu_n(t).$$

This defines a sequence of non-negative measures on $[0, \infty)$. Since $\int t\,d\mu_n(t) = c_1$, given $\varepsilon > 0$ and putting $T = c_1/\varepsilon$, we have:

$$\int_T^\infty d\mu_n(t) < \varepsilon, \qquad n = 1, 2, \ldots$$

Now it follows from the theorems of Helly and Prohorov (see e.g. Billingsley), that a subsequence of μ_n converges to a measure μ and that

$$\int_0^\infty t^k\,d\mu(t) = \lim_{n_i\to\infty} \int_0^\infty t^k\,d\mu_{n_i}(t) = c_k \quad \text{for } k = 0, 1, \ldots$$

In the same way, $f(z) = \lim_{n_i\to\infty} u_{n_i}(-z)$ for $\operatorname{Re} z \geq 0$ is established. Since each derivative of $f(z)$ and $u_n(-z)$ is bounded by the value at $x = 0$, the sequences are equicontinuous and the interpolation conditions (4.11) are also satisfied. □

A function f is called a *Stieltjes function* if it admits a representation of the form (4.10).

The solution of the moment problem, even if it exists, is not always unique. Let

$$c_k = \int_0^\infty t^k \cdot t^{-\log t}\,dt, \qquad k = 0, 1, \ldots \tag{4.12}$$

By elementary calculations one easily checks that

$$\int_0^\infty t^k t^{-\log t} \sin(2\pi \log t)\,dt = 0 \quad \text{for } k = 0, 1, \ldots$$

Therefore, given the moments $c_0, c_1, \ldots$ from (4.12), any measure given by $d\mu(t) = [1 + \lambda \sin(2\pi \log t)]t^{-\log t}\,dt$ where $0 \leq \lambda \leq 1$, provides a solution. Non-uniqueness is strongly connected with the fact (see Akhieser 1965) that the polynomials may not be dense in the Hilbert space $L_{2,\mu}$ equipped with the norm

$$\|\varphi\| := \left[\int_0^\infty |\varphi(t)|^2\,d\mu(t)\right]^{1/2}$$

(cf. Exercise 4.12).

Because of the nonuniqueness, it is natural to ask which one of the solutions of a moment problem is a limit of the Padé approximants. In getting the answer, the following lemma is an essential tool.

4.5 Lemma (Wynn 1968). *Let u_{mn} denote the (m,n) Padé approximant to a Stieltjes function f. Then*

$$u_{mn}(x) < u_{m+1,n+1}(x) < f(x) \quad \text{for } x > 0 \text{ if } m+1-n \text{ is even,}$$

and (4.13)

$$u_{mn}(x) > u_{m+1,n+1}(x) > f(x) \quad \text{for } x > 0 \text{ if } m+1-n \text{ is odd.}$$

Proof. First we consider the case where $m = n - 1$. Let $u_{n-1,n} = p/q$. From Theorem 4.3, we know that we may choose the denominator such that $q(x) > 0$ for $x \geq 0$. Since $M_{n,n+1}(x,f)$ is positive definite, from Proposition 3.4 we may conclude that $h := q(qf - p)$ has at most $n + (n+1) - 1 = 2n$ zeros. For the same reason, we have $h^{(2n)}(0) > 0$. Since $(qf - p)$ has a zero of multiplicity $2n$ at $x = 0$, it follows that $h(x) > 0$ for $x > 0$. Hence $u_{n-1,n}(x) < f(x)$ for $x > 0$.

Since $M_{n,n+1}(x, u_{n,n+1})$ is positive definite and $u_{n-1,n}$ is also the $(n-1,n)$ Padé approximant of $u_{n,n+1}$, the inequality $u_{n-1,n}(x) < u_{n,n+1}(x)$ for $x > 0$ is obtained by the same arguments. This proves the lemma for $l := m + 1 - n = 0$.

If $l > 0$ then the reduced function

$$g(x) = (-1)^l [f(x) - p(x)]/x^l \tag{4.14}$$

(see (3.11)) is a Stieltjes function. Let v be the $(n-1,n)$ Padé approximant to g. Then $p(x) + (-x)^l v(x)$ is the (m,n) Padé approximant to f. Hence the inequalities (4.13) are also true for $l > 0$. □

Now, the convergence of Padé approximants may readily be established.

4.6 Convergence Theorem. *Let f be a Stieltjes function. Then there is a Stieltjes function f_0 such that $f_0(x) \leq g(x)$ holds for each Stieltjes function g with $g^{(k)}(0) = f^{(k)}(0)$, $k = 0, 1, 2, \ldots$ The sequence of (m,n) Padé approximants where $m - n$ is constant, converges monotonely on $[0, \infty)$ to f_0.*

The monotone convergence is a direct consequence of the preceding lemma. From the proof of Theorem 4.4, it is clear that the limit of the sequence is a Stieltjes function. By (4.13), it has the extremal property as stated. □

For other convergence results see Saff (1983).

The *Hamburger moment problem* is related to rational functions which may have poles both on the positive and on the negative real line. Therefore, it will be solved by a superposition of two Stieltjes moment problems.

4.7 Theorem (Hamburger Moment Problem). *The following conditions on a sequence $(c_0, c_1, c_2, \ldots)$ are equivalent:*

1°. *For all $n \geq 1$ the Hankel determinants*

$$\det(c_{i+j-2})_{i,j=1}^{n} \tag{4.15}$$

are positive.

2°. *There is a non-negative Borel measure which is not concentrated on a finite number of points, such that*

$$c_k = \int_{-\infty}^{+\infty} t^k \, d\mu(t) \quad \textit{for } k = 0, 1, 2, \ldots$$

3°. *There are functions g and h analytic for* $\operatorname{Re} z \le 0$ *and* $\operatorname{Re} z \ge 0$, *resp., such that*

$$g(z) = \int_0^{\infty} \frac{d\mu(t)}{1 - tz} \quad \textit{and} \quad h(z) = \int_{-\infty}^{0} \frac{d\mu(t)}{1 - tz}$$

with a measure μ *not concentrated on a finite number of points and* $c_k = \frac{1}{k!}(g^{(k)}(0) + h^{(k)}(0))$ *for* $k = 0, 1, 2\ldots$

Proof. If the determinants (4.15) are positive, then the $(n-1, n)$ Padé approximant has the form

$$u_n(x) = \sum_{k=1}^{n} \frac{a_{nk}}{1 - t_{nk}x}.$$

It is split into two parts

$$v_n(x) = \sum_{(t_{nk} \ge 0)} \frac{a_{nk}}{1 - t_{nk}x}, \qquad w_n(x) = \sum_{(t_{nk} < 0)} \frac{a_{nk}}{1 - t_{nk}x}.$$

We note that $0 \le M_{n-1,n}(0, v_n) \le M_{n-1,n}(0, f)$ and $0 \le M_{n-1,n}(0, w_n) \le M_{n-1,n}(0, f)$. Therefore, one may choose subsequences such that $\lim v_{n'}^{(k)}(0)$ for $k = 0, 1, 2, \ldots$ exists. Now the rest of the proof proceeds as in Theorem 4.4. □

Exercises

4.8. Does there exist an analytic function for which the length of the blocks in the Padé table is not bounded?

4.9. Show the following comparison for Stieltjes functions: Assume that f, g and $g - f$ are Stieltjes functions. Then for any compact interval $[a, b]$, $a \ge 0$, the relation $E_{n-1,n}(f) \le E_{n-1,n}(g)$ holds.

4.10. Let μ be a solution of the moment problem. Verify that the polynomials are contained in the space $L_{2,\mu}$.

4.11. Let μ_1 and μ_2 be two nonnegative measures. Put $v_1 = (2\mu_1 + \mu_2)/3$ and $v_2 = (\mu_1 + 2\mu_2)/3$. Show that the spaces L_{2,v_1} and L_{2,v_2} coincide.

4.12. Prove that the moment problem is not uniquely solvable if and only if there is a solution μ such that the polynomials are not dense in $L_{2,\mu}$, (Achieser 1965).
Hint: Use Exercises 4.10 and 4.11.

§5. The Degree of Rational Approximation

The degree of rational approximation can only be determined in some very rare cases. Then peculiar properties of the underlying functions have to be used, and each case requires a specific technique. Therefore, we will only discuss some illuminating techniques. For example, rational approximation of the function e^x and $\sqrt{x}$ are also important in other fields of mathematics. Rational approximation of e^x is central for understanding implicit methods in the solution of stiff differential equations. When Heron's method (Newton's method) is applied to compute $\sqrt{x}$, a close to best rational approximation is generated.

A. Approximation of e^x on $[-1, +1]$

The degree of the approximation of e^x on the unit real interval can be evaluated from the approximation in the unit disk in $\mathbb{C}$, which in turn may be determined from the Padé approximant.

The following representation for the (m, n) Padé approximant p/q of e^x was probably first given by Perron:

$$\begin{aligned} p(z) &= \int_0^\infty t^n (t+z)^m e^{-t}\,dt, \\ q(z) &= \int_0^\infty (t-z)^n t^m e^{-t}\,dt. \end{aligned} \tag{5.1}$$

To verify this, we note that $q(0) = (n+m)! \neq 0$, and we consider the remainder term

$$\begin{aligned} e^z q(z) - p(z) &= \int_0^\infty (t-z)^n t^m e^{-t+z}\,dt - \int_0^\infty t^n (t+z)^m e^{-t}\,dt \\ &= -\int_0^{-z} t^m (t-z)^n e^{-t}\,dt \\ &= (-1)^n z^{n+m+1} \int_0^1 u^m (1-u)^n e^{uz}\,du. \end{aligned} \tag{5.2}$$

Since the integral in (5.2) is bounded for $|z| \leq 1$, we have $e^z q(z) - p(z) = O(z^{n+m+1})$, and (5.1) provides the Padé approximant. In order to estimate the integral in (5.2), we note that $|e^{uz} - e^{u_0 z}(1 + z(u - u_0))| \leq \frac{1}{2}(u - u_0)^2 |z|^2 e^{|z|}$ whenever $|u - u_0| \leq 1$. By choosing $u_0 = (n+1)/(m+n+2)$, we get

$$\begin{aligned} &\int_0^1 u_n (1-u)^m e^{uz}\,dz \\ &= \frac{m!n!}{(m+n+1)} e^{(n+1)/(m+n+2)z} \left(1 + \frac{\xi}{8(m+n+3)} |z|^2 e^{|z|}\right), \end{aligned} \tag{5.3}$$

where $0 \le \xi \le 1$. Moreover,

$$q(-z) = \int_0^\infty (t+z)^n t^m e^{-t}\,dt$$

$$= \sum_{k=0}^{n} \int_0^\infty \binom{n}{k} z^k t^{m+n-k} e^{-t}\,dt = \sum_{k=0}^{n} \binom{n}{k} (m+n-k)!z^k$$

$$= (m+n)! \sum_{k=0}^{n} \frac{(n-k+1)_k}{(m+n-k+1)_k} \frac{z^k}{k!},$$

where the *Pochhammer symbol* $(a)_k := a(a+1)\ldots(a+k-1)$ appears. Since $|(n-k+1)_k/(m+n-k+1)_k - n^k/(n+m)^k| \le k(k-1)/2(m+n-k+1)$, it follows that

$$q(z) = (m+n)!e^{-nz/(n+m)}(1+o(1)) \tag{5.4}$$

where the o-term may be bounded by $(2m+2n)^{-1}|z|^2e^{2|z|}$ [Perron 1957, p. 248]. From (5.3), (5.4), and $u_0 = \alpha/2 + o(1)$, it follows that

$$e^z - \frac{p(z)}{q(z)} = (-1)^n \frac{m!n!}{(m+n)!(m+n+1)!} z^{m+n+1} e^{\alpha z}(1+o(1)), \tag{5.5}$$

where $\alpha = 2n(n+m)^{-1}$. This error estimate for the Padé approximation is a generalization of the well known remainder term in the Taylor's series.

In contrast to the approximation in the complex plane, a factor of 2^{m+n} is gained for the approximation on the real interval.

5.1 Meinardus' Conjecture. *The degree of approximation on the unit interval* $[-1, +1]$ *is*

$$E_{mn}(e^x) = \frac{2^{-m-n}m!n!}{(m+n)!(m+n+1)!}(1+o(1)) \quad \text{as } m+n \to \infty.$$

Proof. The crucial point is an observation by Newman (1979). Let $p/q \in R_{mn}$. Given $x \in [-1, +1]$, set $z = (x+iy)/2$, where $x^2 + y^2 = 1$. Then the expression $u(x) = p(z)p(\bar{z})/q(z)q(\bar{z})$ is also an (m, n) degree rational function in the variable x. To see this, consider the product of linear polynomials

$$(az+b)(a\bar{z}+b) = abx + (b^2 + a^2/4) \quad \text{if } |z| = \tfrac{1}{2}.$$

Since $e^x = e^z \cdot e^{\bar{z}}$, from the formula for complex numbers $a\bar{a} - b\bar{b} = 2\,\mathrm{Re}\,\bar{a}(a-b) - |a-b|^2$, we obtain

$$e^x - u(x) = \mathrm{Re}\left\{e^{\bar{z}}\left(e^z - \frac{p(z)}{q(z)}\right)\right\} - \left|e^z - \frac{p(z)}{q(z)}\right|^2. \tag{5.6}$$

In the following, we only consider cases where the second term is small when compared with the first one. The following lemma deals with functions which admit a decomposition similar to $e^x = e^z \cdot e^{\bar{z}}$.

5.2 Lemma. *Let $r > 0$. Assume that f is a real analytic function in the disk $|z| < r$ and that $qf - p$ with $p/q \in R_{mn}$ has $n + m + 1$ zeros in the disk while q and f have none. Moreover, let $F(x) = f(z)f(\bar{z})$ where $|z| = r$, $\operatorname{Re} z = rx$. Then*

$$2 \min_{|z|=r} \left| \bar{f}\left(f - \frac{p}{q}\right) \right| \le E_{mn}(F)(1 + o(1)) \le 2 \max_{|z|=r} \left| \bar{f}\left(f - \frac{p}{q}\right) \right|. \tag{5.7}$$

Proof of Lemma 5.2. The upper bound is clear, since a treatment of $f(z)f(\bar{z}) - p(z)p(\bar{z})/q(z)q(\bar{z})$ as in (5.6) shows that the right hand side of (5.7) equals $\|F - u\|$ apart from terms of higher order.

The lower bound will be derived by using de la Vallée-Poussin's theorem (Braess 1984a). Note that

$$\operatorname{Re} \bar{f}\left(f - \frac{p}{q}\right) = \begin{cases} +\left| f\left(f - \dfrac{p}{q}\right) \right| & \text{if } \arg \bar{f}\left(f - \dfrac{p}{q}\right) \equiv 0 (\bmod 2\pi) \\ -\left| f\left(f - \dfrac{p}{q}\right) \right| & \text{if } \arg \bar{f}\left(f - \dfrac{p}{q}\right) \equiv \pi (\bmod 2\pi) \end{cases} \tag{5.8}$$

By assumption $f^{-1}q^{-1}(qf - p)$ has $m + n + 1$ zeros counting multiplicities but no pole in the unit disk. The winding number of this function is $n + m + 1$. The argument of $f^{-1}q^{-1}(qf - p) = \bar{f}(f - p/q)|f|^{-2}$ is increased by $(m + n + 1)2\pi$ when an entire circuit is performed. The argument is increased by $(m + n + 1)\pi$ as z transverses the upper half of the circle. Since the function is real for $z = +1$ and $z = -1$, we get an alternant of the desired length $m + n + 2$ and the lower bound follows from (5.8) and de la Vallée-Poussin's theorem. □

Proof of Meinardus' conjecture (continued). In order to get sharp estimates from (5.7), we will modify the Padé approximant. From (5.5) we have

$$e^z(e^z - p/q) = \text{const. } z^{n+m+1} e^{\beta z}(1 + o(1))$$

with $\beta = 1 + 2n(n + m)^{-1}$, $1 \le \beta \le 3$. The modulus of this expression is not constant on the circle $|z| = 1/2$ mainly because $e^{\beta z} z^{n+m+1}$ is not. We will see that, by choosing z_0 appropriately, the modified expression

$$|e^{\beta z}(z - z_0)^{n+m+1}| \tag{5.9}$$

deviates very little from a constant on the circle.

The Taylor's series for the logarithmic function yields the inequalities:

$$e^{-3/2N} \le \left| e^{\beta z}\left(1 - \frac{\beta z}{N}\right)^N \right| \le e^{3/2N}, \quad \text{whenever } |\beta z| \le \frac{3}{2},\ N \ge 5.$$

Therefore, for $|z|^2 = z\bar{z} = 1/4$, we get

$$e^{-3/2N} \le 2^N \left| e^{\beta z}\left(\bar{z} - \frac{\beta}{4N}\right)^N \right| \le e^{3/2N} \quad \text{for } |z| = \frac{1}{2},\ |\beta| \le 3, \text{ and } N \ge 5.$$

Consequently, when putting $N = n + m + 1$, $z_0 = \beta/[4(n + m + 1)]$, we obtain a close to circularity property for (5.9).

Let z_0 be as above. Then $\tilde{p}(z)/\tilde{q}(z) = e^{z_0}p(z-z_0)/q(z-z_0)$ is the Padé-approximant to e^z at z_0. From the circularity property of

$$e^z\left(e^z - \frac{\tilde{p}(z)}{\tilde{q}(z)}\right) = \frac{(-1)^n m!n!}{(m+n)!(m+n+1)!}(z-z_0)^{n+m+1}e^{\beta z}(1+o(1))$$

and Lemma 5.2, it follows that Meinardus' conjecture is true. □

B. Approximation of e^{-x} on $[0,\infty)$ by Inverses of Polynomials

Let $q_n(x) = \sum_{k=0}^n x^k/k!$ be the truncated power series to e^x. Then

$$\left\| e^{-x} - \frac{1}{q_n} \right\|_{[0,\infty)} \le 2^{-n}. \tag{5.10}$$

Since $1/q_n$ is the Padé approximant of degree $(0,n)$, the representation for the error of the preceding section can be used for proving (5.10):

$$e^{-x} - \frac{1}{q_n(x)} = -\frac{\displaystyle\int_0^x t^n e^{-t}\,dt}{\displaystyle\int_0^\infty (t+x)^n e^{-t}\,dt}.$$

Now, we have for $x \ge 0$:

$$0 \le \int_0^x t^n e^{-t}\,dt \le 2^{-n}\int_0^x (2t)^n e^{-t}\,dt$$
$$\le 2^{-n}\int_0^x (t+x)^n e^{-t}\,dt \le 2^{-n}\int_0^\infty (t+x)^n e^{-t}\,dt.$$

With this inequality, the quotient above can obviously be estimated by 2^{-n} and (5.10) is obtained.

The truncated power series, however, is not optimal:

5.3 Theorem (Schönhage 1973). *For the unifrom approximation on* $[0,\infty)$, *one has*

$$\frac{1}{12[n+3/2]^{1/2}} \le 3^n E_{0n}(e^{-x}) \le \sqrt{2}. \tag{5.11}$$

In particular, $\lim_{n\to\infty} E_{0n}(e^{-x})^{1/n} = 1/3$.

Proof. The normalized Laguerre-polynomials

$$L_n(x) = (e^x/n!)(e^{-x}x^n)^{(n)}, \qquad n = 0, 1, \ldots$$

are orthogonal in the sense that

$$\int_0^\infty L_n(x)L_m(x)e^{-x}\,dx = \delta_{nm}, \qquad n, m = 0, 1, \ldots$$

The expansion of $f(t) = e^{-t/4}$ with these orthogonal polynomials has the coefficients

$$c_k = \int_0^\infty e^{-t} e^{t/4} L_k(t)\, dt = \frac{1}{k!} \int_0^\infty (e^{-t} t^{-k})^{(k)} e^{t/4}\, dt = \frac{(-1)^k \cdot 4}{3^{k+1}}.$$

Here, the integrals are evaluated by k integrations by parts. The best polynomial in Π_n for the weighted L_2-problem is

$$g_n(x) = 4 \sum_{k=0}^{n} \frac{(-1)^k}{3^{k+1}} L_k(x). \tag{5.12}$$

Consequently,

$$\min_{p \in \Pi_n} \int e^{-t}(e^{-t/4} - p(t))^2\, dt = \sum_{k=n+1}^{\infty} c_k^2 = \frac{2}{3^{2n+2}}, \tag{5.13}$$

and the minimum is attained for $p = g_n$.

Instead of approximating e^{-x}, one may consider the function $e^{-x/4}$. Let p be the best uniform approximation, i.e.,

$$|e^{-x/4} - p(x)^{-1}| \le \lambda, \qquad x \ge 0,$$

where $\lambda = E_{0n}(e^{-x/4})$. The inequality is rewritten as

$$|p(x) - e^{x/4}| \le \lambda e^{x/4} p(x), \qquad x \ge 0. \tag{5.14}$$

To prove the lower bound, we assume that $\lambda \le \frac{1}{2} e^{-a/4}$, where $a = (4n + 6)\log 3$. Then

$$p_n(x) \le \frac{e^{x/4}}{1 - \lambda e^{x/4}} \le 2e^{x/4}, \qquad 0 \le x \le a,$$

and from (5.14) it follows that

$$|p(x) - e^{x/4}| \le 2\lambda e^{x/2}, \qquad 0 \le x \le a. \tag{5.15}$$

Next, consider the approximation on $[0, a]$ for the weighted supremum-norm with the weight function $e^{-x/2}$. The best approximation q from Π_n satisfies (5.15) and $\varepsilon(x) = q(x) - e^{x/4}$ has an alternant of length $n + 2$. Since $\varepsilon(x)$ cannot have more than $n + 1$ zeros, it has exactly $n + 1$ zeros with sign changes in $[0, a]$. From $\lim_{x\to\infty} \operatorname{sgn} \varepsilon(x) = -1$, it follows that $\lim_{x\to-\infty} \operatorname{sgn} q(x) = \lim_{x\to-\infty} \operatorname{sgn} \varepsilon(x) = (-1)^n$. Since q has the exact degree n, we have $\lim_{x\to+\infty} \operatorname{sgn} q(x) = +1$.

We claim that $q(x) \ge 0$ for $x \ge a$. Suppose to the contrary that $q(z) < 0$ for some $z > a$. Let $x_1 < x_2 < \cdots < x_{n+1}$ be the $n + 1$ zeros of ε. Specifically, $q(x_{n+1}) > 0$. By the mean values theorem, there are points

$$x_j' \in (x_j, x_{j+1}) \quad \text{with } \varepsilon'(x_j') = 0, j = 1, 2, \ldots, n,$$

$$z' \in (x_{n+1} z) \quad \text{with } q'(z') < 0.$$

By induction we get $\xi < \zeta$ such that $(e^{x/4} - q)^{(n)}(\xi) = 0$ and $q^{(n)}(\zeta) < 0$. Since $q^{(n)}$

is constant, this is a contradiction. Therefore,

$$0 \le q(x) \le e^{x/4} \quad \text{for } x \ge a,$$

which, together with (5.13) and (5.14) yields

$$\frac{2}{3^{2n+2}} \le \int_0^\infty e^{-t}(q(t) - e^{t/4})^2\,dt \le \int_0^a e^{-t}(2\lambda e^{t/2})^2\,dt + \int_a^\infty e^{-t/2}\,dt$$
$$= 4a\lambda^2 + 2e^{-a/2} = 4a\lambda^2 + 2\cdot 3^{-2n-3}.$$

With this, the lower bound for λ in (5.11) is established.

In order to obtain a good approximation, we note that

$$\frac{3}{4}e^x \int_x^\infty e^{t/4}e^{-t}\,dt = e^{x/4},$$

and consider the polynomial which is analogously constructed with g_n defined in (5.12):

$$p_n(x) = \frac{3}{4}e^x \int_x^\infty g_n(t)e^{-t}\,dt.$$

By the Cauchy-Schwarz inequality one gets

$$\begin{aligned} |e^{x/4} - p_n(x)| &\le \frac{3}{4}e^x \int_x^\infty |e^{t/4} - g_n(t)|\,e^{-t}\,dt \\ &\le \frac{3}{4}e^x \left(\int_x^\infty e^{-t}\,dt\right)^{1/2} \left(\int_x^\infty |e^{t/4} - g_n(t)|^2 e^{-t}\,dt\right)^{1/2} \qquad (5.16) \\ &\le \frac{2^{1/2}}{4\cdot 3^n}e^{x/2}, \qquad x \ge 0. \end{aligned}$$

Again let q_n be the best approximation from Π_n to $e^{x/4}$ on $[0, b]$, where $b = 4n\log 3 + \log 4$, for the weight function $e^{-x/2}$. Then (5.16) remains true for $0 \le x \le b$ if p_n is replaced by q_n. Hence,

$$q_n(x) \ge e^{x/4}\left(1 - \frac{2^{1/2}}{4\cdot 3^n}e^{x/4}\right) \ge \frac{1}{2}e^{x/4}.$$

Combining this and (5.16), we have

$$|e^{-x/4} - 1/q_n(x)| \le \frac{1}{2^{1/2}\cdot 3^n} \text{ for } 0 \le x \le b. \qquad (5.17)$$

With the same arguments as in the first part of the proof, q' has no zero for $x > b$, and $q(x) - e^{x/4}$ is negative:

$$\tfrac{1}{2}e^{b/4} < q_n(x) < e^x \quad \text{for } x > b,$$

which in turn yields

$$0 < 1/p_n(x) - e^{-x} < 2e^{-b/4} \quad \text{for } x > b.$$

With this and (5.17), the proof of the upper bound is complete. □

C. Rational Approximation of e^{-x} on $[0, \infty)$

The asymptotic behaviour of $E_{nn}(e^{-x})$ for the approximation on the infinite interval $[0, \infty)$ has generated much interest. Computations of Cody, Meinardus and Varga (1969) and Schönhage's result for $E_{0n}(e^{-x})$ gave rise to the 1/9-*conjecture*, namely that

$$\Lambda := \lim_{n\to\infty} E_{nn}(e^{-x})^{1/n}$$

exists, and is equal to 1/9.

The conjecture was recently disproved. Opitz and Scherer (1985) showed that $\bar{\Lambda} < 1/9.037$ (if the denotations $\bar{\Lambda}$ and $\underline{\Lambda}$ are used for the limes superior and limes inferior, resp.). Extended numerical computations of $E_{nn}(e^{-x})$ for $n \leq 30$ by Carpenter, Ruttan, and Varga (1985) and extrapolation indicate that

$$\Lambda = 9.289\,025\,491\,.. \tag{5.18}$$

Quite recently, Magnus (1986/7) announced an estimate of $\bar{\Lambda}$ via the Carathéodory-Fejer approximation (see Trefethen and Gutknecht 1983b). He stated that $\Lambda = \exp[-\pi \mathsf{K}'(k)/\mathsf{K}(k)]$ where the modulus k is determined by $\mathsf{K}(k) = 2\mathsf{E}(k)$. Here K and E are complete elliptic integrals, cf. Theorem 5.5. This theory leads to the number given in (5.18). On the other hand, there is a gap to the rigorous lower bound $\Lambda \geq 1/12.93$ given by Schönhage (1982).

Here, we will only present some rough bounds. First, we consider the (m, n) Padé approximant p/q to e^{-x}, where

$$p(x) = \int_0^\infty (t - x)^m t^n e^{-t}\, dt,$$

$$q(x) = \int_0^\infty t^m (t + x)^n e^{-t}\, dt.$$

The following error bound is easily obtained:

$$\|e^{-x} - p/q\|_{[0,\infty)} \leq 3^{-n}, \text{ if } n = 3m. \tag{5.19}$$

To verify this, we use the representation of the error function which was employed in the preceding sections:

$$e^{-x} - \frac{p(x)}{q(x)} = -\frac{\displaystyle\int_0^x (t - x)^m t^n e^{-t}\, dt}{\displaystyle\int_0^\infty t^m (t + x)^n e^{-t}\, dt}.$$

Note that for $x, t \geq 0$:

$$t(x+t)^3 = 27t^3(x-t) + t(x+7t)(x-2t)^2 \geq 27t^3(x-t).$$

Therefore,

$$0 \leq \int_0^x [t^3(x-t)]^m e^{-t}\,dt \leq 3^{-n} \int_0^x [t(x+t)^3]^m e^{-t}\,dt$$

$$\leq 3^{-n} \int_0^\infty t^m (x+t)^{3m} e^{-t}\,dt = 3^{-n} q(x)$$

and (5.18) is immediate. □

For the construction of better approximations, Opitz and Scherer (1985) used a more general ansatz:

$$p(x) = \int_0^\infty P(t-x) Q(t) e^{-t}\,dt,$$

$$q(x) = \int_0^\infty P(t) Q(t+x) e^{-t}\,dt.$$

Specifically, they chose $P(t) = (p_k)^{n/k}$, $Q(t) = (q_k)^{n/k}$, where the polynomials p_k, $q_k \in \Pi_k$ were computed such that the bound became as good as possible within their concept. In particular, for $k = 60$ they obtained the rigorous bound which disproved the 1/9-conjecture.

The first lower bound $E_{nn}(e^{-x}) \geq \frac{1}{2}1280^{-n}$ was established by Newman (1974). The following variant of this proof is due to Blatt and Braess (1980) and yields the estimate

$$E_{nn}(e^{-x}) \geq \tfrac{1}{2} 54^{-n}, \quad n = 1, 2, \ldots \tag{5.20}$$

First we observe that

$$E_{nn}(e^{-x}) \leq E_{n-1,n}(e^{-x}) \leq 2E_{nn}(e^{-x}). \tag{5.21}$$

To verify this, let p/q be the best approximation from R_{nn}. Then $p_1 = p - aq \in \Pi_{n-1}$ where $a = \lim_{x\to\infty} p(x)/q(x)$. Since $|a| \leq \|e^{-x} - p/q\|$, it follows that $\|e^{-x} - p_1/q\| \leq 2\|e^{-x} - p/q\|$, which proves (5.21).

We may consider the approximation of a^{-x} instead of e^{-x}, where $a > 2$ will be fixed later. Let $p \in \Pi_{n-1}$, $q \in \Pi_n$ and

$$\left| a^{-x} - \frac{p(x)}{q(x)} \right| \leq \lambda \quad \text{for } 0 \leq x \leq 2n.$$

We normalize q on the integers between 0 and $2n$ such that

$$\max_{0 \leq \nu \leq 2n} 2^{-\nu} |q(\nu)| = 1. \tag{5.22}$$

Thus, the function $r(x) := q(x) - a^x p(x)$ is bounded

$$|r(x)| \leq \lambda a^x |q(x)| \leq \lambda (2a)^x \quad \text{for } x = 0, 1, \ldots, 2n. \tag{5.23}$$

Let E denote the shift operator, which is defined by $Ef(x) = f(x+1)$. Moreover, let I be the identity and $\Delta = E - I$ the forward difference operator. From (5.23), one obtains for $x = 0, 1, 2, \ldots, n$:

$$\begin{aligned} |(E - aI)^n q(x)| = |(E - aI)^n \{q(x) - a^x p(x)\}| &\le (E + aI)^n |r(x)| \\ &\le \lambda (3a)^n (2a)^x. \end{aligned} \tag{5.24}$$

Note that $S := (E - aI)^n q \in \Pi_n$. Therefore, the series for the inversion of $(E - aI)^n$, which is obtained from Taylor's series of $(1-x)^{-n}$, contains only finitely many nonzero terms

$$\begin{aligned} q(x) = (E - aI)^{-n} S(x) &= (1-a)^{-n} \left(I - \frac{1}{a-1} \Delta \right)^{-n} S(x) \\ &= (1-a)^{-n} \sum_{j=0}^{n} \binom{n+j-1}{j} \frac{\Delta^j S(x)}{(a-1)^j}. \end{aligned}$$

If Δ^k is applied, then k more terms vanish:

$$\Delta^k q(x) = (1-a)^{-n} \sum_{j=0}^{n-k} \binom{n+j-1}{j} \frac{\Delta^{j+k} S(x)}{(a-1)^j}.$$

From (5.24), it follows that $|\Delta^m S(0)| \le \lambda (3a)^n (1+2a)^m$, and for $k = 0, 1, 2, \ldots, n$:

$$\begin{aligned} |\Delta^k q(0)| &\le \lambda \left(\frac{3a}{a-1} \right)^n \sum_{j=0}^{n-k} \binom{n+j-1}{j} \frac{(1+2a)^{j+k}}{(a-1)^j} \\ &\le \lambda \left(\frac{3a}{a-1} \right)^n (1+2a)^n \sum_{j=0}^{\infty} \binom{n+j-1}{j} \frac{1}{(a-1)^j} \\ &= \lambda \left(\frac{3a}{a-1} \right)^n (1+2a)^n \left[1 - \frac{1}{a-1} \right]^n = \lambda \left[\frac{3a(1+2a)}{a-2} \right]^n. \end{aligned}$$

Since $\Delta^k q = 0$ for $k > n$, we obtain for any $v \ge 0$ the estimate

$$|q(v)| = |(I + \Delta)^v q(0)| \le \sum_{k=0}^{v} \binom{v}{k} |\Delta^k q(0)| < \lambda \left[\frac{3a(1+2a)}{a-2} \right]^n 2^v. \tag{5.25}$$

When setting $a = 4$, we finally obtain $|q(v)| < \lambda \cdot 54^n \cdot 2^v$. In view of the normalization (5.22), it follows that

$$\lambda > 54^{-n},$$

and the proof of (5.20) is complete. □

Schönhage (1982) obtained the sharper bound $E_{nn}(e^{-x}) \ge 13.92^{-n}$ by a very involved analysis with suitable orthogonal polynomials.

D. Rational Approximation of $\sqrt{x}$

Formulas for best rational approximation to $\sqrt{x}$ on a positive interval $[\alpha, \beta]$ were already known to Zolotarov (1877) in the last century. He has provided the best uniform approximations with weight $x^{-1/2}$ in terms of elliptic functions (see also Achieser 1930). The asymptotic behaviour of the error has been otherwise determined with tools from the theory of harmonic functions and from the logarithmic capacity. The square root function is a Stieltjes functions and the $\lim E_{n,n}^{1/2n}$ essentially depends on the location of the support of the representing measure (see Theorem 3.11).

Here two different and more elementary methods will be combined. Firstly, estimates which are sharp for $\beta/\alpha \approx 1$, will be established via Newman's trick. Secondly, we analyse Heron's method, i.e. a very old method for the computation of square roots, from the viewpoint of rational approximation. This shows a strong connection with Gauss' arithmetic-geometric mean, (Braess 1984b) which makes it possible to determine the asymptotical behaviour of the error for any $\beta > \alpha$.

Throughout this section, the weighted Chebyshev approximation on an interval $[\alpha, \beta] \subset (0, \infty)$ with the weight function $w(x) = x^{-1/2}$ will be considered. Suppose that we know the best approximation u from R_{mn}. Then $\sqrt{c}u(x/c)$ is the solution for the best approximation problem on the interval $[c\alpha, c\beta]$, if $c > 0$. Moreover, the weighted error has the same norm. Thus, the solution for the interval $[\alpha, \beta]$ mainly depends on the ratio

$$\frac{\alpha}{\beta}. \tag{5.26}$$

When we shift the interval and consider $\sqrt{\rho + x}$ with $\rho > 1$ on $[-1, +1]$, then the argument of the square root lies between $\alpha = \rho - 1$ and $\beta = \rho + 1$, and $k := (\rho - 1)/(\rho + 1)$ is the characteristic parameter (5.26).

For the decomposition of the square root function in the sense of Lemma 5.2, we observe that

$$\left.\begin{aligned}(\rho + z)(\rho + \bar{z}) &= \rho^2 + 1 + 2\rho x\\ &= 2\rho(a + x)\end{aligned}\right\} \quad \text{for } |z| = 1,\ x = \operatorname{Re} z, \tag{5.27}$$

where $a = \frac{1}{2}(\rho + \rho^{-1})$. Therefore, when setting $f(z) = \sqrt{\rho + z}$, the induced function in the sense of Lemma 5.2 is $F(x) = \text{const.} \sqrt{a + x}$. The associated parameter is

$$k^2 = \frac{a - 1}{a + 1} = \left(\frac{\rho - 1}{\rho + 1}\right)^2. \tag{5.28}$$

Furthermore, ρ equals the sum of the semiaxes of that ellipse in $\mathbb{C}$ with foci $+1$ and -1 in which $F(x) = (a + x)^{1/2}$ is an analytic function.

The *stair case* elements of the Padé table, and more generally, the solutions of the Cauchy interpolation problem can be given in closed form. Let $x_1, x_2, \ldots, x_k$ be positive numbers (which need not be distinct). Set

$$h(w) = \sum_{i=1}^{k} (\sqrt{x_i} + w).$$

Then we define a rational function by $p(x) = \{h(\sqrt{x}) + h(-\sqrt{x})\}/2$, and $q(x) = \{h(\sqrt{x}) - h(-\sqrt{x})\}/2\sqrt{x}$, which interpolates $\sqrt{x}$ at $x_1, x_2, \ldots, x_k$. This follows from

$$\sqrt{x} - \frac{p(x)}{q(x)} = -\frac{1}{q(x)h(\sqrt{x})} \prod_{i=1}^{k} (x_i - x). \tag{5.29}$$

Obviously, $p/q \in R_{m,m}$ if $k = 2m + 1$ is odd, and $p/q \in R_{m,m-1}$ if $k = 2m$ is even.

Let $m = n$ or $m = n + 1$ and $k = m + n + 1$. The (m, n) Padé approximant to $f(z) = (\rho + z)^{1/2}$ is given by

$$\begin{aligned} p(z) &= \frac{1}{2}\{(\sqrt{\rho} + \sqrt{\rho + z})^k + (\sqrt{\rho} - \sqrt{\rho + z})^k\}, \\ q(z) &= \frac{1}{2\sqrt{\rho + z}}\{(\sqrt{\rho} + \sqrt{\rho + z})^k - (\sqrt{\rho} - \sqrt{\rho + z})^k\}, \end{aligned} \tag{5.30}$$

and the error may be written in the form

$$\sqrt{\rho + z} - \frac{p}{q} = -\frac{(\sqrt{\rho} - \sqrt{\rho + z})^k}{q(z)}.$$

Following Borwein (1982) we apply Newman's trick (see Section 5.A). Recalling (5.27) we obtain

$$\begin{aligned} \sqrt{a + x} - \frac{p(\bar{z})p(z)}{q(\bar{z})q(z)} &= -2\,\mathrm{Re}\sqrt{\rho + z}\,\frac{(\sqrt{\rho} - \sqrt{\rho + z})^k}{q(z)}(1 + o(1)) \\ &= 4\sqrt{a + x}\,\mathrm{Re}\frac{z^k}{(\sqrt{\rho} + \sqrt{\rho + z})^{2k} - z^k}(1 + o(1)). \end{aligned}$$

The rational function of x on the left hand side will be denoted by u_k. Since $\rho > 1$, we have $|z^k| = o(\rho^k)$ for $|z| = 1$ and

$$\sqrt{a + x} - u_k(x) = 2^{-2k}\rho^{-k}\sqrt{a + x}\,\mathrm{Re}\frac{z^k}{\varphi(z)^{2k}}(1 + o(1)),$$

where $\varphi(z) = (1 + \sqrt{1 + z/\rho})/2 = 1 + o(\rho^{-1})$. From this estimate and Lemma 5.2, we obtain

$$E_{mn}(\sqrt{x + a}) = \frac{4}{[(4 - \delta)\rho]^{m+n+1}}(1 + o(1)) \quad \text{for } n \le m \le n + 1, \tag{5.31}$$

where $|\delta| = |\delta(m, n, \rho)| \le 3\rho^{-1}$.

The accuracy of the best polynomial approximation is well known from Bernstein's results for analytic functions: $E_{m+n,0}(\sqrt{x + a}) \approx \text{const}\,\rho^{-m-n}$. Therefore, rational approximation with the same number of coefficients is better by a factor of $[4 - \delta]^{m+n}$.

The gap which is left in (5.31) is larger than the gap in the corresponding estimates for the exponential function. Here, it becomes small for $\rho \to \infty$, i.e., when the singularity is far away from the interval $[-1, +1]$.

At an early age, Gauss became enamored of a sequential procedure that has come to be known as the *arithmetic-geometric process*. Given two numbers $0 < y_0 < z_0$, one successively takes the arithmetic mean and the geometric mean:

$$y_{l+1} = \sqrt{y_l z_l}, \qquad z_{l+1} = \tfrac{1}{2}(y_l + z_l). \tag{5.32}$$

The common limit $\lim_{l\to\infty} y_l = \lim_{l\to\infty} z_l$ is called the *arithmetic-geometric mean* of (y_0, z_0). It can be expressed in terms of a complete elliptic integral

$$I(a,b) = \int_0^\infty \frac{dt}{\sqrt{(a^2+t^2)(b^2+t^2)}} = \int_0^{\pi/2} \frac{d\varphi}{\sqrt{a^2\cos^2\varphi + b^2\sin^2\varphi}}. \tag{5.33}$$

By the substitution $t = (x + ab/x)/2$ and some simple calculations, one sees that $I(y_l, z_l) = I(y_{l+1}, z_{l+1})$. From this invariance and the obvious formula $I(a, a) = \pi/2a$ Gauss deduced that the arithmetic-geometric mean of a and b equals $\pi/2I(a, b)$.

Set $\lambda_l = z_l/y_l$. From

$$\lambda_{l+1} = \frac{1}{2}\left(\sqrt{\lambda_l} + \frac{1}{\sqrt{\lambda_l}}\right), \qquad \lambda_l = (\lambda_{l+1} + \sqrt{\lambda_{l+1}^2 - 1})^2 \tag{5.34}$$

the fast convergence of the sequences (y_l) and (z_l) is obvious (cf. the table below). The associated quantitites μ_l and λ_l' with $\mu_l = (\lambda_l + 1)/\lambda_l - 1)$ and $\lambda_l^{-2} + (\lambda_l')^{-2} = 1$ are transformed as the λ's in (5.34), but in the reverse direction.

Now we turn to the iterative method often used for the computation of the square root of a given positive number x:

$$v_{l+1} = \frac{1}{2}\left(v_l + \frac{x}{v_l}\right), \qquad l = 0, 1, 2, \ldots \tag{5.35}$$

Table 2. Sequence of arithmetic-geometric process with $\lambda_0 = 1 + \sqrt{2}$ and $\lambda_l^{-2} + (\lambda_l')^{-2} = 1$

l	λ_l	$\dfrac{\lambda_l + 1}{\lambda_l - 1}$	λ_l'
-4	$6.825 \cdot 10^{14}$	$1 + 2.930 \cdot 10^{-15}$	$1 + 1.07 \cdot 10^{-30}$
-3	$1.306 \cdot 10^{7}$	$1 + 1.531 \cdot 10^{-7}$	$1 + 2.930 \cdot 10^{-15}$
-2	1807.08	1.001107	$1 + 1.531 \cdot 10^{-7}$
-1	21.26	1.099	1.001107
0	2.414	2.414	1.099
1	1.099	21.26	2.414
2	1.001107	1807.08	21.26
3	$1 + 1.53 \cdot 10^{-7}$	$1.306 \cdot 10^{7}$	1807.08
4	$1 + 2.930 \cdot 10^{-15}$	$6.825 \cdot 10^{-14}$	$1.306 \cdot 10^{7}$

The procedure is called *Heron's method* or sometimes the *Babylonian method*. Actually, successively better rational approximations of $\sqrt{x}$ are generated by the iteration. Indeed, $v_l \in R_{m,m-1}$ implies that $v_{l+1} \in R_{2m,2m-1}$. A start with a constant v_0 leads to $v_1 \in R_{10}$ and generates $v_l \in R_{m,m-1}$ with $m = m(l) = 2^{l-1}$ for $l \geq 1$.

A simple calculation shows that Heron's method combined with a normalization yields best rational approximations. This was possibly first discovered by Rutishauser (1963) and it was afterwards often used for the construction of best starting values for the iteration (see Taylor 1972). The starting point is the observation that from (5.35) we obtain the recursion

$$\frac{v_{l+1}(x) - \sqrt{x}}{v_{l+1}(x) + \sqrt{x}} = \left[\frac{v_l(x) - \sqrt{x}}{v_l(x) + \sqrt{x}}\right]^2. \tag{5.36}$$

Consider again the weighted Chebyshev approximation. Assume that

$$-\lambda \leq x^{-1/2}[u(x) - \sqrt{x}] \leq \lambda \quad \text{for } x \in [\alpha, \beta] \tag{5.37}$$

with some $\lambda < 1$. A simple calculation shows that $v = (1 - \lambda^2)^{-1/2} u$ satisfies

$$-\mu \leq \frac{v(x) - \sqrt{x}}{v(x) + \sqrt{x}} \leq \mu, \tag{5.38}$$

where

$$\frac{1 + \lambda}{1 - \lambda} = \left(\frac{1 + \mu}{1 - \mu}\right)^2. \tag{5.39}$$

Similarly, $w = (1 - \lambda)^{-1} u$ provides a one-sided approximation

$$0 \leq \frac{w(x) - \sqrt{x}}{w(x) + \sqrt{x}} \leq \lambda. \tag{5.40}$$

Moreover, the lower or the upper bound in (5.37) is attained at some $x \in [\alpha, \beta]$, if and only if the same holds for (5.38) and (5.40).

Now assume that u is the best weighted approximation to $\sqrt{x}$ from $R_{m,m-1}$. We know from Exercise 2.11 that $\sqrt{x}$ is normal and that $x^{-1/2}(u(x) - \sqrt{x})$ has an alternant of length $2m + 1$. Let v be the associated function in the sense of (5.38). One step of Heron's method yields a one-sided approximation $\tilde{w} = \frac{1}{2}(v + x/v) \in R_{2m,2m-1}$.

Recalling (5.40), we obtain a balanced approximation $\tilde{u}$ from $\tilde{w}$ with an alternant of length $4m + 1$. Specifically, the alternant of $\tilde{u}$ consists of the alternant of u and the zeros of $v(x) - \sqrt{x}$. Hence, $\tilde{u}$ is the best weighted approximation from $R_{2m,2m-1}$. From (5.36), (5.39), and (5.40) we conlude that

$$\frac{1 + E_{m,m-1}}{1 - E_{m,m-1}} = \left(\frac{1 + E_{2m,2m-1}^{1/2}}{1 - E_{2m,2m-1}^{1/2}}\right)^2.$$

This formula may be rewritten

$$E_{m,m-1}^{-1} = (E_{2m,2m-1}^{1/2} + E_{2m,2m-1}^{-1/2})/2, \tag{5.41}$$

$$E_{2m,2m-1} = \left[\frac{E_{m,m-1}}{1 + \sqrt{1 - E_{m,m-1}^2}}\right]^2. \tag{5.42}$$

Obviously, $E_{m,m-1}^{-1}$ behaves like the parameter λ' in the arithmetic-geometric process.

Note that $w_0 = \beta^{1/2}$ is the best constant for the one-sided approximation of $\sqrt{x}$ on the interval $[\alpha, \beta]$. Moreover, $0 \le (w_0 - \sqrt{x})/(w_0 + \sqrt{x}) \le (\sqrt{\beta} - \sqrt{\alpha})/(\sqrt{\beta} + \sqrt{\alpha})$. Thus, from (5.28) and (5.40), we calculate

$$E_{00}(\sqrt{x + a}) = \frac{\sqrt{\beta} - \sqrt{\alpha}}{\sqrt{\beta} + \sqrt{\alpha}} = \frac{1}{\rho}, \tag{5.43}$$

and $\rho = E_{00}^{-1}, E_{10}^{-1}, E_{21}^{-1}, E_{43}^{-1}, \ldots E_{2^l, 2^{l-1}}^{-1}, \ldots$ is a sequence which can be evaluated by the arithmetic-geometric process.

A similar method yields a useful transformation of the interval. It enables us to derive sharp estimates from (5.31), i.e. from a formula which originally is good only for large ρ.

5.4 Lemma. *Let $m \ge 1$ and (y_l), (z_l) be sequences according to the arithmetic-geometric process* (5.32). *Then*

$$E_{m,m-1}(\sqrt{x}, [y_{l+1}^2, z_{l+1}^2]) = E_{2m,2m-1}(\sqrt{x}, [y_l^2, z_l^2]). \tag{5.44}$$

Proof. Given a positive interval $[\alpha, \beta] = [y_l^2, z_l^2]$, the function

$$u_1(x) = \frac{x + y_l z_l}{y_l + z_l} \frac{2}{y_l z_l}$$

is a multiple of the best relative approximation to $\sqrt{x}$ from $\Pi_1 = R_{10}$. Some simple calculations show that

$$y_{l+1} \le \sqrt{\frac{x}{u_1^2(x)}} \le z_{l+1} \quad \text{for } y_l \le \sqrt{x} \le z_l.$$

Next, note that $\sqrt{x} = u_1(x)\sqrt{x/u_1^2(x)} = u_1(x) \cdot \sqrt{\xi}$, where $\xi = x/u_1^2(x)$. Let $p/q \in R_{m,m-1}$ be a best relative approximation to $\sqrt{x}$ for the interval $[y_{l+1}^2, z_{l+1}^2]$. Then

$$\frac{P(x)}{Q(x)} = u_1(x) \frac{p(x/u_1^2(x))}{q(x/u_1^2(x))} \in R_{2m,2m-1}$$

provides an approximation for the original interval with the same norm for the relative error as p/q on the smaller interval. An alternant of length $4m + 1$ is also generated from the alternant of length $2m + 1$, since $x \mapsto x/u_1^2(x)$ is monotone on the subintervals $[y_l^2, y_l z_l]$ and $[y_l z_l, z_l^2]$, respectively. Hence P/Q is the best approximation from $R_{2m,2m-1}$, and the proof is complete. □

The asymptotic rate of convergence will be established in terms of the elliptic integrals $\mathsf{K}(k)$, see (3.18). From the connection with the symmetric integrals (5.33):

$$I(a, b) = a^{-1}\mathsf{K}'\left(\frac{b}{a}\right), \quad \text{for } 0 < b < a,$$

the transformation rules for the arithmetic-geometric process are easily calculated

$$\mathsf{K}(k_1) = (1 + k)\mathsf{K}(k_1), \qquad \mathsf{K}'(k_1) = \frac{1 + k}{2}\mathsf{K}'(k), \tag{5.45}$$

where $k_1 = 2\sqrt{k}/(1 + k)$, $k_0 = k$, and $\lambda_l = 1/k_l$ $(l = 0, 1)$ provide the connection with (5.34). The asymptotic behaviour as $k \to 0$ and $k \to 1$ is well known:

$$\lim_{k\to 0} \mathsf{K}(k) = \frac{\pi}{2}, \qquad \lim_{k\to 1} \frac{\mathsf{K}(k)}{\log\frac{1}{k}} = 1. \tag{5.46}$$

Now we are in a position to establish sharp bounds. Note that they are in accordance with Theorem 3.11.

5.5 Theorem. *Let* $0 < k < 1$, *then*

$$(1 + \sqrt{1 - 16\omega^{-4m}})^2\omega^{-2m} \le E_{m,m-1}(\sqrt{x}, [k^2, 1]) \le 4\omega^{-2m}, \; m = 1, 2, \ldots \tag{5.47}$$

where

$$\omega = \omega(k) := \exp\left[\frac{\pi\mathsf{K}(k)}{\mathsf{K}'(k)}\right] \tag{5.48}$$

is the Riemann modulus of the doubly-connected domain $\mathbb{C}\setminus((-\infty, 0] \cup [k^2, 1])$ *and* $\sqrt{1 - 16\omega^{-4m}}$ *is to be replaced by* 0 *whenever* $\omega^m \le 2$.

Proof. From (5.31) it follows that

$$\varphi(k) := \liminf_{m\to\infty} E_{m,m-1}^{-1/2m}(\sqrt{x}, [k^2, 1])$$

exists and is not smaller than ρ, where ρ is given by (5.28). Furthermore, observe that $\varphi(k)$ does not change if we take the limes inferior only for even m, since $\rho^{1/m} \to 1$ as $m \to \infty$. Now we conclude from the preceding lemma that $\varphi(k_1) = \varphi^2(k)$ where as above, $k_1 = 2\sqrt{k}/(1 + k)$. Next, recalling (5.28) and (5.31), the asymptotic behaviour is immediate

$$\lim_{k\to 1} \frac{\varphi(k)}{4\dfrac{1 + k}{1 - k}} = 1$$

We claim that $\varphi(k)$ is already uniqueley determined by these properties and equals $\omega(k)$.

Indeed, the transformation rules (5.45) give rise to the functional equation $\mathsf{K}(k_1)/\mathsf{K}'(k_1) = 2\mathsf{K}(k)/\mathsf{K}'(k)$. Therefore, the auxiliary function $\psi(k) := \log\varphi(k)/\log\omega(k) = \log\varphi(k)\cdot\mathsf{K}'(k)/\pi\mathsf{K}(k)$ satisfies the functional equation

$$\psi(k_1) = \psi(k).$$

Next, the formulas (5.46) yield $\pi \mathsf{K}(k)/\mathsf{K}'(k) \to 2\log(1/k') = \log[1/(1-k^2)]$ as $k \to 1$. From this and the asymptotics of $\varphi(k)$ we have

$$\lim_{k\to 1} \psi(k) = 1.$$

Since ψ is invariant under the arithmetic-geometric process, given $k \in (0, 1)$ there is a $\tilde{k}$ which is arbitrarily close to 1 such that $\psi(k) = \psi(\tilde{k})$. Consequently,

$$\psi(k) = 1 \quad \text{for } 0 < k < 1.$$

Since the same arguments apply to the limes superior, we obtain from $\psi \equiv 1$:

$$\lim_{m\to\infty} E_{m,m-1}^{-1/2m}(\sqrt{x}, [k^2, 1]) = \omega(k).$$

From the result for the asymptotic rates, the estimates (5.47) for finite m will be derived. First, (5.42) implies that

$$\tfrac{1}{4}E_{2m,2m-1} \geq (\tfrac{1}{4}E_{m,m-1})^2. \tag{5.49}$$

Therefore, when considering $(\frac{1}{4}E_{M,M-1})^{-1/2M}$ for $M = m$, $2m$, $4m$, $8m$, $\ldots$ we get a monotone decreasing sequence with limit ω and $(\frac{1}{4}E_{m,m-1})^{-1/2m} \geq \omega$. This gives the upper bound for $E_{m,m-1}$ of (5.47). Similar, dividing (5.42) by $(1 + \sqrt{1 - E_{m,m-1}^2})^2$ and noting that $E_{m,m-1} \geq E_{2m,2m-1}$ yields

$$\frac{E_{2m,2m-1}}{(1 + \sqrt{1 - E_{2m,2m-1}^2})^2} \leq \left[\frac{E_{m,m-1}}{(1 + \sqrt{1 - E_{m,m-1}^2})^2}\right]^2, \tag{5.50}$$

Therefore, when considering $[(1 + \sqrt{1 - E_{M,M-1}^2})^{-2}E_{M,M-1}]^{-1/2M}$ for $M = m$, $2m$, $4m$, $8m$, $\ldots$, we get a monotone increasing sequence with limit ω. This, together with the fact that $E_{m,m-1} \leq \min\{4\omega^{-2m}, 1\}$, yields the lower bound. □

A fast computation of $\omega(k)$ is possible due to the asymptotic behaviour

$$\omega(k) = 4\frac{k+1}{k-1}(1 + o(1)) \quad \text{as } k \to 1,$$

in connection with the Landen transformation considered above. A variant is considered in Exercise 5.12.

E. Rational Approximation of $|x|$

Though the square root function has a singularity at $z = 0$, the rational approximation on an interval of the form $[0, b]$, say on $[0, 1]$, is still very good. Instead of looking for n-th degree rational approximations to $\sqrt{x}$, one may approximate x by expressions of the form $p(x^2)/q(x^2)$ on $[0, 1]$. By the invariance principle, this is equivalent to the approximation of the even function $|x|$ on $[-1, +1]$.

The first step was done by Newman (1964) who showed that $\frac{1}{2}e^{-9\sqrt{n}} \le E_{n+1,n}(|x|) \le e^{-\sqrt{n}}$. Later, Bulanov (1968) und Vjačeslavov (1975) improved the bounds:

$$e^{-\pi\sqrt{n+1}} \le E_{n,n}(|x|) \le ce^{\pi\sqrt{n}}. \tag{5.51}$$

Gončar's more general conjecture, that $\lim\{n^{-1/2}\log E_{n,n}(x^\alpha,[0,1])\} = -\pi\sqrt{2\alpha}$ for any positive non-integer α, was proved by Ganelius (1979).

In order to avoid some technical difficulties we will confine ourselves to the weaker inequalities

$$e^{-\pi\sqrt{n+2}} \le E_{n+1,n}(|x|) \le 6ne^{-\pi\sqrt{n+7}} \quad \text{for } n \ge 3, \tag{5.52}$$

which provide the correct constant in the exponent.

To prove the lower bound, let v be the nearest point to $|x|$ from $R_{n-1,n-2}$. Set $\lambda := \||x| - v\|$. After multplying the numerator and the dominator by x^2, we may write $v(x) = xg(x)/h(x)$ for $x \ne 0$ with $g, h \in \Pi_n$. From the symmetry properties $g(-x) = -g(x)$ and $h(-x) = h(x)$ one gets the representation

$$v(x) = x\frac{p(x) - p(-x)}{p(x) + p(-x)} \tag{5.53}$$

with $p = g + h$. The error may be written in the form

$$|x| - v(x) = x\frac{2p(-x)}{p(x) + p(-x)} = 2x\frac{p(-x)}{p(x)}\cdot\frac{1}{1 + \dfrac{p(-x)}{p(x)}} \quad \text{for } x > 0. \tag{5.54}$$

From the quotient in the expression in the middle, it follows that $|p(-x)| \le |p(x)|$ for $x \ge \lambda$. Therefore $|p(x) + p(-x)| \le 2|p(x)|$ and

$$\lambda \ge ||x| - v(x)| \ge \left|x\frac{p(-x)}{p(+x)}\right| \quad \text{for } \lambda \le x \le 1. \tag{5.55}$$

5.6 Lemma. *Given $p \in \Pi_n$, $p \ne 0$, there is a point in $[e^{-\pi\sqrt{n}}, 1]$ where*

$$\left|x\frac{p(-x)}{p(x)}\right| > e^{-\pi\sqrt{n}}.$$

Proof. Let $\xi = u + iv$ by any complex number and $0 \le a < b$. Then

$$\int_a^b \log\left|\frac{t+\xi}{t-\xi}\right|\frac{dt}{t} > -\frac{\pi^2}{2}. \tag{5.56}$$

Indeed, we have for $t \ge 0$:

$$\left|\frac{t+\xi}{t-\xi}\right| = \left(\frac{(t+u)^2 + v^2}{(t-u)^2 + v^2}\right)^{1/2} \ge \left(\frac{(t-|u|)^2}{(t+|u|)^2}\right)^{1/2} = \left|\frac{t-|u|}{t+|u|}\right|.$$

We may clearly assume that $u \ne 0$. Then

$$\int_a^b \log\left|\frac{t-\xi}{t+\xi}\right|\frac{dt}{t} \geq \int_a^b \log\left|\frac{t-|u|}{t+|u|}\right|\frac{dt}{t} = \int_{a/|u|}^{b/|u|} \log\left|\frac{t-1}{t+1}\right|\frac{dt}{t}$$

$$\geq \int_0^\infty \log\left|\frac{t-1}{t+1}\right|\frac{dt}{t} = -\frac{\pi^2}{2}.$$

Next set $\delta = \exp(-\pi\sqrt{n})$, and assume that $|xp(-x)/p(x)| \leq \exp(-a\sqrt{n})$ for $\delta \leq x \leq 1$, $a > 0$. Then

$$\int_\delta^1 \log\left|t\frac{p(-t)}{p(t)}\right|\frac{dt}{t} \leq -a\sqrt{n}\int_\delta^1 \frac{dt}{t} = -a\pi n. \tag{5.57}$$

On the other hand, one has

$$\log\left|t\frac{p(-t)}{p(t)}\right| = \log t + \sum_\xi \log\left|\frac{t+\xi}{t-\xi}\right|,$$

where ξ runs through the zeros of p. Noting that $\int_\delta^1 t^{-1}\log t\,dt = -n\pi^2/2$ and applying (5.56) to each term in the sum, we get

$$\int_\delta^1 \log\left|t\frac{p(-t)}{p(t)}\right|\frac{dt}{t} > -\frac{n}{2}(\pi^2 + \pi^2).$$

The comparison of this inequality with (5.57) gives $a < \pi$. □

From the lemma above, we conclude that λ in (5.55) must be larger than $\exp(-\pi\sqrt{n})$ and the lower bound in (5.52) is established.

In order to get rational functions which are close to the optimal one, a polynomial from Π_{2n} of the form

$$P(x) = \prod_{k=0}^{n-1} (x + \xi_k)(x + \xi_k^{-1}) \tag{5.58}$$

with appropraite knots ξ_k will be constructed. Note that $z_k := 2n - 2\sqrt{n(n-k)}$, $k = 0, 1, \ldots, n$, satisfy

$$\int_0^{z_k} \left(1 - \frac{t}{2n}\right) dt = k. \tag{5.59}$$

Now let $a = \pi/2\sqrt{n}$ and

$$\xi_k = e^{-az_k},\ k = 0, 1, \ldots, n.$$

Assume that $\xi_{j-1} < x < \xi_j$ for $1 \leq j \leq n$. Write $x = e^{-ay}$. By making use of the monotonicity of the function

$$h(t) := \frac{e^{-ay} - e^{-at}}{e^{-ay} + e^{-at}},\ t \geq y,$$

and recalling (5.59) we estimate the contribution of $\xi_j, \xi_{j+1}, \ldots, \xi_n$ to $P(-x)/P(x)$:

$$\begin{aligned}\log \prod_{k=j}^{n-1} \frac{x-\xi_k}{x+\xi_k} &= \log \prod_{k=j}^{n-1} h(z_k)\\ &= \sum_{k=j}^{n-1} \log h(z_k)\cdot \int_{z_k}^{z_{k+1}} \left(1-\frac{t}{2n}\right) dt \qquad (5.60)\\ &\le \int_{z_j}^{z_n} \log h(t)\left(1-\frac{t}{2n}\right) dt = \int_y^{z_n} \dots - \int_y^{z_j} \dots .\end{aligned}$$

To obtain a bound for the second integral, note that $\delta := \int_y^{z_j} (1 - t/2n)\, dt \le 1$. Since the weight function $(1 - t/2n)$ is smaller than 1, the monotonicity of h implies that

$$\begin{aligned}\int_y^{z_j} \log h(t)(1-t/2n)\, dt &\ge \int_y^{y+\delta} \log h(t)\, dt \ge \int_y^{y+1} \log h(t)\, dt\\ &= \int_0^1 \log \frac{1-e^{-at}}{1+e^{-at}}\, dt \ge -1.6 - \frac{1}{2}\log n.\end{aligned}$$

The last inequality is obtained by the following integral inequality, which is derived from $(1-e^{-t})/(1+e^{-t}) > t/e$ for $0 < t < 1$:

$$\int_{e^{-x}}^1 \log \frac{1-t}{1+t}\frac{dt}{t} = \int_0^x \log \frac{1-e^{-t}}{1+e^{-t}}\, dt > x(2-\log x), \qquad 0 < x \le 1. \quad (5.61)$$

Another inequality needed below is obtained by making use of the power series for $\log[(1-t)/(1+t)]$:

$$\int_0^x \log \frac{1-t}{1+t}\frac{dt}{t} \ge -\frac{\pi^2}{4}x, \qquad 0 < x \le 1, \qquad (5.62)$$

with equality if $x = 1$.

Now we proceed with estimating (5.50):

$$\log \prod_{k=j}^{n-1} \frac{x-\xi_k}{x+\xi_k} \le \int_y^{z_n} \log \frac{e^{-ay}-e^{-at}}{e^{-ay}+e^{-at}}\left(1-\frac{t}{2n}\right) dt + 1.6 + \frac{1}{2}\log n. \quad (5.63)$$

Since the weight function $(1 - t/2n)$ is negative for $t > z_n = 2n$, we obtain the upper bound for (5.63):

$$\begin{aligned}&\int_y^\infty \log \frac{e^{-ay}-e^{-at}}{e^{-ay}+e^{-at}}\left(1-\frac{t}{2n}\right) dt + 1.6 + \frac{1}{2}\log n\\ &= \int_0^\infty \log \frac{1-e^{-at}}{1+e^{-at}}\left(1-\frac{y+t}{2n}\right) dt + 1.6 + \frac{1}{2}\log n\\ &= \left(1-\frac{y}{2n}\right) a^{-1} \int_0^1 \log \frac{1-t}{1+t}\frac{dt}{t} - \frac{1}{2n}\int_0^\infty \log \frac{1-e^{-at}}{1+e^{-at}}\, t\, dt + 1.6 + \frac{1}{2}\log n\\ &= -\frac{\pi}{2}n^{1/2} - \frac{1}{2}\log x + 2.1 + \frac{1}{2}\log n. \qquad (5.64)\end{aligned}$$

Similarly when the product for $k < j$ is evaluated, we obtain an inequality analogous to (5.63):

$$\log \prod_{k=0}^{j-1} \frac{\xi_k - x}{\xi_k + x} \leq \int_0^y \log \frac{e^{-at} - e^{-ay}}{e^{-at} + e^{-av}} \left(1 - \frac{t}{2n}\right) dt + 1.6 + \frac{1}{2} \log n$$
$$= \int_0^y \log \frac{1 - e^{-at}}{1 + e^{-at}} \left(1 - \frac{y - t}{2n}\right) dt + 1.6 + \frac{1}{2} \log n. \tag{5.65}$$

Moreover, with the same arguments as in (5.60), one gets

$$\log \prod_{k=0}^{n-1} \frac{\xi_k^{-1} - x}{\xi_k^{-1} + x} \leq \int_{-z_n}^{0} \log \frac{e^{-at} - e^{-ay}}{e^{-at} + e^{-ay}} \left(1 - \frac{|t|}{2n}\right) dt$$
$$= \int_y^{y+2n} \log \frac{1 - e^{-at}}{1 + e^{-at}} \left(1 + \frac{y - t}{2n}\right) dt. \tag{5.66}$$

Note that $\left(1 - \dfrac{y + t}{2n}\right)$ is negative for $t > 2n$ and that it is smaller than the weight factors in (5.65) and (5.66). Therefore combining (5.65) and (5.66), we conclude that

$$\log \prod_{k=0}^{j-1} \frac{\xi_k - x}{\xi_k + x} \prod_{k=0}^{n-1} \frac{\xi_k^{-1} - x}{\xi_k^{-1} + x}$$
$$\leq \int_0^\infty \log \frac{1 - e^{-at}}{1 + e^{-at}} \left(1 - \frac{y + t}{2n}\right) dt + 1.6 + \frac{1}{2} \log n \tag{5.67}$$
$$= -\frac{\pi}{n} \sqrt{n} - \frac{1}{2} \log x + 2.1 + \frac{1}{2} \log n.$$

Combining (5.64) and (5.67) we have $|P(-x)/P(x)| \leq \dfrac{n}{x} \exp(-\pi\sqrt{n} + 4.2)$ for $\xi_n \leq x \leq 1$. From this and the fact that $P(x) \geq P(-x) > 0$ for $0 \leq x < \xi_n$, it follows that

$$\left| x \frac{P(-x)}{P(x)} \right| \leq n e^{-\pi\sqrt{n} + 4.2} \quad \text{for } 0 \leq x \leq 1. \tag{5.68}$$

Now we transform the variable $z = 2x(1 + x^2)^{-1}$ and the knots $\zeta_k = 2\xi_k(1 + \xi_k^2)^{-1}$ and introduce the polynomial:

$$p(z) = \prod_{k=0}^{n-1} (z + \zeta_k).$$

Noting that

$$\frac{-z + 2(1 + \zeta^2)^{-1}}{z + 2(1 + \zeta^2)^{-1}} = \frac{(-x + \xi)(-x + \xi^{-1})}{(x + \xi)(x + \xi^{-1})} \quad \text{for } \xi > 0,$$

it follows that

$$\max_{0\le z\le 1}\left|z\frac{p(-z)}{p(z)}\right| = \max_{0\le x\le 1}\left|\frac{2x}{1+x^2}\,\frac{P(-x)}{P(x)}\right| < \lambda := ne^{-\pi\sqrt{n+5}}.$$

Let $z_0 = 3\lambda$. The inequality

$$\left|z\frac{p(-z)}{p(z)}\right| < \lambda \quad \text{for } z_0 \le z \le 1, \tag{5.69}$$

remains true if we replace the knots by $\zeta_k := \max\{(z_0, 2\xi_k(1+\xi_k^2)^{-1}\}$ for $k = 0, 1, \ldots, n-1$, without changing the notation. Then we have $|p(-z)/p(z)| \le 1/3$ for $z_0 \le z \le 1$ and $p(z) \ge p(-z) > 0$ for $0 \le z < z_0$. Recalling (5.54), the proof of the upper bound in (5.52) is complete. □

Exercises

5.7. In order to compare rational and polynomial approximation for the square root function, derive upper and lower bounds for $E_n(\sqrt{x+a})$ from the well known result for $E_n(1/(x+a))$, via Bernstein's comparison theorem.

5.8. How good is the result for $E_n(1/(x+a))$ in polynomial approximation which is obtained via Newman's trick?

5.9. Let $\lim m_i = \infty$, $\lim m_i/n_i = \beta (0 < \beta < 1)$, and p_i/q_i be the (m_i, n_i) Padé approximant to e^{-x}. Show that

$$\lim_{i\to\infty} \sqrt[n]{\|e^{-x} - p/q\|_{[0,\infty)}} = \frac{\beta^\beta(1-\beta)^{1-\beta}}{2^{1-\beta}}$$

(cf. Saff, Varga, and Ni 1976). Moreover, show that $\|e^{-x} - p/q\|_{[0,\infty)} \le 8^{-n/2}$, if p/q is the (m, n) Padé approximant and $n = 2m$.

5.10. Let $\lambda_n = \inf\{\|e^{-x} - 1/q^2(x)\|_{[0,\infty)}: q \in \Pi_n\}$. Show that $\lim \lambda_n^{1/n} = 1/5$.

5.11. Determine $E_{3m,3m-1}(\sqrt{x}, [\alpha, \beta])$ from $E_{m,m-1}(\sqrt{x}, [\alpha_1, \beta_1])$, where β_1/α_1 is given in terms of $E_{11}(\sqrt{x}, [\alpha, \beta])$.

5.12. Let ω be as in Theorem 5.5. Show that $(1 + \sqrt{1 - E_{m,m-1}^2})^2/E_{m,m-1} \le \omega^{2m} \le 4/E_{m,m-1}$ for $m \ge 1$ and that $E_{10}^{-1} = \rho$ implies that

$$(1 + \sqrt{1-\rho^{-2}})^2 \cdot \rho \le \omega \le 2(\rho + \sqrt{\rho^2 - 1}) < 4\rho.$$

How many transformations from the arithmetic-geometric process are necessary when $\lim_{m\to\infty} E_{m,m-1}^{1/m}(\sqrt{2+x})$ is to be computed with 8 significant digits?

5.13. Let $v \in R_{n,n-1}$ be the best approximation to $|x|$ on the interval $[-1, +1]$. Multiply numerator and denominator only by x. This yields the representation $v(x) = x[p(x) + p(-x)]/[p(x) - p(-x)]$. Prove the lower bound in (5.51) with this observation.

The next three Exercises establish the connection of the theorems 3.11 and 5.5 with the ideas of Zolotarev (1877), see also Gončar (1967):

5.14. Let $E = [a, b]$ and $F = [c, d]$ be disjoint real intervals. Show that the quantity $k^2 := (d-a)(c-b)(d-b)^{-1}(c-a)^{-1}$ is invariant under Möbius transformations $z \mapsto (\alpha z + \beta)/(\gamma z + \delta)$ whenever $\alpha\delta - \beta\gamma \ne 0$.

5.15. There is a Möbius transformation which sends the intervals $(-\infty, 0)$ and $[k^2, 1]$ to the intervals $(-1, -k_1^2)$ and $[k_1^2, 1]$. Express k_1 in terms of k.

5.16. Let $0 < a < b$ and let $\omega > 1$. Show that the following are equivalent:
1°. There exists a sequence (u_n) with $u_n \in R_{nn}$ such that

$$|u_n(x)| \le 1 \quad \text{for } x \in [-b, -a] \quad \text{and} \quad |u_n(x)| \ge c_1 \omega^{2n} \quad \text{for } x \in [a, b]$$

with some $c_1 > 0$.
2°. There exists a sequence (p_n) with $p_n \in \Pi_n$ such that

$$\left|\frac{p(x)}{p(-x)}\right| \le c_2 \omega^{-n} \quad \text{for } x \in [a, b]$$

with some $c_2 > 0$ (and $|p(z)/p(-z)| = 1$ for $\operatorname{Re} z = 0$).
3°. For the Chebyshev approximation with the weight function $w(x) = x^{-1/2}$, one has

$$E_{n+1,n}(\sqrt{x}, [a^2, b^2] \le c_3 \omega^{-2n},$$

for some $c_3 > 0$, cf. Theorem VI.3.4.
4°. There exists a sequence (B_n), where B_n is a *Blasche product* of order n such that

$$|B_n(x)| \le c_4 \omega^{-n} \quad \text{for } x \in \left[\frac{a-1}{a+1}, \frac{b-1}{b+1}\right]$$

with some $c_4 > 0$. By definition, a Blaschke product of order n is a function of the form

$$B_n(z) = \prod_{k=1}^{n} \frac{z - \bar{\alpha}_k}{1 - \alpha_k z} \quad \text{with } |\alpha_k| < 1 \ (k = 1, 2, \ldots, n).$$

Obviously, $|B_n(z)| = 1$ for $|z| = 1$. These functions play an important role for rational approximation in Hardy spaces (Loeb and Werner 1974, Newman 1979, Andersen and Bojanov 1984). Hint: Consider the Möbius transformation $w = (z + 1)/(z - 1)$.

5.17. Show that Meinardus' conjecture is also true for relative rational approximation of e^x, i.e. if the error is weighted with e^{-x} (Nemeth 1977).

5.18. Let $\mu_0 > 1$ and $\mu_{l+1} = (\mu_l + \sqrt{\mu_l^2 - 1})^2$ for $l = 0, 1, \ldots$. Determine $\lim_{l \to \infty} (\mu_l)^{1/2^l}$.

5.19. Let p/q be the best approximation to $\sqrt{x}$ from $R_{m,m-1}$ on an interval on the positive real line. Show that the zeros and poles of p/q lie on the negative real line. Hint: The zeros and poles of p/q are the poles of the best approximation from $R_{2m,2m-1}$.

5.20. Assume that $m \ge n$ and $p/q \in R_{mn}$. Given $x = (z + z^{-1})/2 \in [-1, +1]$, let $u(x) = p(z)/q(z) + p(z^{-1})/q(z^{-1})$. Show that u is an (m, n) degree rational function and compare this well known construction with Newman's trick.

§ 6. The Computation of Best Rational Approximations

Best rational approximations are often used for the representation of special functions. Therefore, with the advent of high-speed digital computers came an interest in methods of computing best rational approximations with respect to the supremum norm. Two algorithms have turned out to be favorable, the *differential correction algorithm* and the (nonlinear) *Remes algorithm*. Specifically, the differential correction algorithm (properly initialized), possesses global convergence properties. It is more robust than the Remes algorithm, although the latter is faster if it converges. For a systematic comparison of several algorithms, we refer to Lee and Roberts (1973).

A. The Differential Correction Algorithm

The differential correction algorithm has a history which is interesting in terms of the relation between practical use and the existence of convergence proofs. The algorithm was first described by Cheney and Loeb (1961). A modified version of the method was subsequently considered by the same authors in 1962, and this was shown to have sure convergence properties. The modified version was more popular than the original one, although it was known to be slower, until Barrodale, Powell and Roberts (1972) showed that the method in its original form is globally convergent. Moreover, it is quadratically convergent whenever the given function is normal.

In the following, X will be a finite set or an interval and $\|f\| = \sup_{x\in X}|f(x)|$. We note that some caution is required, since there may be no best approximation on a finite set. A rational function P/Q with $P \in \Pi_m$ and $Q \in \Pi_n$ is *admissable* if

$$Q(x) > 0 \quad \text{for } x \in X. \tag{6.1}$$

Moreover, we will use the normalization

$$\|Q\| = 1. \tag{6.2}$$

6.1 The Differential Correction Algorithm (Original Version).
Step 1 (Initialization). Choose an initial approximation $p_0/q_0 \in R_{mn}$. Set $k = 0$.
Step 2. Determine polynomials $P \in \Pi_m$, $Q \in \Pi_n$ with $\|Q\| = 1$, to minimize

$$\max_{x\in X}\left\{\frac{|f(x)Q(x) - P(x)| - \Delta_k Q(x)}{q_k(x)}\right\} \tag{6.3}$$

where

$$\Delta_k = \|f - p_k/q_k\|. \tag{6.4}$$

Step 3. Unless the approximation is good enough, set $p_{k+1} = P$, $q_{k+1} = Q$, $k = k + 1$ and return to Step 2.

We note that in the k-th step of the modified version, the expression

$$\max_{x\in X}\{|f(x)Q(x) - P(x)| - \Delta_k Q(x)\} \tag{6.5}$$

is minimized instead of the quantity (6.3). The auxiliary minization problems can be solved by linear programming techniques if X is finite. The extension to infinite X is straightforward in view of the well known Remes algorithm (see Cheney and Loeb 1962, Watson 1980).

The motivation for the method comes about by expressing a product of two quantities by the linear terms in a Taylor's series expansion and is as follows:

For any choice of admissable polynomials P and Q, let Δ be a number such that

$$|f(x) - P(x)/Q/x)| \leq \Delta \tag{6.6}$$

holds for all $x \in X$. This relation can be rewritten as

$$|f(x)Q(x) - P(x)| \leq \Delta Q(x) \approx \Delta_k Q(x) + (\Delta - \Delta_k) q_k(x),$$

if we neglect the second order term $(\Delta - \Delta_k)(Q - q_k)$. This may be rewritten as

$$\frac{|f(x)Q(x) - P(x)| - \Delta_k Q(x)}{q_k(x)} + \Delta_k \lesssim \Delta. \tag{6.7}$$

Thus, from (6.7), the quantity Δ is minimized approximately by computing the minimum of (6.3).

It is an essential feature of the differential correction algorithm that on every iteration the condition (6.1) is maintained automatically. Roughly speaking, when minimizing the expression (6.3), due to the last term in the numerator, there is a tendency to generate a Q which attains large values on X.

6.2 Lemma. *If q_k satisfies Condition* (6.1) *and if $\Delta_k > E_{mn}(f)$, then $\Delta_{k+1} < \Delta_k$ and $q_{k+1}(x) > 0$ for $x \in X$.*

Proof. Since $\Delta_k > E_{mn}(f)$, there exists some $\bar{p}/\bar{q} \in R_{mn}$ with $\|\bar{q}\| = 1$ such that $\|f - \bar{p}/\bar{q}\| = \bar{\Delta} < \Delta_k$. Set $\eta := \min_{x \in X} |\bar{q}(x)|$. Now by the definition of $p_{k+1} = P$ and $q_{k+1} = Q$,

$$\begin{aligned}
&\max_{x \in X} \left\{ \frac{|f(x) q_{k+1}(x) - p_{k+1}(x)| - \Delta_k q_{k+1}(x)}{q_k(x)} \right\} \\
&\leq \max_{x \in X} \left\{ \frac{|f(x)\bar{q}(x) - \bar{p}(x)| - \Delta_k \bar{q}(x)}{q_k(x)} \right\} \\
&= \max_{x \in X} \left\{ \left[\left| f(x) - \frac{\bar{p}(x)}{\bar{q}(x)} \right| - \Delta_k \right] \frac{\bar{q}(x)}{q_k(x)} \right\} \\
&\leq -\min_{x \in X} \left\{ [\Delta_k - \bar{\Delta}] \frac{\bar{q}(x)}{q_k(x)} \right\} \leq -\eta[\Delta_k - \bar{\Delta}] < 0.
\end{aligned} \tag{6.8}$$

The expression in the first line of this sequence can only be less than (or equal to) $-\eta[\Delta_k - \bar{\Delta}]$, if

$$q_{k+1}(x)/q_k(x) \geq \eta[\Delta_k - \bar{\Delta}]/\Delta_k. \tag{6.9}$$

Hence, $q_{k+1}(x) > 0$. Finally, from

$$\begin{aligned}
\left| f - \frac{p_{k+1}}{q_{k+1}} \right| &= \Delta_k + \frac{q_k}{q_{k+1}} \frac{|f q_{k+1} - p_{k+1}| - \Delta_k q_{k+1}}{q_k} \\
&\leq \Delta_k - \frac{q_k}{q_{k+1}} \cdot \eta[\Delta_k - \bar{\Delta}],
\end{aligned} \tag{6.10}$$

it follows that also $\Delta_{k+1} < \Delta_k$ is true. □

6.3 Theorem (Barrodale, Powell and Roberts 1972). *If X is a finite set, then* $\lim_{k\to\infty} \|f - p_k/q_k\| = E_{mn}(f)$.

Proof. The lemma above shows that the sequence Δ_k decreases monotonically and is bounded below, so it has a limit $\tilde{\Delta}$. Suppose that $\tilde{\Delta} > E_{mn}(f)$. Then we can choose $\bar{P}/\bar{Q} \in R_{mn}$ such that $\bar{\Delta} := \|f - \bar{P}/\bar{Q}\| < \tilde{\Delta}$. Following the notation of the proof of Lemma 6.2 and setting $c = \min\{\frac{1}{2}, \eta(1 - \bar{\Delta}/\tilde{\Delta})\}$, we deduce from (6.10) that

$$\left| f(x) - \frac{p_{k+1}(x)}{q_{k+1}(x)} \right| \leq \Delta_k \left(1 - c\frac{q_k(x)}{q_{k+1}(x)} \right). \tag{6.11}$$

Since the sequence (Δ_k) converges, $\min\{q_k/q_{k+1}\}$ tends to zero as $k \to \infty$ and we find $\xi_k \in X$, $k = 1, 2, \ldots$ such that

$$\lim_{k\to\infty} \frac{q_k(\xi_k)}{q_{k+1}(\xi_k)} = 0. \tag{6.12}$$

Now we assume that X consists of N points, i.e. $X = \{x_1, x_2, \ldots, x_N\}$. By (6.12), there is an integer K such that

$$q_{k+1}(\xi_k) \geq c^{-N} q_k(\xi_k)$$

holds for all $k \geq K$. Moreover, we rewrite (6.9) as $q_{k+1}(x) \geq c q_k(x)$ and use this relation for all $x \in X$, $x \neq \xi_k$, to obtain

$$\prod_{i=1}^{N} q_{k+i}(x_i) \geq \frac{c^{N-1}}{c^N} \prod_{i=1}^{N} q_k(x_i) \geq 2 \prod_{i=1}^{N} q_k(x_i).$$

It follows that $\prod q_k(x_i)$ diverges as $k \to \infty$, contradicting $\|q_k\| = 1$. □

If the approximation problem is considered on an infinite set X, an extra condition can provide sure convergence. Let X be compact, and assume that

$$\|f - p_0/q_0\| \leq E_{m-1,n-1}(f). \tag{6.13}$$

In this case, one has $\|f - p_k/q_k\| \leq d$ for $k = 1, 2, \ldots$ with $d = \|f - p_1/q_1\| < E_{m-1,n-1}(f)$. By Lemma 1.1, no subsequence of (q_k) converges to a polynomial with a zero in X. Therefore, $\min_{x\in X}\{q_k(x)\}$ is bounded from below, say by $\eta^* > 0$. Now from (6.10), we obtain $\Delta_{k+1} \leq \Delta_k - \eta^* \cdot \eta[\Delta_k - \bar{\Delta}]$, and $\Delta_{k+1} - \bar{\Delta} \leq (1 - \eta^*\eta)[\Delta_k - \bar{\Delta}]$. Therefore, $\lim \Delta_k = \bar{\Delta}$ and (p_k/q_k) converges to the best approximation.

The starting condition (6.13) is no serious restriction, since one may start with the solution for $R_{m-1,n-1}$. More generally, the best approximations from $R_{m-r,n-r}$, $R_{m-r+1,n-r+1}$, $\ldots$, $R_{m,n}$ may be successively computed with $r = \min\{m, n\}$. A sequence of approximation problems is thus solved with sure convergence in each step.

The rate of convergence is quadratic, if f is (m,n)-normal. As in the Gauss-Newton method, quadratic convergence is strongly connected with strong uniqueness. Here, a strong uniqueness formula is required which is suitable for the normalization (6.2).

6.4 Lemma. *Let X contain at least $m + n + 2$ points and let the best approximation $u^* = p^*/q^*$ from R_{mn} be non-degenerate. Then there is a constant $c > 0$ such that*

$$\|f - p/q\| \geq \|f - p^*/q^*\| + c\,\|q - q^*\|, \tag{6.14}$$

whenever $p/q \in R_{mn}$ and $\max_{x\in X}\{q(x)\} = 1$.

Proof. Recall that a neighborhood of u^* in R_{mn} is an $(m + n + 1)$-dimensional Haar sub-manifold of $C(X)$. We will use a C^1-parametrization from $\mathbb{R}^{m+n+1}$ into R_{mn} which is consistent with this structure. We restrict ourselves to the case where $q^* \notin \Pi_{n-1}$, since the case where $p^* \notin \Pi_{m-1}$ and $q^* \in \Pi_{n-1}$ can be treated analogously. In a first step a different normalization is used, namely the highest coefficient is fixed: $q(x) = a_0 + a_1 x + \cdots + a_{n-1}x^{n-1} \pm x^n$ and $p(x) = a_n + a_{n+1}x + \cdots + a_{n+m}x^m$. The mapping $F\colon \mathbb{R}^{m+n+1} \to R_{mn}$

$$(a_0, \ldots, a_{n+m}) \mapsto u = p/q \in R_{mn}$$

provides a C^1-parametrization such that the derivatives span the tangent space (cf. (2.3)). Thus, we have strong uniqueness in the form

$$\begin{aligned} \|f - F(a)\| &\geq \|f - F(a^*)\| + c_1\|a - a^*\| \qquad (6.15)\\ &\geq \|f - F(a^*)\| + c_2\|q(a) - q(a^*)\|, \end{aligned}$$

since all norms on finite dimensional spaces are equivalent. Rewrite (6.15) as an estimate for $\|q - q^*\|$ where $q = q(a)$ and $q^* = q(a^*)$. Moreover, $|\max q(x) - \max q^*(x)| \leq \|q - q^*\|$, and by some elementary calculations, strong uniqueness for the normalized denominators is established. □

Now we are in a position to prove

6.5 Theorem (Barrodale, Powell and Roberts 1972, Dua and Loeb 1973). *Let X contain at least $m + n + 2$ points. Assume that f is (m, n)-normal. Then the rate of convergence is at least quadratic.*

Proof. Since the best approximation $u^* = p^*/q^*$ exists, we may choose $\bar{p}/\bar{q} = u^*$ in (6.8). From strong uniqueness, we deduce that

$$\min_{x\in X}\left\{\frac{q^*(x)}{q_k(x)}\right\} \geq \min_{x\in X}\left\{\frac{q^*(x)}{q^*(x) + \|q^* - q_k\|}\right\} \geq \frac{\eta}{\eta + [\Delta_k - \Delta^*]/c}. \tag{6.16}$$

Noting that $\Delta_{k+1} \leq \Delta_k$ with similar reasoning, we have that

$$\min_{x\in X}\left\{\frac{q_{k+1}(x)}{q_k(x)}\right\} \geq \frac{\eta - (\Delta_k - \Delta^*)/c}{\eta + (\Delta_k - \Delta^*)/c}.$$

We now substitute (6.16) into the third line of (6.8), and obtain the bound

$$\left[\left|f - \frac{p_{k+1}}{q_{k+1}}\right| - \Delta_k\right]\frac{q_{k+1}(x)}{q_k(x)} \le -(\Delta_k - \Delta^*)\frac{\eta^*}{\eta^* + [\Delta_k - \Delta^*]/c}.$$

Multiplying both sides of this expression by q_k/q_{k+1} and using (6.14), we deduce the bound

$$(\Delta_{k+1} - \Delta_k) \le -(\Delta_k - \Delta^*)\frac{\eta^*}{\eta + [\Delta_k - \Delta^*]/c}\cdot\frac{\eta^* - [\Delta_k - \Delta^*]/c}{\eta + [\Delta_k - \Delta^*]/c}.$$

Finally we add $\Delta_k - \Delta^*$ to both sides:

$$(\Delta_{k+1} - \Delta^*) \le (\Delta_k - \Delta^*)^2\frac{3\eta^*/c + [\Delta_k - \Delta^*]/c^2}{(\eta^* + [\Delta_k - \Delta^*]/c)^2} \le (\Delta_k - \Delta^*)^2\frac{3}{\eta c},$$

and the proof of quadratic convergence is complete. □

Here, we want to recall that there is some freedom in the choice of the functionals when a best uniform approximation is characterized via the Kolmogorov criterion. This may be considered as a reason for the fact that the minimization of (6.3) turns out to be successful. In the L_p-approximation problem there is no such freedom. Therefore, analogous approches for rational L_p-approximation are not reasonable (cf. Dunham 1974).

B. The Remes Algorithm

The differential correction algorithm and the Remes algorithm are generally viewed as being the most effective ones for computing the best rational approximation on an interval $X = [\alpha, \beta]$. The latter algorithm is faster, since the auxiliary problems refer to equations in only $m + n + 2$ points from X. This method is used, for instance, when best approximations are computed with an extremely high accuracy, as in the computations by Carpenter. Ruttan, and Varga (1984) in connection with the 1/9-conjecture. The lack of robustness can often be overcome by starting the iteration with a good rational function p_0/q_0, or by similar devices.

The idea of the Remes algorithm is the determination of the alternant of the best approximation. One obvious difficulty is that the length of the alternant is not known in advance. However, we know from the discussion of the previous algorithm that it is no drawback to assume that f is (m, n)-normal. Then there is an alternating set which consists of $l := m + n + 2$ points. The following algorithm was modelled by Werner (1962) and Maehly (1963) after Remes' famous algorithm of 1934 for the computation of best polynomials.

6.6 Remes Algorithm for Rational Approximation.

Step 1 (Initialization). Choose an initial approximation $p_0/q_0 \in R_{mn}$ and a reference set of l points $\{x_i^1\}_{i=1}^l$. Set $k = 1$.

Step 2. Determine $p_k/q_k \in R_{mn}$ and $\eta_k \in \mathbb{R}$ such that

$$f(x_i^k) - p_k(x_i^k)/q_k(x_i^k) = (-1)^i \eta_k \quad \text{for } i = 1, 2, \ldots, l. \tag{6.17}$$

Step 3. Determine a new reference set $x_1^{k+1} < x_2^{k+1} < \cdots < x_l^{k+1}$ such that

$$s(-1)^i (f - p_k/q_k)(x_i^{k+1}) \geq |\eta_k| \quad \text{for } i = 1, 2, \ldots, l, \tag{6.18}$$

and that for one $i \in \{1, 2, \ldots, l\}$ the left hand side of (6.18) equals $\|f - p_k/q_k\|$. Here, s is either $+1$ or -1.

Step 4. (Note that $|\eta_k| \leq E_{mn}(f) \leq \|f - p_k/q_k\|$.) Unless the approximation is not good enough, replace k by $k + 1$ and return to Step 2. □

For the study of the algorithm, we will assume for the moment that the nonlinear interpolation problem (6.17) always has a solution with a nonnegative denominator.

The system of equations (6.17) may be rewritten (after suppressing the iteration count k), as

$$f(x_i)q(x_i) - p(x_i) = (-1)^i \eta q(x_i). \tag{6.19}$$

The homogeneity of the (unnormalized) form shows the equivalence to an algebraic eigenvalue problem, where η plays the role of an eigenvalue. To get some more insight, we define σ by $\sigma(x_i) = (-1)^i$, multiply (6.19) by x^j for $j = 0, 1, 2, \ldots, n$ and take divided differences of order $l - 1$ in order to eliminate p (see Werner 1964).

$$(fx^j q)[x_1, \ldots, x_l] = \eta(\sigma x^j q)[x_1, \ldots, x_l] \text{ for } j = 0, 1, \ldots, n.$$

Therefore, letting $q(x) = \sum_{k=0}^n b_k x^k$ yields an eigenvalue problem in matrix-vector form:

$$Ab = \eta Bb, \tag{6.20}$$

where $A_{jk} = (fx^{j+k})[x_1, \ldots, x_l]$, $B_{jk} = (\sigma x^{j+k})[x_1, \ldots, x_l]$, and $b = (b_0, b_1, \ldots, b_n)$. Obviously, A and B are real, symmetric matrices. Furthermore, given any polynomial $q \in \Pi_n$, $q \neq 0$, we see that σq^2 has $l - 1$ weak sign changes and cannot be interpolated by a polynomial from Π_{l-2}. Hence, $b^T Bb = (\sigma q^2)[x_1, \ldots, x_l] \neq 0$ and B is definite. Consequently, (6.19) and (6.20) have $n + 1$ real eigenvalues.

At most one eigenvector belongs to a polynomial q which is positive on $\{x_i\}_{i=1}^l$. Indeed, given two eigenvectors $b \neq \tilde{b}$ we have the orthogonality relation $0 = \tilde{b}^T Bb = (\sigma q\tilde{q})[x_1, \ldots, x_l]$. Now with the same arguments as above, we deduce that $\sigma q\tilde{q}$ cannot alternate and $q\tilde{q}$ cannot be positive on all l points.

A pole-free solution does not always exist.

Example (Maehly 1963). Let $f(x) = x$ in $[-1, +1]$ and $x_1 = 1$, $x_2 = 0$, $x_3 = 1$. The two solutions in R_{01}: $q^{(1)} = 2x - \sqrt{2}$ and $q^{(2)} = 2x + \sqrt{2}$ change their sign between x_1 and x_3.

For the situation which is observed in most cases, a convergence proof (in the classical sense) can be established.

6.7 Theorem. *Let $f \in R_{mn}$ be (m, n)-normal. Assume that there exists a $c > 0$ such that $c \le q_k(x) \le 1$ for all $k \ge 1$ and all $x \in [\alpha, \beta]$. Then (p_k/q_k) converges to the nearest point for f from R_{mn}.*

Proof. Since Π_{m+n} is a Haar subspace of $C[\alpha, \beta]$, there exists a nontrivial vector $\lambda^{(k+1)} \in \mathbb{R}^l$ such that

$$\sum_{i=1}^{l} \lambda_i^{(k+1)} P(x_i^{k+1}) = 0 \quad \text{for } P \in \Pi_{m+n}.$$

Specifically, the components of $\lambda^{(k+1)}$ alternate in sign, i.e. $(-1)^i \lambda_i^{(k+1)} > 0$. Set

$$\gamma_i^{(k+1)} := \lambda_i^{(k+1)} q_k(x_i^{k+1}) q_{k+1}(x_i^{k+1})$$

and assume that $\lambda^{(k+1)}$ has been normalized such that $\sum_i |\gamma_i^{(k+1)}| = 1$. It follows that

$$\sum_{i=1}^{l} \gamma_i^{(k+1)} (u_k - u_{k+1})(x_i^{k+1}) = 0, \tag{6.21}$$

using the abbreviation $u_k := p_k/q_k$. Now, by (6.17) and (6.18)

$$(f - u_{k+1})(x_i^{k+1}) = (-1)^i \eta_{k+1},$$
$$(f - u_k)(x_i^{k+1}) = s(-1)^i \zeta_i^k,$$

where $\zeta_i^k \ge |\eta^k|$ for $i = 1, 2, \ldots, l$ and s is either $+1$ or -1. Thus, using (6.21), $|\eta_{k+1}| = \sum_i |\gamma_i^{(k+1)}| \zeta_i^k$ and so

$$|\eta_{k+1}| - |\eta_k| = \sum_{i=1}^{l} |\gamma_i^{(k+1)}| (\zeta_i^k - |\eta_k|) > 0. \tag{6.22}$$

It follows that $(|\eta_k|)$ is an increasing sequence, which by the de la Vallée-Poussin Theorem is bounded above, and hence convergent.

We claim that there is a $c_1 > 0$ such that

$$|\gamma_i^{(k+1)}| \ge c_1 > 0 \quad \text{for all } i, k. \tag{6.23}$$

Indeed, since the denominators are bounded from below, the rational functions u_k are equicontinuous. So are the error functions $\varepsilon_k := f - u_k$. By (6.18), ε_k changes its value between x_i^{k+1} and x_{i+1}^{k+1} by at least $2|\eta_1| > 0$. Hence, there is some $\delta > 0$ such that $x_{i+1}^{k+1} - x_i^k \ge \delta$ for all i and k. By a standard compactness argument, lower bounds for $|\lambda^{(k+1)}|$ and $|\gamma^{(k+1)}|$ as postulated in (6.23) are established.

Finally, we note that one ζ_i^k in (6.22) equals $\|f - u_k\|$ and

$$|\eta_{k+1}| - |\eta_k| \ge |\eta_k| + c_1(\|f - u_k\| - |\eta_k|).$$

Since the sequence $(|\eta_k|)$ converges, it follows that $\|f - u_k\|$ and $|\eta_k|$ tend to $E_{mn}(f)$. The strong uniqueness ensures that u_k tends to the best approximation. □

Up to now, we have always assumed that Step 2 of the algorithm can always be executed. For the practical use of the Remes algorithm, one needs to know under which conditions a break-down can be excluded or what extra devices are able to overcome a break-down.

1. If f is (m, n)-normal and if u_0 is sufficiently close to the best approximation u^*, then the Remes algorithm converges. Indeed, if we rewrite (6.17) as $(f - v)(x_i^k) = (-1)^i\eta$ for $i = 1, 2, \ldots, l$, then there is a solution v in the linear manifold $u^* + T_{u^*}R_{mn}$. By the Implicit Function Theorem, we deduce that there is also a solution in the family of rational functions. Problem 6.10 asserts that the constructed rational function is again close to the best approximation. Therefore, Theorem 6.7 is applicable and yields convergence.

2. In most computer programs, the nonlinear equations (6.17) are not solved via the eigenvalue problem, but via Newton's method. More precisely, one is content with the approximate solution which one gets after 2 or 3 steps of the Newton iteration. Moreover, if the Newton iteration turns out to yield denominators with negative values on $\{x_i^k\}_{i=1}^l$, then the Newton corrections are damped and only a small portion of the correction is added to p_k and q_k respectively. [The philosophy is the same as for the damped Gauss-Newton method. The linear problem predicts some increase of the lower bound $|\eta|$ and the iteration is controlled by the postulate that the gain with the new constructed function is at least $(1/4)\cdot$damping factor$\cdot$gain in the linearized problem.]

3. Obviously, if a solution of (6.17) exists, it provides a best approximation to f for the domain $\{x_i^k\}_{i=1}^l$. More generally, a solution for this approximation problem may be regarded as a *weak solution* of (6.17). For this reason, there are variants of the Remes algorithm which employ the differential correction algorithm when performing Step 2 in the cases with a threat of a break-down.

4. In all versions of the Remes algorithm, there arise some variants of rational interpolation problems. These are to be solved via continued fractions, since the associated equations in matrix vector form usually have a large condition number (see e.g. Werner 1972, Cuyt and Wuytack 1986). In this way, one achieves the same numerical stability as one gets in polynomial interpolation by making use of Newton's interpolation formula.

For this reason, the Remes algorithm is less sensitive to rounding errors than the differential correction (unless the results of the linear programs are recalculated by using continued fractions). Therefore, which of the two competing algorithms is more robust also depends on the question of whether or not rounding errors are taken into consideration.

Exercises

6.8. Provide a characterization of the solution (P, Q) which minimizes (6.3), and is analogous to Theorem I.2.9 for linear Chebyshev approximation.

6.9. Let u^* be the best uniform approximation to f from an n dimensional linear Haar subspace $M \subset C[\alpha, \beta]$. Show that there is a strong uniqueness constant $c > 0$ for a de la Vallée-Poussin type inequality in the following sense: For any $u \in M$, one has

$$\|u - u^*\| \le c[d(f, M) - \eta], \tag{6.24}$$

whenever there are $n + 1$ points $\alpha \le x_0 < x_1 < \cdots < x_n \le \beta$ such that

$$(-1)^i s(f - u)(x_i) \ge \eta \quad \text{for } i = 0, 1, \ldots, n,$$

where s is either $+1$ or -1.

6.10. Let u^* be the non-degenerate best uniform approximation to f from R_{mn}. Show that a *local* strong uniqueness property of the form (6.24) holds.

Chapter VI. Approximation by Exponential Sums

The rational functions and exponential sums belong to those concrete families of functions which are the most frequently studied in nonlinear approximation theory. The starting point in the consideration of exponential sums is an approximation problem often encountered for the analysis of decay processes in natural sciences. A given empirical function on a real interval is to be approximated by sums of the form

$$\sum_{\nu=1}^{n} \alpha_\nu e^{t_\nu x}$$

where the parameters α_ν and t_ν are to be determined, while n is fixed.

The first systematic study, in the sense that one minimizes the norm of the error function, was done by Rice in 1960 for $n = 1$ and in 1962 for $n > 1$. In the latter investigation, it was described how the families of exponential sums have to be closed to get a mathematically reasonable theory. Central in the theory of exponential sums is Rice's concept of varisolvency, although it later turned out (Braess 1967) that a best approximation is not always unique, and that varisolvency alone does not provide a complete theory. Today, exponential approximation is considered as interesting, since one has families for which non-trivial bounds for the number of solutions are known.

In this chapter, the basic properties of exponential sums are derived and the existence problem is treated. Here, the existence proof is completely different from the existence proofs for other families, since the central argument involves an à priori estimate for derivatives. The uniqueness problem, as well as the characterization of (local) solutions will be treated in the next chapter. There, exponential sums are considered as special γ-polynomials, i.e., functions of the form

$$\sum_{\nu=1}^{n} \alpha_\nu \gamma(t_\nu, x)$$

with the kernel $\gamma(t, x) = e^{tx}$.

§ 1. Basic Facts

A. Proper and Extended Exponential Sums

In the analysis of decay processes, one often wants to approximate a given function on a real interval by exponential sums of the form

$$\sum_{\nu=1}^{n} \alpha_\nu e^{t_\nu x} \tag{1.1}$$

where the parameters α_ν and t_ν are to be determined. The parameters α_ν and t_ν of these *proper exponential sums* are interpreted as mass concentrations and as inverse time constants, resp., while n is considered fixed. Now it is easily seen that the families

$$E_n^0 := \left\{u\colon u = \sum_{\nu=1}^{k} \alpha_\nu e^{t_\nu x}, \alpha_\nu, t_\nu \in \mathbb{R}, k \leq n\right\} \tag{1.2}$$

are not existence sets whenever $n \geq 2$. The function xe^x can be arbitrarily well approximated by exponential sums of the form

$$\delta^{-1}[e^{(1+\delta)x} - e^x], \qquad \delta \neq 0.$$

Hence, there is no nearest point to $f(x) = xe^x$ from E_2^0.

In order to guarantee existence, one has to consider the closure of E_n^0 for $n > 1$, i.e., the set of *extended exponential sums*:

$$E_n = \left\{u\colon u = \sum_{\nu=1}^{l} p_\nu(x) e^{t_\nu x}, p_\nu \in \Pi_n, t_\nu \in \mathbb{R}, k := \sum_{\nu=1}^{l} (1 + \partial p_\nu) \leq n\right\}. \tag{1.3}$$

The extended exponential sums are solutions of linear differential equations of order k with constant coefficients, whose characteristic polynomials have only real zeros. The term $(1 + \partial p_\nu)$ can be interpreted as the multiplicity of the characteristic number t_ν.

Generally, we may assume that

$$t_1 < t_2 < \cdots < t_l. \tag{1.4}$$

The parameters k and l, which are associated with each exponential sum in (1.3) will be important in the sequel. Referring to the parametrization (1.3), we call $k = k(u)$ the *order* of the exponential sum and $l = l(u)$ its *length*. Obviously, $k(u) \geq l(u)$, and the proper exponential sums are characterized by $k(u) = l(u)$.

Moreover, we will consider the set of positive exponential sums for $n \geq 1$,

$$E_n^+ = \left\{u\colon u(x) = \sum_{\nu=1}^{n} \alpha_\nu e^{t_\nu x}, \alpha_\nu \geq 0, t_\nu \in \mathbb{R}\right\} \tag{1.5}$$

(though the notation of nonnegative exponential sums would be more precise).

B. The Descartes' Rule of Signs

As will be shown in the next section, the families E_n^0 are varisolvent. To prove this, we need information on the number of zeros.

1.1 Lemma. *Each exponential sum of order $\leq k$ has at most $k - 1$ zeros, or vanishes identically.*

Proof. If $k(u) = 1$, then $u(x) = \alpha_1 e^{t_1 x}$ and the statement of the lemma is obvious.

Assume that u is an exponential sum of order $k > 1$, and that the lemma has already been proved for $k-1$. Let u be given as in (1.3) and evaluate:

$$\frac{d}{dx}(e^{-t_1 x}u(x)) = p_1'(x) + \sum_{\nu=2}^{l} [p_\nu' + (t_\nu - t_1)p_\nu]e^{(t_\nu - t_1)x}. \tag{1.6}$$

The right hand side is an exponential sum of order $\leq k-1$. From the induction hypothesis, we know that this function has at most $k-2$ zeros or that it vanishes identically. By Rolle's theorem u has at most $k-1$ zeros or $u = 0$. □

For proper exponential sums, an improved bound for the number of zeros has long been known (cf. Laguerre 1898).

1.2 Descartes' Rule of Signs. *Let u be a proper exponential sum of the form* (1.1) *with r zeros $x_1 < x_2 < \cdots < x_r$. Moreover, assume that $t_1 < t_2 < \cdots < t_n$ and that $\alpha_\nu \neq 0$ for $\nu = 1, 2, \ldots, n$. Then there are at least r sign changes in the sequence $\alpha_1, \alpha_2, \ldots, \alpha_n$. If the number of sign changes equals r, then*

$$\operatorname{sgn} u(x) = \operatorname{sgn} \alpha_n \quad \text{for } x > x_r. \tag{1.7}$$

Proof. The proof proceeds by induction on $\tilde{r}$, the number of sign changes in the sequence $\alpha_1, \alpha_2, \ldots, \alpha_n$. If $\tilde{r} = 0$, then the statement is obvious. If $\tilde{r} > 0$, we have $\alpha_j \alpha_{j+1} < 0$ for some $j < n$. Choose t in (t_j, t_{j+1}) and note that

$$\frac{d}{dx}(e^{-tx}u(x)) = \sum_{\nu=1}^{n} \alpha_\nu (t_\nu - t)e^{(t_\nu - t)x}$$

is an exponential sum such that the sequence with the induced factors $\beta_\nu = \alpha_\nu(t_\nu - t)$ has one sign change less than the given sequence. From Rolle's theorem we get the statement on the zeros of u. □

For an extension, see Theorem VII.1.10 and Exercise VII.1.18.

Exercises

1.3. Give a sequence in E_n^0 which converges to $x^{n-1}e^{tx}$.

1.4. Give a representation for those exponential sums in E_n which satisfy $u(x) = -u(x)$.

1.5. Assume that $u_i \in E_n^0 \setminus E_{n-1}^0$, $i = 1, 2, \ldots, N$ are such that u_i and u_j have no common spectral value if $i \neq j$. Show that there is an $f \in L_2[a, b]$ to which $u_1, u_2, \ldots u_N$, are strict local best approximations from E_n^0.
Hint: The tangent space at $u = F(\alpha_1, \ldots, \alpha_n, t_1, \ldots, t_n) := \sum_\nu \alpha_\nu e^{t_\nu x}$ is spanned by the derivatives of F to the $2n$ parameters.

§2. Existence of Best Approximations

There are many similarities between rational functions and exponential sums when the existence problem for best approximations is considered. For instance, by

$$u_j(x) = e^{-jx}, \qquad j = 1, 2, \ldots$$

a sequence in $C[0, 1]$ is given such that

$$\lim_{j\to\infty} u_j(x) = \begin{cases} 1 & \text{if } x = 0, \\ 0 & \text{if } x > 0. \end{cases}$$

The first rigorous proof that $E_n (n \in \mathbb{N})$ is an existence set, was given by Werner (1969). The investigations were simplified by E. Schmidt (1970). We will follow his proof which provides more convenient bounds for derivatives of exponential sums. A proof which is very close to Werner's concept and proceeds without Lemma 2.1 and Theorem 2.2, is indicated in Exercise 2.15. A different approach was used by Kammler (1973).

A. A Bound for the Derivatives of Exponential Sums

In order to estimate the derivatives of exponential sums we need some à priori information.

2.1 Lemma (Werner 1969). *Let X be a real closed interval of length $d > 0$ and let $f \in C^m(X)$, $m \geq 1$. Then there is a point $z \in X$ such that*

$$|f^{(m)}(z)| \leq \frac{2^{2m-1} m!}{d^m} \|f\|_\infty. \tag{2.1}$$

Proof. First assume that $X = [-1, +1]$. For the degree of polynomial approximaton, there is the obvious upper bound: $E_{m-1}(f) \leq \|f\|_\infty$. By Bernstein's comparison theorem, there is a point $z \in [-1, +1]$ such that $E_{m-1}(f) = (2^{-m+1}/m!)|f^{(m)}(z)|$. Since the length of the standard interval $[-1, +1]$ equals 2, the lemma is proven for this case.

The general case can be reduced to the case treated above. To this end, the given interval is mapped onto $[-1, +1]$ by a linear transformation. □

In the sequel $\|\cdot\|_{[\alpha,\beta]}$ will refer to the supremum norm for the interval $[\alpha, \beta]$.

2.2 Theorem (E. Schmidt 1970). *There is a constant c which only depends on n such that for any $u \in E_n$:*

$$\|u'\|_{[a+d,b-d]} \leq \frac{c}{d} \|u\|_{[a,b]} \tag{2.2}$$

holds whenever $0 < d < (b - a)/2$.

Remark. The estimate (2.2) excludes the boundaries of the given interval. This is necessary: Consider the function $u(x) = \exp(-x/d) \in E_1$. We have $\|u\|_{[0,\infty)} = 1$ but $|u'(d)| = e^{-1} d^{-1} \|u\|_{[0,\infty)}$. The shrinkage of intervals means that there is a difference between (2.2) and the well known inequalities of Markov and Bernstein for polynomials. On the other hand, the shrinkage is more specific than in a generalization which also applies to rational functions (see Exercise 2.14).

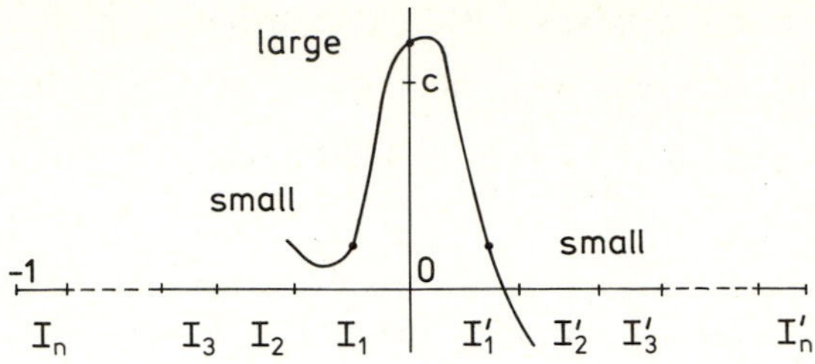

Fig. 27. Division of $[-1, +1]$ into $2n$ subintervals for the proof of Theorem 2.2 and graph of u'

Proof of Theorem 2.2. It is sufficient to estimate $|u'(0)|$ by the supremum of $|u|$ on the interval $[-1, +1]$ for $u \in E_n$. Moreover, after multiplying u by an appropriate factor if necessary, we may assume that $\|u\|_{[-1,+1]} = 1$.

Choose n pairs of subintervals in the ordering $I_n, I_{n-1}, \ldots, I_1, x = 0, I'_1, I'_2, \ldots, I'_n$, (see Fig. 27). Let d_j denote the length of I_j and I'_j for $j = 1, 2, \ldots, n$. Then $\sum_{j=1}^n d_j = 1$. Finally, we put

$$r_j = 2^{2j-1} j!(d_j)^{-j}, \qquad C_j = \sum_{\nu=j}^{n} r_\nu, \qquad j = 1, 2, \ldots, n, \tag{2.3}$$

and $C_{n+1} = 0$. It is our aim to show that (2.2) holds with $c = C_1$.

We may restrict ourselves to the case where $u \in E_n$ is not a polynomial. Suppose that $u'(0) > C_1$. By the preceding lemma, there are points in I_1 and I'_1 at which $|u'| \le r_1$. Now we make use of the fact that $u'(0)$ is large. By the Mean Value Theorem there are points $\zeta_1 \in I_1$ and $\zeta'_1 \in I'_1$ with

$$u''(\zeta_1) > +C_1 - r_1 = +C_2,$$
$$u''(\zeta'_1) < -C_1 + r_1 = -C_2.$$

Moreover, u'' has a zero between ζ_1 and ζ'_1.

Let $2 \le j \le n$. Assume that there are points $\zeta_{j-1} < 0 < \zeta'_{j-1}$ such that $\max\{|\zeta_{j-1}|, \zeta'_{j-1}\} \le d_1 + d_2 + \cdots d_{j-1}$, and

$$u^{(j)}(\zeta_{j-1}) > C_j,$$
$$(-1)^{j-1} u^{(j)}(\zeta'_{j-1}) > C_j.$$

Moreover, let $u^{(j)}$ have $j-1$ zeros $x_1 < x_2 < \cdots x_{j-1}$ in $(\zeta_{j-1}, \zeta'_{j-1})$. Without loss of generality, we may assume that x_1 is the first zero in the interval, and that x_{j-1} is the last one. By the preceding lemma, there are points $z_j \in I_j$ and $z'_j \in I'_j$ at which $|u^{(j)}| \le r_j$. From the Mean Value Theorem it follows that there are points $\zeta_j \in (z_j, \zeta_{j-1})$ and $\zeta'_j \in (\zeta'_{j-1}, z'_j)$ such that

$$u^{(j+1)}(\zeta_j) > C_j - r_j = C_{j+1},$$
$$(-1)^j u^{(j+1)}(\zeta'_j) > C_j - r_j = C_{j+1}.$$

Since $u^{(j+1)}(x_1) \le 0$ and $(-1)^j u^{(j+1)}(x_{j-1}) \le 0$, from the Intermediate Value Theorem we conclude that two zeros are in $(\zeta_j, x_1] \cup [x_{j-1}, \zeta_j')$. By Rolle's theorem, $u^{(j+1)}$ has $j-2$ more zeros in (x_1, x_{j-1}). Hence, there are at least j zeros in (ζ_j, ζ_j').

By induction, we obtain n zeros for $u^{(n+1)}$. Since $u^{(n+1)} \in E_n$ this contradicts Lemma 1.1. Therefore, $|u'(0)|$ cannot be larger than C_1. □

B. Existence

From the bound on the derivatives, a compactness result is obtained in which the boundaries are again excluded.

2.3 Theorem (E. Schmidt 1970). *Let (u_j) be a sequence in E_n, which is bounded in the supremum norm on a non-degenerate interval $[a, b]$. Then there is a subsequence which converges uniformly to some $u^* \in E_n$ on each subinterval $[\alpha, \beta]$ with $a < \alpha < \beta < b$.*

Proof. Choose $2n+2$ points $a < a_1 < a_2 < \cdots < a_{n+1} < b_{n+1} < b_n \cdots < b_1 < b$. By applying Theorem 2.2 repeatedly, we conclude that sequences of derivatives $(u_j^{(i)})$, $i = 1, 2, \ldots, n+1$, are uniformly bounded in $[a_i, b_i]$. Therefore, the sequences $(u_j^{(i)})$, $i \leq n$, are equicontinuous in $[a_{n+1}, b_{n+1}]$. By the Arzelà-Ascoli theorem, one may choose a subsequence such that the limits

$$\lim_{j\to\infty} u_j^{(i)} = g_i, \qquad i = 0, 1, \ldots, n,$$

exist. (Without loss of generality we may assume that this is already the given sequence). The functions $g_0, g_1, \ldots, g_n$ are defined in $[a_{n+1}, b_{n+1}]$. Moreover, we know that $g_i = g_0^{(i)}$ holds for $i = 1, 2, \ldots, n$. We will prove that $g_0 \in E_n$.

Let $t_i^{(j)}$, $i = 1, 2, \ldots, n$, be the characteristic numbers of u_j. Then we have

$$\prod_{i=1}^{n} \left(\frac{d}{dx} - t_i^{(j)}\right) u_j = 0. \tag{2.4}$$

After passing to a subsequence, if necessary, we deduce that any sequence of characteristic numbers either converges or tends to $\pm\infty$. This means that, after a relabelling we have, for some k between 0 and n:

$$\lim_{j\to\infty} t_i^{(j)} = t_i \quad \text{for } i = 1, 2, \ldots, k,$$

$$\lim_{j\to\infty} 1/t_i^{(j)} = 0 \quad \text{for } i = k+1, \ldots, n.$$

Now the differential equation (2.4) is divided by the factor $\prod_{i>k}(-t_i^{(j)})$:

$$\prod_{i\leq k} \left(\frac{d}{dx} - t_i^{(j)}\right) \prod_{i>k} \left(1 - \frac{1}{t_i^{(j)}} \frac{d}{dx}\right) u_j = 0.$$

Passing to the limit for $j \to \infty$ yields

$$\prod_{i=1}^{k} \left(\frac{d}{dx} - t_i\right) g_0 = 0.$$

Hence, g_0 is an extended exponential sum, and the subsequence of (u_j) converges uniformly to g_0 on $[a_{n+1}, b_{n+1}]$.

Finally, we have to verify that the subsequence does not only converge in $[a_{n+1}, b_{n+1}]$ but converges uniformly on each subinterval $[\alpha, \beta]$ whenever $a < \alpha < \beta < b$. Suppose that this is not true. By repeating the arguments above for $[\tilde{a}_{n+1}, \tilde{b}_{n+1}] = [\alpha, \beta]$, it would be possible to select a sequence, which converges to another exponential sum on $[\alpha, \beta]$. This, however, is impossible. □

By the preceding theorem, each $\|\cdot\|_\infty$-bounded set in E_n contains a sequence, which converges pointwise to an element of E_n in the open interval. Now the general existence theorem II.1.4 guarantees a nearest point to each f from E_n.

2.4 Existence Theorem (Werner 1969). *Let $n \geq 1$ and X be a compact real interval. Then E_n is an existence set in $C(X)$.*

The L_p-approximation problem can be treated with the following lemma.

2.5 Lemma. *For any $u \in E_{n-1}$:*

$$\|u\|_{[a+d, b-d]} \leq \frac{c}{d} \|u\|_{L_1[a,b]} \tag{2.5}$$

holds whenever $0 < d < (b-a)/2$. Here c refers to the constant for E_n from (2.2).

Proof. Given $u \in E_{n-1}$ define

$$v(x) = \int_a^x u(t)\,dt$$

Obviously, $v \in E_n$ and $\|v\|_{[a,b]} \leq \|u\|_{L_1[a,b]}$. By applying Theorem 2.2 to v we get (2.5). □

Finally, it follows from the Hölder Inequality that

$$\|u\|_{L_1(a,b)} \leq (b-a)^{(p-1)/p} \|u\|_{L_p(a,b)}, \tag{2.6}$$

whenever $1 \leq p < \infty$. Therefore, any L_p-bounded set in E_n is L_1-bounded. By the lemma above, it contains a sequence which converges in the sense of Theorem 2.2. Therefore, with the same arguments as for rational L_p-approximation we get existence.

2.6 Theorem. *Let $1 \leq p < \infty, n \geq 1$ and X be a compact real interval. Then E_n is an existence set in $L_p(X)$.*

Exercises

2.7. Show that one has a rough bound for the constant in (2.2) given by $c(n) \leq 2^{3n} \cdot n!$

2.8. For the treatment of the L_2-approximation problem verify that

$$\|u\|_{[a+d,b-d]} \leq c' d^{-1/2} \|u\|_{L_2[a,b]}$$

where $c' = c(n^2+1)$. (This improves the bound one obtains by combining (2.5) and (2.6).)

2.9. Let (u_j) be a sequence in E_n which converges to u^* in the sense of Theorem 2.3. Moreover, let $\lim_{j\to\infty} u_j(x) = u^*(x)$ for $x = a$ and for $x = b$. Why is it nevertheless still possible that the sequence does not converge uniformly?

2.10. Is there a function $f \in C^m(X)$ for which (2.1) is optimal, i.e., for which (2.1) cannot be replaced by a strict inequality?

2.11. Let (u_j) be a sequence in E_n^+ which is bounded in the supremum norm. Show that a subsequence converges in the sense of Theorem 2.3 to some $u^* \in E_n^+$.
Hint: Let $u = \sum_\nu \alpha_\nu e^{t_\nu x} \in E_n^+$. Then each summand $\alpha_\nu e^{t_\nu x}$ is bounded by $\|u\|$.

2.12. Let $1 < p < \infty$ and $n \geq 1$. Show that no best L_p-approximation to f from E_n belongs to E_{n-1} whenever $f \notin E_{n-1}$.
Hints: Recall Example III.2.2 and show that 0 is not a best approximation from E_1 if $f \neq 0$.

2.13. Construct functions with two best L_p-approximations from E_n, and show that each $(n+1)$-dimensional subspace of L_p contains a function with more than one best approximation.

2.14. Let $n > 0$ and $u \in E_n$ or $u \in R_{nn}$. Assume that $|u(x)| \leq M$ holds for $x \in X \subset \mathbb{R}$. Given $\delta > 0$, there is a set $X_\delta \subset \mathbb{R}$ with $\mathrm{mes}(X_\delta) \leq \delta$ such that

$$|u'(x)| \leq 2mM/\delta \quad \text{for } x \in X \backslash X_\delta. \tag{2.7}$$

Here $m = n$ if $u \in E_n$, and $m = 2n$ if $u \in R_{nn}$. Prove (2.7).
Hint: Since u' has at most $m - 1$ sign changes, we have

$$\int_X |u'(x)|\,dx \leq m \cdot 2M.$$

We note that Dolženko (1961) has given the formula (2.7) with $2nM/\delta$ on the right hand side for rational functions, although he verifies it by referring to the (at most) $2n$ sign changes of u'.

2.15. Werner (1969) proved that E_n is an existence set before Schmidt (1970) derived a predecessor of Theorem 2.2. He used the idea from the preceding exercise. In order to follow his proof, perform the following steps:

Let X be an interval and $u \in E_n$ be as in Exercise 2.14.

(1) Show that (2.7) holds with X_δ being the union of (at most) $n + 1$ intervals.

(2) Show that given $\delta > 0$, there is a union of $\leq (n+1)^2$ intervals denoted as Y_δ with $\mathrm{mes}(Y_\delta) \leq (n+1)\delta$ such that

$$|u^{(i)}(x)| \leq (2n\delta^{-1})^i M \text{ for } x \in X \backslash Y_\delta, \qquad i = 1, 2, \ldots, n+1. \tag{2.8}$$

(3) Assume that (u_j) is a sequence in E_n such that

$$|u_j(x)| \leq M \quad \text{for } x \in X, j = 1, \ldots$$

Then there is a union of $\leq (n+1)^2$ intervals denoted as Y_δ with $\mathrm{mes}(Y_\delta) \leq 2(n+1)^3\delta$ such that the elements of a subsequence satisfy (2.8).

(4) Each bounded set in E_n contains a subsequence which pointwise converges to some $u^* \in E_n$ with possibly $(n+1)^2$ points excluded. □

§ 3. Some Facts on Interpolation and Approximation

A. Interpolation by Exponential Sums

The interpolation by exponential sums shares many properties with rational interpolation, when considered from a formal viewpoint. The interpolation in E_1 is easily understood. A function f can be interpolated at two nodes $z_1, z_2 \in \mathbb{R}$ where $z_1 \neq z_2$ by a function in E_1, if and only if $f(z_1) = f(z_2) = 0$ or if $f(z_1) \cdot f(z_2) > 0$.

For higher order n the problem is settled only for the interpolation in the class E_n^+. Here, the Hankel matrices (cf. Karlin 1968, p. 70)

$$H_n(x, f) := \begin{pmatrix} f(x) & f'(x) & \dots & f^{(n-1)}(x) \\ f'(x) & f''(x) & \dots & f^{(n)}(x) \\ \dots & & & \\ f^{(n-1)}(x) & f^{(n)}(x) & \dots & f^{(2n-2)}(x) \end{pmatrix}, n = 1, 2, \dots \tag{3.1}$$

play the role which the matrices M_{mn} had done in rational interpolation. We note that the derivatives in (V.3.8) carry different factors.

3.1 Lemma. *If $f \in C^{2n-2}[a, b]$ and if $H_n(x, f)$ is positive definite for $a < x < b$, then $f - u$ has at most $2n - 2$ zeros whenever $u \in E_{n-1}$.*

Proof. For any $t \in \mathbb{R}$ we have $(d/dx)(e^{tx} f) = [f' + tf]e^{tx}$. Therefore,

$$\det H_n(x, e^{tx} \cdot f) = e^{ntx} \cdot \det H_n(x, f), \tag{3.2}$$

since the determinant does not change its value if we subtract t times the i-th row from the $(i + 1)$-st one successively for $i = n - 1, n - 2, \dots, 1$, and do the same with the columns.

The lemma is true for $n = 1$, since $f - u$ has no zero if $f > 0$ and $u \in E_0$, i.e., $u = 0$.

Assume that $n \geq 2$ and $H_n(x, f)$ is positive definite. Let $u \in E_{n-1}$ have the characteristic number t. Put $g = e^{-tx} f$, and recall (3.2). Since $H_n(x, g)$ is definite, the submatrix $H_{n-1}(x, g'')$ is also definite. Therefore, by the induction hypothesis $g'' - u_1$ has at most $2(n - 1) - 2$ zeros whenever $u_1 \in E_{n-2}$. Note that $(d^2/dx^2)[e^{-tx}(f - u)] = g'' - u_1$, where indeed $u_1 \in E_{n-2}$. By Rolle's theorem, $f - u$ cannot have more than $2n - 2$ zeros. □

3.2 Theorem. *Assume that $f \in C^{2n-1}[a, b]$ and $H_n(x, f)$ is positive definite for $a < x < b$. Then, given $a \leq z_1 < \dots < z_{2n} \leq b$, there is a unique exponential sum*

$$u(x) = \sum_{\nu=1}^{n} \alpha_\nu e^{t_\nu x}, \qquad \alpha_\nu > 0, \qquad t_1 < t_2 < \dots < t_n, \tag{3.3}$$

which interpolates f at $z_1, z_2, \dots, z_{2n}$. If moreover, $H_n(x, f')$ is positive definite (negative definite, resp.) in (a, b), then all characteristic numbers of the interpolant are positive (negative resp.).

We will only sketch the proof. It follows the lines of the path lifting lemma IV.1.7., only the compactness argument has to be adapted to the situation for the exponential sums.

Choose an exponential sum u_0 of the form (3.3) and note that $H_n(x, u_0)$ is positive definite. Put $f_\lambda = (1 - \lambda)u_0 + \lambda f$ for $0 \leq \lambda \leq 1$. Define a subset Λ of $[0, 1]$ as follows: If $\lambda \in \Lambda$, then there exists an interpolant u of the form (3.3) to f_λ

for the given points. The set Λ is open because the derivatives

$$\frac{\partial u}{\partial \alpha_\nu}, \frac{\partial u}{\partial t_\nu}, \nu = 1, 2, \ldots, n,$$

span a $2n$-dimensional Haar subspace, and one has local solvability of degree $2n$ (see III §4B). The set Λ is also closed. Indeed let (λ_k) be a sequence in Λ which converges to λ^*. Then $|u_{\lambda_k}|$ is bounded in $[z_1, z_{2n}]$ by $\max\left\{\|f\|, \|u_0\|\right\}$. By Theorem 2.3, a subsequence converges uniformly on $[z_2, z_{2n-1}]$ to an exponential sum u^*. Hence u^* interpolates at least at $z_2, z_3, \ldots, z_{2n-1}$. If the characteristic numbers in the sequence are bounded from below (or from above, resp.), then u^* also interpolates at z_1 (at z_{2n}, resp.). From Lemma 3.1, it follows that u^* cannot be in E_{n-2} or E_{n-1}. Therefore, the sequence converges even in $[z_1, z_{2n}]$ and $\lambda^* \in \Lambda$.

The set Λ contains $\lambda = 0$, it is closed and open in $[0, 1]$. Hence, $\Lambda = [0, 1]$ and f is interpolated by u_1. By Lemma 1.1, the existence of two interpolants is excluded.

Finally, note that the derivative of a positive exponential sum is again in E_n^+ if the characteristic numbers are positive. Therefore, the statements on the spectral values can be proven as for Theorem V.3.5. □

The analogue to the result on Stieltjes functions is a well-known theorem (see e.g. Donoghue 1974).

3.3 Big Bernstein Theorem. *If f is completely monotone for $x \geq 0$, i.e.,*

$$(-1)^n f^{(n)}(x) \geq 0, \qquad n \geq 0, \qquad x \geq 0,$$

then it is the restriction to the half-axis of the Laplace transform of a nonnegative measure:

$$f(z) = \int_0^\infty e^{-tz}\, d\mu(t). \tag{3.4}$$

B. The Speed of Approximation of Completely Monotone Functions

Let f be a completely monotone function. Given $2n$ points, by Theorem 3.2 there is an exponential sum which interpolates f. For this reason the degree of approximation can be estimated from above by following Gonćar's method for proving Theorem V.3.11.

3.4 Theorem (Braess and Saff 1987). *Assume that f is completely monotone and analytic for* $\operatorname{Re} z > 0$, *and let* $0 < a < b$. *Then for the uniform approximation on the interval* $[a, b]$,

$$\varlimsup_{n\to\infty} d(f, E_n)^{1/n} \leq \frac{1}{\omega^2}, \tag{3.5}$$

where

$$\omega = \exp\frac{\pi \mathrm{K}(k)}{\mathrm{K}'(k)} \quad \text{and} \quad k = \frac{a}{b}. \tag{3.6}$$

Proof. (1) We will assume that f is analytic for $\operatorname{Re} z > -\varepsilon$ with some $\varepsilon > 0$. This is possible since the given problem is equivalent with the approximation of $f(x+\varepsilon)$ on $[a-\varepsilon, b-\varepsilon]$ and ω is a continuous function of a and b.

(2) Let $a < x_1 < x_2 < \cdots < x_{2n} < b$ be points which will be fixed later. By Theorem 3.2 there is an exponential sum $u_n \in E_n$ which interpolates f at the given points. Since f is completely monotone, $f - u_n$ has no other zero on $(0, b)$ and thus

$$f(x) - u_n(x) > 0 \quad \text{for } 0 < x < x_1.$$

From (3.4) and the positivity of the coefficients α_i of u_n, it follows that

$$|f(z)| \le f(x) \quad \text{and} \quad |u_n(z)| \le u_n(x) \quad \text{for } x = \operatorname{Re} z \ge 0.$$

By combining this observation with the monotonicity of f and u_n, we conclude that

$$|f(z) - u_n(z)| \le |f(z)| + |u_n(z)| \le 2f(0) \quad \text{for } \operatorname{Re} z \ge 0. \tag{3.7}$$

(3) By Theorem V.5.5, there are polynomials $p \in \Pi_n$ and $q \in \Pi_{n-1}$ such that

$$|\sqrt{x} - p(x)/q(x)| \le 4\omega^{-2n}\sqrt{x} \quad \text{for } a^2 \le x \le b^2,$$

with ω given by (3.6). We rewrite this inequality as $|1 - p(z^2)/zq(z^2)| \le 4\omega^{-2n}$ for $a \le z \le b$. In view of the alternation property of best approximations, all $2n$ zeros of $P(z) := p(z^2) - zq(z^2)$ lie in (a, b). Let $h(z) = P(z)/P(-z)$. For sufficiently large n we have $4\omega^{-2n} < 1$ and

$$|h(z)| \le 4\omega^{-2n} \quad \text{for } a \le z \le b,$$

$$|h(z)| = 1 \quad \text{if } \operatorname{Re} z = 0.$$

Since P is a polynomial, we may choose $R > 0$ large such that

$$|h(z)| > \tfrac{1}{2} \quad \text{for } |z| = R.$$

(4) Now assume that u_n actually interpolates f at the $2n$ zeros of the polynomial P. Then $(f - u_n)/h$ is analytic in $D = \{z \in \mathbb{C}\colon \operatorname{Re} z > 0, |z| < R\}$ and

$$|(f - u_n)/h| \le 4f(0) \quad \text{on } \partial D. \tag{3.8}$$

By the maximum prinicple, the right hand side of (3.8) provides a bound of $|(f - u_n)/h|$ for $[a, b] \subset D$. Consequently,

$$|(f - u_n)(z)| = |h(z)| \cdot \left|\frac{(f - u_n)(z)}{h(z)}\right| \le 16 f(0)\omega^{-2n}$$

for $a \le z \le b$, and (3.5) follows from $d(f, E_n) \le \|f - u_n\|_{[a,b]}$. □

The following table contains the approximations to $f(x) = (1 + x)^{-1} = \int_0^\infty e^{-xt}e^{-t}\,dt$ from E_n $(n \leq 6)$ for the interval $[0, 1]$. Theorem 3.4 yields $\lim_{n\to\infty} d((1 + x)^{-1}, E_n)^{1/n} \leq 1/135$. The numerical results support the conjecture that the bound is sharp in this case.

If f is an entire function, then $\Pi_n \subset E_{n+1}$ implies that $\lim d(f, E_n)^{1/n} = 0$. The converse is not true. The function

$$f(z) = \sum_{n=1}^{\infty} e^{-n^2 x}$$

is completely monotone and by considering the partial sums we see that

$$d(f, E_n) < 2e^{-n^2 a} \quad \text{if } n > a^{-1},$$

although f has a singularity at $z = 0$.

Table 3. Approximation of $(1 + x)^{-1}$ by exponential sums on $[0, 1]$

n	Best approximation in E_n	$d(f, E_n)$
1	$0.977 \cdot e^{-0.715x}$	$2.13 \cdot 10^{-2}$
2	$0.714 \cdot e^{-0.407x} + 0.286 \cdot e^{-2.443x}$	$2.08 \cdot 10^{-4}$
3	$0.459 \cdot e^{-4.507x} + 0.349 \cdot e^{-1.601x} + 0.560 \cdot e^{-0.287x}$	$1.83 \cdot 10^{-16}$
4	$0.00572 \cdot e^{-6.740x} + 0.1074 \cdot e^{-3.187x} + 0.4264 \cdot e^{-1.212x}$	
	$+ 0.4605 \cdot e^{-0.223x}$	$1.54 \cdot 10^{-8}$
5	$6.16 \cdot 10^{-4} \cdot e^{-9.078x} + 0.0211 \cdot e^{-4.996x} + 0.159 \cdot e^{-2.506x}$	
	$+ 0.428 \cdot e^{-0.978x} + 0.391 \cdot e^{-0.182x}$	$1.26 \cdot 10^{-10}$

Exercises

3.5. For which exponential sums is $H_n(x, u')$ positive (semi-) definite?

3.6. Assume that $H_1(x, f') = f'$ is positive. What can be said about the interpolation by functions of the form $\alpha_0 + \alpha_1 e^{t_1 x}$?

3.7. Prove that the definiteness of $M_{n-1,n}(x, f)$ implies definiteness of $H_n(x, f)$.

3.8. Show that $H_n(x, (x + a)^{-1})$ is definite for $n \geq 1$, $a > 0$, $x \geq 0$.
Hint: Use the Big Bernstein Theorem.

3.9. Prove that the interpolation problem with $f(x) = \cos(x - x_0)$, is not solvable in E_2 for any choice of 4 distinct points in $(-\pi/2, +\pi/2)$.

3.10. Prove that the interpolation problem for $f(x) = 1 - x^2$ and E_2 is not solvable for any choice of 4 symmetrically located points in $[-1, +1]$.

3.11. Show that a $u \in E_n$ solves the Hermite-type interpolation problem $u^{(j)}(0) = f^{(j)}(0)$ for $j = 0, 1, \ldots, 2n - 1$ whenever $H_n(0, f)$ is definite.

We anticipate a result on the characterization of best approximations which is useful in the solution of the next two exercises. By Theorem VII.2.8, the exponential sum u is the unique best uniform approximation to a completely monotone function f from E_n, if and only if $f - u$ has a positive alternant of length $2n + 1$.

3.12. Prove the following extension of Bernstein's comparison theorem: (Braess 1974c, Borwein 1983). If f, g and $g - f$ are completely monotone, then

$$d(f, E_n) \leq d(g, E_n).$$

3.13. Assume that f is completely monotone. Show that

$$d(f, E_n) \leq d(f, \Pi_{2n-1}) \tag{3.9}$$

(see Borwein 1983) by using a de la Vallée-Poussin type argument and Rolle's theorem. Compare Bernstein's result for $d(f, \Pi_{2n-1})$ with the result (3.5) for $d(f, E_n)$.

Chapter VII. Chebyshev Approximation by γ-Polynomials

The approximation by sums of exponentials shares some of the properties of rational approximation. But there is an essential difference: the best approximation is not always unique. There may be more than one isolated solution, which means that phenomena arise which are not met in the linear theory. Fortunately, it is possible to establish explicit bounds for the number of solutions. In order to get them, the results for Haar embedded manifold (derived with methods of global analysis), are applied.

The sums of exponentials are γ-polynomials with the special kernel $\gamma(t, x) = e^{tx}$. Therefore, the theory will be treated in the more general framework of γ-polynomials. γ-polynomials are also found in the theory of *monosplines for totally positive* kernels although in that theory they are usually written with the kernel K.

In this chapter, approximation in the sense of Chebyshev will be considered, while some questions in connection with L_p-approximation are studied in the next chapter.

§ 1. Descartes Families

A. γ-Polynomials

Let X be an interval on the real axis and let $\gamma \in C(T \times X)$ where T is a subset of $\mathbb{R}$. Due to Hobby and Rice (1967), the function

$$u(x) = \sum_{\nu=1}^{k} \alpha_\nu \gamma(t_\nu, x), \qquad \alpha_\nu \in \mathbb{R}, \qquad t_\nu \in T, \tag{1.1}$$

is called a *γ-polynomial* and its *order* is k if it cannot be expressed by a sum with $k - 1$ terms. The interest in γ-polynomials stems from approximation by exponentials and by spline functions, where the kernel γ is e^{tx} and $(x - t)_+^n$, resp., although splines require some extra formulations. Other interesting kernels are $\cosh tx$, x^t, $(1 + xt)^{-1}$, $\arctan tx$ and $\log(1 + tx)$, see p. 189.

A unified theory for approximation by γ-polynomials can be formulated, if the functions

$$\gamma(t_1, .), \gamma(t_2, .), \ldots, \gamma(t_k, .)$$

form a Chebyshev system for distinct t_i. If moreover, T is an interval, then even a Descartes' rule on sign holds.

1.1 Definition. A *Descartes' rule* holds for the γ-polynomials of order $\leq k$, if the following implication is true: Let the γ-polynomial u of the form (1.1) with $t_1 < t_2 < \cdots < t_n$ satisfy

$$(-1)^{r-i}u(x_i) > 0, \qquad i = 1, 2, \ldots, r, \tag{1.2}$$

where $x_1 < x_2 < \cdots < x_r$. Then there are at least $r - 1$ sign changes in the sequence $\alpha_1, \alpha_2, \ldots, \alpha_k$. If the number of changes equals $r - 1$, then

$$\operatorname{sgn} \alpha_1 = \tilde{\varepsilon}_r(-1)^{r-1}, \qquad \operatorname{sign} \alpha_k = \tilde{\varepsilon}_r. \tag{1.3}$$

Here $\tilde{\varepsilon}_1, \tilde{\varepsilon}_2, \ldots, \tilde{\varepsilon}_k$ are constants, each of them is either $+1$ or -1.

The Descartes' rule of signs for exponential sums (see VI.1.2) was already known to Laguerre (1898), and the simple proof via Rolle's theorem is possibly due to him. He also observed the connection with the more famous rule of signs for polynomials which has various applications in approximation theory, see e.g. Bernstein's old estimate of $E_n(|x|)$ from below (1926), The general theory for this rule will be discussed in Section B.

1.2 Definition. Let $\gamma \in C(T \times X)$. The set

$$G_n^0 := \left\{u\colon u = \sum_{\nu=1}^{k} \alpha_\nu \gamma(t_\nu, x),\, \alpha_\nu \in \mathbb{R},\, t_\nu \in T,\, k \leq n\right\}$$

is called a *Descartes family*, if a Descartes' rule holds for the γ-polynomials of order $\leq 2n$.

When Hobby and Rice (1967) introduced the γ-polynomials, they had already noticed that one has to consider the closure of the families of γ-polynomials. If the derivatives $\gamma^{(\mu)} := (\partial^\mu/\partial t^\mu)\gamma$ exist and are continuous in $T \times X$, one has to adjoin the *extended γ-polynomials*

$$u(x) = \sum_{\nu=1}^{l} \sum_{\mu=0}^{m_\nu} \alpha_{\nu\mu} \gamma^{(\mu)}(t_\nu, x) \tag{1.4}$$

of order $\leq n$. Specifically, if $t_1 < t_2 < \cdots < t_l$ and $\alpha_{\nu m_\nu} \neq 0$ for $\nu = 1, 2, \ldots, l$, then

$$k = k(u) := \sum_{\nu=1}^{l} (m_\nu + 1)$$

is the *order* of u. The order coincides with the *length* of the γ-polynomial $l = l(u)$, if u has the special form (1.1). The parameters t_ν in (1.4) are called the *characteristic numbers* of u and $m_\nu + 1$ their multiplicities. The set $\operatorname{spect}(u) := \{t_\nu\colon \nu = 1, 2, \ldots, l(u)\}$ is said to be the *spectrum*.

1.3 Definition. Let the kernel γ be $(2n - 1)$-times differentiable in t and $\gamma^{(\mu)} \in C(T \times X)$ for $\mu = 1, 2, \ldots, 2n - 1$. Moreover, assume that G_n^0 is a Descartes family. If each γ-polynomial of order $k \leq 2n$ has at most $k-1$ zeros, then

$$G_n = \left\{ u = \sum_{\nu=1}^{l} \sum_{\mu=0}^{m_\nu} \alpha_{\nu\mu} \gamma^{(\mu)}(t_\nu, .): \alpha_{\nu\mu} \in \mathbb{R}, t_\nu \in T, \sum_{\nu=1}^{l} (m_\nu + 1) \leq n \right\}$$

is called an *(extended) Descartes family.*

A Descartes' rule for the extended γ-polynomials will be established in Section C (see also Exercise 1.18).

B. Sign-Regular and Totally Positive Kernels

When γ is the exponential kernel and in other cases of interest (cf. Section 1D, below), the Descartes' rule can be directly proven. Another approach which was established in particular by Karlin (1968), is via determinants. In this framework, the Descartes' rule is considered as a variation-diminishing property and is deduced from determinants which are found with the given kernels.

Let T and X be subsets of $\mathbb{R}$ and $\gamma \in C(T \times X)$. We consider the following determinants:

$$\gamma\begin{pmatrix} t_1, t_2, \ldots, t_r \\ x_1, x_2, \ldots, x_r \end{pmatrix} := \begin{vmatrix} \gamma(t_1, x_1) & \gamma(t_1, x_2) & \ldots & \gamma(t_1, x_r) \\ \gamma(t_2, x_1) & \gamma(t_2, x_2) & \ldots & \gamma(t_2, x_r) \\ & & \cdots & \\ \gamma(t_r, x_1) & \gamma(t_r, x_2) & \ldots & \gamma(t_r, x_r) \end{vmatrix} \tag{1.5}$$

1.4 Definition. Let $\gamma \in C(T \cdot X)$. If there exist $\varepsilon_1, \varepsilon_2, \ldots, \varepsilon_k$, each either $+1$ or -1, such for all

$$\begin{matrix} x_1 < x_2 < \cdots < x_r, & x_i \in X, \\ t_1 < t_2 < \cdots < t_r, & t_i \in T, \end{matrix} \qquad (1 \leq r \leq k) \tag{1.6}$$

the relation

$$\varepsilon_r \gamma \begin{pmatrix} t_1, t_2, \ldots, t_r \\ x_1, x_2, \ldots, x_r \end{pmatrix} \geq 0 \tag{1.7}$$

holds, then the kernel γ is called *sign-regular* of order k(short: SR_k). If in addition $\varepsilon_1 = \varepsilon_2 = \cdots = \varepsilon_k = +1$, then γ is *totally positive* (short: TP_k). If strict inequality holds in (1.6), then we say that γ is *strictly sign-regular* (short SSR_k) or *strictly totally positive* (short STP_k), respectively.

For connected sets T and X, being SSR_r is obviously equivalent to $\gamma(t_1, .)$, $\gamma(t_2, .), \ldots, \gamma(t_r, .)$ forming a Chebyshev system for distinct t_i's.

The theory can be generalized to the extended γ-polynomials. For this purpose, Karlin has made the following convention for the determinants: If t_ν, $t_{\nu+1}, \ldots, t_{\nu+m}$ is a block of coincident arguments, then the corresponding elements in the i-th row in (1.5) are to be replaced by $\gamma(t_\nu, x_i), \gamma^{(1)}(t_\nu, x_i), \ldots, \gamma^{(m)}(t_\nu, x_i)$, i.e., we replace (1.4) by

$$\gamma^*\begin{pmatrix} t_1, t_2, \ldots, t_r \\ x_1, x_2, \ldots, x_r \end{pmatrix} := \begin{vmatrix} \gamma(t_1, x_1) & \gamma(t_1, x_2) & \ldots & \gamma(t_1, x_r) \\ & & \ldots & \\ \gamma(t_\nu, x_1) & \gamma(t_\nu, x_2) & \ldots & \gamma(t_\nu, x_r) \\ \gamma^{(1)}(t_\nu, x_1) & \gamma^{(1)}(t_\nu, x_2) & \ldots & \gamma^{(1)}(t_\nu, x_r) \\ & & \ldots & \\ \gamma^{(m)}(t_\nu, x_1) & \gamma^{(m)}(t_\nu, x_2) & \ldots & \gamma^{(m)}(t_\nu, x_r) \\ & & \ldots & \end{vmatrix} \tag{1.8}$$

It is admitted that several blocks of different lengths occur in the determinant.

1.5 Definition. Let $\gamma(t, x)$ be $(r - 1)$-times differentiable in t, suppose that $(\partial^{r-1}/\partial t^{r-1})\gamma \in C(T \times X)$, and let each $\varepsilon_1, \varepsilon_2, \ldots, \varepsilon_r$ be $+1$ or -1. Assume that

$$\varepsilon_k \gamma^*\begin{pmatrix} t_1, t_2, \ldots, t_k \\ x_1, x_2, \ldots, x_k \end{pmatrix} > 0 \qquad (1 \le k \le r)$$

for all $t_1 \le t_2 \le \cdots \le t_k (t_i \in T)$ and for all $x_1 < x_2 < \cdots < x_k (x_i \in X)$. Then $\gamma(t, x)$ is an *extended sign-regular* kernel of order r in the t-variable $(ESR_r(t))$. If all of $\varepsilon_1, \varepsilon_2, \ldots, \varepsilon_r$ are $+1$, $\gamma(t, x)$ is an *extended totally positive* kernel $(ETP_r(t))$.

As usually, we will write SSR, STP, $ESR(t)$ or $ETP(t)$ and suppress the specification of the order, if the corresponding property holds for any order.

1.6 Lemma. *If γ is strictly sign-regular of order k, then a Descartes' rule holds for the γ-polynomials of order $\le k$ with $\tilde{\varepsilon}_1 = \varepsilon_1$ and $\tilde{\varepsilon}_j = \varepsilon_{j-1}/\varepsilon_j, j = 2, 3, \ldots, k$.*

Proof. Let u be a γ-polynomial with at most $r - 1$ sign changes in the sequence $\alpha_1, \alpha_2, \ldots, \alpha_k$. Then we may split the sum (1.1) into r parts,

$$u(x) := \sum_{\nu=1}^{r} s_\nu w_\nu(x), \tag{1.9}$$

with partial sums $w_\nu := \sum_\mu |\alpha_\mu| \gamma(t_\mu, .)$, such that all characteristic numbers of w_ν are larger than those of $w_{\nu-1}, 2 \le \nu \le r$. From Laplace's theorem on determinants and (1.7), it follows that

$$\varepsilon_r \begin{vmatrix} w_1(x_1) & w_1(x_2) & \ldots & w_1(x_r) \\ w_2(x_1) & w_2(x_2) & \ldots & w_2(x_r) \\ & & \ldots & \\ w_r(x_1) & w_r(x_2) & \ldots & w_r(x_r) \end{vmatrix} > 0$$

whenever $x_1 < x_2 < \cdots < x_r$. Hence, $w_1, w_2, \ldots, w_r$ span a Haar subspace and no nontrivial linear combination has r zeros. Given $u(x_1), \ldots, u(x_r)$, we obtain a linear system of equations for $s_1, s_2, \ldots, s_r$ by means of (1.9). From (1.2) and Cramer's rule, (1.3) follows immediately, with the factors $\tilde{\varepsilon}_r = \varepsilon_{r-1}/\varepsilon_r$. □

We want to point out that actually strict sign-regularity only of the order r has been used.

The consequences for the case where T and X are intervals, are clear. Then a Descartes' rule of signs holds for the γ-polynomials or order $\leq k$ if each γ-polynomial of order $r \leq k$ has at most $r - 1$ zeros or vanishes identically. Only an investigation of the number of zeros of γ-polynomials is needed when a Descartes' rule is to be proven.

C. The Generalized Descartes' Rule

The Descartes' rule may be generalized to include the extended γ-polynomials. To this end, for every γ-polynomial of order k, we recursively define a sign vector $(s_1, s_2, \ldots, s_k)$ with k components.

1.7 Definition. (1) To the special γ-polynomial having a single characteristic number of multiplicity $m + 1$:

$$u = \sum_{\mu=0}^{m} \alpha_\mu \gamma^{(\mu)}(t, .), \qquad \alpha_m \neq 0, \qquad \operatorname{sgn} \alpha_m = \sigma,$$

we associate the *sign vector* with $m + 1$ components

$$\operatorname{sign} u = ((-1)^m \sigma, (-1)^{m-1} \sigma, \ldots, \sigma, -\sigma, \sigma).$$

(2) If all characteristic numbers of u_1 are smaller than those of u_2, the following composition rule holds:

$$\operatorname{sign}(u_1 + u_2) = (\operatorname{sign} u_1, \operatorname{sign} u_2).$$

Note that here $k(u_1 + u_2) = k(u_1) + k(u_2)$.

The components of $\operatorname{sign} u$ are called *generalized signs* of u. The number of positive (negative, resp.) signs is denoted by $k^+(u)$, ($k^-(u)$, resp.). Clearly,

$$k^+(u) + k^-(u) = k(u). \tag{1.10}$$

1.8 Example. Let $u = (3x + 4)e^{-4x} - 2e^x + (-7x^2 - 5x + 2)e^{2x}$. Then

$$\operatorname{sign}((3x + 4)e^{-4x}) = (-, +), \qquad \operatorname{sign}(-2e^x) = (-)$$

and

$$\operatorname{sign}((-7x^2 - 5x + 2)e^{2x}) = (-, +, -).$$

Hence,

$$\operatorname{sign} u = (-, +, -, -, +, -).$$

Fig. 28. Generalized sign pattern for γ-polynomial in Example 1.8

If u is a proper γ-polynomial of the form (1.1), then $\operatorname{sgn}\alpha_j$ is the j-th component of sign u ($j = 1, 2, \ldots, k$), provided that $t_1 < t_2 < \cdots < t_k$. To show that Definition 1.7 is reasonable, we consider a limiting process. For sufficiently differentiable kernels

$$\sum_{\mu=0}^{m} \beta_\mu \gamma^{(\mu)}(t, .) = \lim_{\substack{t_i \to t \\ t_i \neq t_j}} \sum_{\mu=0}^{m} \mu!\, \beta_\mu \gamma(t_1, t_2, \ldots, t_{\mu+1}; x). \tag{1.11}$$

The divided differences of a function $h(t)$ are defined as usual (see Chap. I). Equality (1.11) follows from the existence of a mean value τ with $h(t_1, t_2, \ldots, t_{\mu+1}) = h^{(\mu)}(\tau)/\mu!$. Using the formula

$$h(t_1, t_2, \ldots, t_{\mu+1}) = \sum_{i=1}^{\mu+1} h(t_i) \prod_{\substack{j=1 \\ j \neq i}}^{\mu+1} \frac{1}{t_i - t_j}$$

(t_i distinct), we get

$$\begin{aligned} \sum_{\mu=0}^{m} \beta_\mu \gamma^{(\mu)}(t, x) &= \lim_{\substack{t_i \to t \\ t_i \neq t_j}} \sum_{i=1}^{m+1} \gamma(t_i, x) \prod_{j \neq i} \frac{1}{t_i - t_j} \\ &\quad \times \left\{ m!\, \beta_m + \sum_{\mu=i}^{m} (\mu - 1)!\, \beta_{\mu-1} \prod_{j=\mu+1}^{m+1} (t_i - t_j) \right\}. \end{aligned} \tag{1.12}$$

Assume that $\beta_m \neq 0$. If the characteristic numbers are sufficiently close to each other, then the value in the curly brackets takes on the same sign as β_m. We also get

$$\operatorname{sgn} \prod_{j \neq i} \frac{1}{t_i - t_j} = (-1)^{p_i}$$

with

$$p_i = \text{number of characteristic numbers } s_j \text{ larger than } t_i.$$

On the right side of (1.12), the coefficients of $\gamma(t_i, x)$ are alternatingly positive and negative, and the coefficient of γ with the largest t_i has the same sign as β_m. This corresponds exactly to Definition 1.7. Defining $\tilde{\varepsilon}_j$ as in Lemma 1.6, we get a generalized Descartes' rule.

1.9 Lemma. *Let $\gamma(t, x)$ be $SSR_{\min}\{k, r\}$ and let the extended γ-polynomial* (1.4) *satisfy*

$$(-1)^{r-i} u(x_i) > 0, \qquad i = 1, 2, \ldots, r. \tag{1.13}$$

where $x_1 < x_2 < \cdots < x_r$. Then there are at least $r - 1$ sign changes in the sequence $s_1, s_2, \ldots, s_k$ of generalized signs of u. If the number of changes equals $r - 1$, then

$$s_1 = \tilde{\varepsilon}_r \operatorname{sgn} u(x_1), \qquad s_k = \tilde{\varepsilon}_r \operatorname{sgn} u(x_r). \tag{1.14}$$

Proof. Since u has $r - 1$ zeros, we have $k \geq r$. If u is a proper γ-polynomial, the statement is a consequence of Lemma 1.6. In the general case, let $\delta > 0$ and consider the proper γ-polynomial

$$u_\delta(x) = \sum_{\nu=1}^{l} \sum_{\mu=0}^{m_\nu} \alpha_{\nu\mu}\mu!\,\gamma(t_\nu, t_\nu + \delta, \ldots, t_\nu + \mu\delta; x).$$

Relations (1.11) and (1.13) yield, for a sufficiently small δ,

$$(-1)^{r-i}u_\delta(x_i) > 0, \qquad i = 1, 2, \ldots, r.$$

Thus, the lemma holds for $u_\delta(x)$. Since the generalized signs of u and of u_δ coincide for sufficiently small δ, the lemma holds for extended polynomials. □

If the kernel γ is *ESR*, a stronger result holds:

1.10 Descartes's Rule for Extended γ-Polynomials (Braess 1973b). *Let the extended γ-polynomial $u(x)$ of order k satisfy*

$$(-1)^{r-i}u(x_i) \geq 0, \qquad i = 1, 2, \ldots, r. \tag{1.15}$$

where $x_1 < x_2 < \cdots < x_r$, and assume the kernel to be $ESR_{\min(k,r)}(t)$. Then, if $u \neq 0$, there are at least $r - 1$ sign changes in the sequence $s_1, s_2, \ldots, s_k$ of generalized signs of u. If the number of changes equals $r - 1$, then

$$s_1 = \tilde{\varepsilon}_r(-1)^{r-1}, \qquad s_k = \tilde{\varepsilon}_r.$$

Proof. We distinguish two cases.

(1) Suppose that $k \geq r$. Choose numbers $m'_\nu \leq m_\nu$, $\nu = 1, 2, \ldots, l$, such that $\sum(m'_\nu + 1) = \min(k, r)$. Since γ is $ESR_r(t)$, there exists a γ-polynomial

$$v = \sum_{\nu=1}^{l} \sum_{\mu=0}^{m_\nu} \beta_{\nu\mu}\gamma^{(\mu)}(t_\nu, x)$$

(terms with $m'_\nu = -1$ are ignored), which solves the interpolation problem

$$v(x_i) = (-1)^{r-i}, \qquad i = 1, 2, \ldots, \min(k, r).$$

Hence, for every positive δ, we have $(-1)^{r-i}(u + \delta v)(x_i) > 0$. This, together with the preceding lemma, proves our conclusion for sign $(u + \delta v)$. For sufficiently small δ, the latter equals sign u.

(2) Suppose that $k < r$. If $u(x_i) = 0$ for $i = 1, 2, \ldots, k + 1$, then $u = 0$. Otherwise, we would have $u(x_j) \neq 0$ for some $j \leq k + 1$. If we repeat the arguments with v specified by

$$v(x_i) = (-1)^{r-i}, \qquad 1 \leq i \leq k + 1, \qquad i \neq j,$$

we obtain $u + \delta v = 0$, for δ sufficiently small.

Hence, (1.15) implies that $u = 0$ whenever $k < r$. □

In the study of approximation by γ-polynomials, Descartes' rule and bounds on the number of zeros will often be employed. They will be used for different orders, both for proper γ-polynomials and for extended families. A brief and precise characterization is then made possible by means of the properties SSR_r and $ESR_r(t)$.

The difference between the sign-regular and totally positive kernels is the fact that a sign factor $\tilde{\varepsilon}_r$ is often found in formulas with sign-regular kernels. Later, total positivity will be assumed to eliminate this burden.

Finally, we introduce the subsets of γ-polynomials having a specified sign vector. Let $s \in \{-1, +1\}^n$, then

$$G_n(s) := \{u \in G_n \backslash G_{n-1}\colon \text{sign } u = s\}. \tag{1.16}$$

D. Further Generalizations

In Chapter VIII, further generalizations of Descartes' rule will be of interest.

1.11 Remark (Barrar and Loeb 1978). Let

$$M(x) = \int_T \gamma(t, x) w(t)\, dt - v, \tag{1.17}$$

where v is an (extended) γ-polynomial and w is a piecewise continuous function. Choose a partition $\tau_0 < \tau_1 < \cdots < \tau_r$ of T such that each subinterval (τ_{j-1}, τ_j) contains no characteristic number of v and no point where w changes its sign. We introduce the *formal* γ-polynomial

$$\tilde{M} = \sum_{j=1}^{r} \tilde{\alpha}_j(\tilde{t}_j, .) - v$$

where $\tau_{j-1} < \tilde{t}_j < \tau_j$ and $\tilde{\alpha}_j \cdot w(t) \geq 0$ for $t \in (\tau_{j-1}, \tau_j)$. Finally, a generalized sign for M is defined by

$$\text{sign } M := \text{sign } \tilde{M}.$$

With this sign, the generalized Descartes' rule holds for the function M of the form (1.17).

To verify this extension, we first consider the corresponding generalization of Lemma 1.9. Assume that (1.13) holds if u is replaced by M. We may approximate the integral in (1.17) by a Riemann sum M_0 such that also (1.13) holds. Since sign $\tilde{M}$ has as many sign changes as sign M_0, the statement on the signs is clear in this case.

Next, we observe that the linear space which contains the γ-polynomials with the same spectrum as v and the functions $\int_{\tau_{j-1}}^{\tau_j} \gamma(t, .) w(t)\, dt$, is a Haar subspace of $C(X)$ spanned by a Chebyshev system. Therefore, we obtain the Descartes' rule 1.10 again by a perturbation argument from the statement, in the sense of Lemma 1.9. □

Up to now, we have only distinguished *nodal* and *nonnodal* zeros, i.e., we have counted multiplicities only up to order 2. One may consider zeros with the full multiplicity if $\gamma(t, x)$ is sufficiently often differentiable in the x variable. For this purpose, the definition of the determinants (1.8) will be generalized to also include blocks of coincident arguments in the x variable. The modification is analogous to the modification for coincident arguments in the t variable (see Karlin 1968).

For kernels which are extended sign-regular in both variables, the Descartes' rule of signs can be stated, counting zeros with their full multiplicities.

E. Examples

1.12. $\gamma(t,x) = e^{tx}$, $T = X = (-\infty, +\infty)$. From Lemma VI.1.1 and Rule VI.1.2, it follows that the kernel is *ETP*. Moreover, the generalized Descartes' rule for the exponential kernel may be directly proven by Rolle's theorem (see Exercise 1.18). The Descartes families for this kernel are denoted by E_n^0 and E_n.

1.13. $\gamma(t,x) = \cosh tx$, $T = X = (0, \infty)$. Each γ-polynomial of order k for this kernel belongs to E_{2k}, and therefore has at most $2k - 1$ zeros in $(-\infty, +\infty)$. There are at most $k - 1$ zeros in $(0, \infty)$, since $\gamma(t, -x) = \gamma(t, x)$. Hence γ is *ESR*. Since the behaviour for $x \to \infty$ is determined by the signs of the terms with the largest characteristic number, it follows that γ is *ETP*. The Descartes' rule for this kernel may also be easily deduced from the results for exponential sums.

In order to get an existence set, it is necessary to use the similar kernel $\tilde{\gamma}(t,x) = \cosh xt^{1/2}$, which is $ETP(t)$ on $T = X = [0, \infty)$. Since T is not an open interval, the characterization of a best approximation is more involved (cf. Section 4C). Finally, we emphasize that approximation by γ-polynomials with this kernel is not equivalent to approximation of even functions by exponential sums of twice the order (cf. Example 3.1).

1.14. $\gamma(t,x) = (1 - tx)^{-1}$, $T = (-1, +1)$, $X = [-1, +1]$. The rational functions from $R_{n-1,n}^p$ are γ-polynomials with this Cauchy kernel. The extended γ-polynomials of order k belong to $R_{k-1,k}$ and thus have at most $k - 1$ zeros. Hence, the kernel is *ESR*. The Descartes' rule is also easily proven. Let

$$u(x) = \sum_{\nu=1}^{k} \frac{\alpha_\nu}{1 - t_\nu x} = \frac{p(x)}{q(x)},$$

where $t_1 < t_2 < \cdots < t_k$. If $\operatorname{sgn} \alpha_\nu = \operatorname{sgn} \alpha_{\nu+1}$, then $u(t_{\nu+1}^{-1} + 0)$ and $u(t_\nu^{-1} - 0)$ have opposite signs. It follows that to each missing sign change in the sequence (α_ν), there is a zero of u outside $[-1, +1]$. Since p has at most $k - 1$ zeros on the whole real line, the proof of the Descartes' rule is complete.

The similar kernel $\gamma(t,x) = (t - x)^{-1}$ is obtained via the transformation $t \to t^{-1}$.

1.15. $\gamma(t,x) = \arctan tx$, $T = X = (0, \infty)$. For any extended γ-polynomial u of order k, the derivative u' is an even rational function and has at most $k - 1$ zeros in $(0, \infty)$. Since $u(0) = 0$, u also has at most $k - 1$ zeros in $(0, \infty)$. Hence γ is $ESR(t)$. However, this kernel does not generate existence sets (cf. Exercise 1.20).

1.16. $\gamma(t,x) = \log(1 + tx)$, $T = (-a^{-1}, \infty)$, $X = [0, a]$ where $a > 0$. As in the preceding example, the kernel is seen to be $ESR(t)$ by considering the derivatives of the γ-polynomials. The kernels also do not generate existence sets, (cf. Exercise 1.19 and Schmidt 1979).

1.17. $\gamma(t,x) = \sin tx$, $T = (0,a)$, $X = (0,\pi/a)$, $a > 0$. This kernel is $ESR(t)$. Indeed, each γ-polynomial of order k has at most $k-1$ zeros. For an inductive proof, an idea of Meinardus (1971) may be used. Let t_1 be in the spectrum of u. Consider $(1/\sin t_1 x)(d/dx)[\sin^2 t_1 x(d/dx)(u(x)/\sin t_1 x)]$, and apply Rolle's theorem twice.

Exercises

1.18. Prove the generalized Descartes' rule 1.10 for exponential sums directly.
Hint: Let u be an extended γ-polynomial and t_ρ be a characteristic number of multiplicity ≥ 2, consider $(d/dx)(e^{-t_\rho x}u)$ and apply Rolle's theorem.

1.19. Prove the generalized Descartes' rule 1.10 for the special case $r = k$ directly via Cramer's rule.

1.20. Consider the family with the functions of the form

$$\alpha_0 + \sum_{v=1}^{k} \alpha_k \log(1 + t_k x), \qquad \alpha_k \in \mathbb{R},\ t_k \in (-1, \infty). \tag{1.18}$$

Show that the closure in $C[0,1]$ consists of those functions for which the derivative is in $R_{k-1,k}$, and has only poles in $\mathbb{R}\backslash[0,1]$. Moreover, prove that the closure is an existence set (Schmidt 1979).

1.21. Consider the kernel from Example 1.15. Show that u is an extended γ-polynomial of order k if $u(0) = 0$ and $u'(\sqrt{x})$ is a rational function in $R_{k-1,k}$ with all poles on the negative real line.

1.22. Show that $\gamma(t,x) = x^t$ is ETP for $T = [0,\infty)$ and $X = (0,\infty)$.

1.23. Show that the kernel $\gamma(t,x) = \cos xt^{1/2}$, $T = [0,a^2)$, $X = [0,\pi/2a]$, $a > 0$ is $ESR(t)$. Why is the similar kernel $\cos xt$ not considered?

1.24. Assume that $\gamma(t,x)$ is totally positive. Verify that $\tilde{\gamma}(t,x) = \gamma(t,-x)$ is sign-regular (on a transformed domain) and that $\tilde{\varepsilon}_r = (-1)^r$ for $r = 1, 2, \ldots$.

1.25. Let $\gamma(t,x)$ be $ESR_r(t)$ on $T \times X$, where T is an interval. Assume that $\Phi\colon \mathbb{R} \to X$ is onto X, sufficiently often differentiable, and $\Phi'(t) > 0$ for all $t \in \mathbb{R}$. Show that $\tilde{\gamma}(t,x) := \gamma(\Phi(t),x)$ is $ESR_r(t)$ on $\mathbb{R} \times X$.

§2. Approximation by Proper γ-Polynomials

A. Varisolvency

Throughout the rest of the chapter unless otherwise stated, X will be a compact interval and T will be an open set in $\mathbb{R}$.

2.1 Theorem (Hobby and Rice 1967). *Every Descartes family G_n^0 is varisolvent of degree $n + k(u)$ at u and has the density property.*

Proof. (1) For any $u_1 \in G_n^0$, the difference $u_1 - u$ is a γ-polynomial of order $\leq n + k(u)$. Thus Property Z of degree $n + k(u)$ is a consequence of γ being SSR_{2n}.

(2) Assume u to be a γ-polynomial of the form (1.1) with the order k. Let $t_1 < t_2 < \cdots < t_k$. Choose numbers t_v $(v = k+1, k+2, \ldots, n)$ with $t_k < t_{k+1} < \cdots < t_n$. Furthermore put $m = n + k$ and

$A := \{b \in \mathbb{R}^m: b_\nu \neq 0, b_{n+\nu} \in T\ (\nu = 1, 2, \ldots, k), b_{n+1} < b_{n+2} < \cdots < b_{n+k} < t_{k+1}\}$.

Then the mapping $F: A \to G_n^0$,

$$F(b) = \sum_{\nu=1}^{k} b_\nu \gamma(b_{n+\nu}, .) + \sum_{\nu=k+1}^{n} b_\nu \gamma(t_\nu, .)$$

is continuous and one-one. Moreover, A is open and the difference of two elements of $F(A)$ is a γ-polynomial of order $\leq n + k$. Since $u \in F(A)$, by Lemma III.3.3, G_n^0 is locally solvent of degree $m = n + k$ at u.

(3) Let $t_1 \in \operatorname{spect}(u)$. We have $u \pm \varepsilon\gamma(t_1, .) \in G_n^0$ and the density property holds. □

Now the theory of varisolvent families may be applied.

2.2 Uniqueness and Characterization Theorem (Hobby and Rice 1967). *To each $f \in C(X)$ there is at most one best approximation from a Descartes family G_n^0, and u is a best approximation of f from G_n^0 if and only if $(f - u)$ has an alternant of length $n + k(u) + 1$.*

We recall that each local best approximation is a global solution.

Finally, we note that u is a regular point in G_n^0 from the viewpoint of varisolvency if and only if $u \notin G_{n-1}^0$.

B. Sign Distribution

The Descartes families G_n^0, $n \geq 1$, are varisolvent, because $\gamma(t_1, .)$, $\gamma(t_2, .)$, $\ldots$, $\gamma(t_r, \ldots)$ form a Chebyshev system for distinct t_ν's. From the Descartes' rule, we get some information about the signs of the factors α_ν.

Henceforth, we will repeatedly use a *de la Vallée-Poussin type argument*: Assume that there is an alternant $x_0 < x_1 < \cdots < x_m$ to $f - u_0$ with sign σ. If u is an equally good approximation, then

$$\sigma(-1)^{m-i}[u(x_i) - u_0(x_i)] \geq 0, \qquad i = 0, 1, \ldots, m. \tag{2.1}$$

If u is a better approximation to f than u_0, then strict inequalities hold in (2.1).

2.3 Theorem. *Let u and v be the best approximations to f from the Descartes families G_m^0 and G_n^0, resp. If $n > m$, then v contains at least as many positive, and at least as many negative factors as u.*

Remark. The conclusion holds even for each proper γ-polynomial v which approximates f as least as well as u, and for (the generalized sign of) each extended γ-polynomial which is a better approximant; cf. Lemma 1.9 for the proof.

Proof of Theorem 2.3. By Theorem 2.1, there is an alternant of length $r = m + k(u) + 1$ for $f - u$. Let $v \neq u$ be at least as good an approximation to f. Then (2.1) holds and the Descartes' rule asserts that the factors of $(v - u)$ change signs at least $m + k(u)$ times, if the terms are reordered according to the size of the characteristic numbers. Hence, $k^+(v - u) \geq (m + k(u))/2$ and $k^-(v - u) \geq (m + k(u))/2$. The positive factors of the difference stem from positive factors of v or from negative factors of u. Therefore, recalling that $k = k^+ + k^-$, we have

$$k^+(v) + k^-(u) \geq k^+(v - u) \geq (m + k(u))/2 = k^+(u) + k^-(u) + (m - k(u))/2.$$

Since $k(u) \leq m$, we obtain

$$k^+(v) \geq k^+(u) + (m - k(u))/2 \geq k^+(u). \tag{2.2}$$

In the same way, we may show that $k^-(v) \geq k^-(u)$. □

If $m > k(u)$, then (2.2) implies $k^+(v) \geq k^+(u) + 1$ and the last theorem can be sharpended.

2.4 Corollary. *Let the best approximation to f from the Descartes family G_n^0 exist and be of order $k < n$. Then each better approximating γ-polynomial has at least one more positive and one more negative factor.*

In the next section another improvement will be required.

2.5 Proposition. *Let u be the best approximation to f from the Descartes family G_n^0. Furthermore, let γ be SSR_{2n+1}. If $f - u$ has an alternant with the sign $+\tilde{\varepsilon}_{2n+1}(-\tilde{\varepsilon}_{2n+1}$, resp.), then each better approximating γ-polynomial has at least one more positive (negative, resp.) factor than u.*

Proof. It is only necessary to consider the case of an alternant of length $2n + 1$, since otherwise Corollary 2.4 can be applied. Assume that v is a better approximation. From (2.1) and the Descartes' rule (Lemma 1.9) it follows that there are at least $2n$ sign changes in the difference $v - u$. Moreover, if the number of sign changes in the factors is exactly $2n$, then by (1.14) the factor of the term with the highest characteristic number is positive. In all cases, the difference contains at least $n + 1$ positive factors. Therefore, as in the proof of Theorem 2.3,

$$k^+(v) + k^-(u) \geq n + 1 \geq k(u) + 1 = k^+(u) + k^-(u) + 1.$$ □

Finally, the positions of the characteristic numbers also depend on signs.

2.6 Proposition. *Assume that the best approximations u and v to $f \in C(X)$ from the Descartes families G_{n-1}^0 and G_n^0 exist and that $u \neq v$. Let t_ν and $t_{\nu+1}$ be two consecutive characteristic numbers of u. If the associated factors α_ν and $\alpha_{\nu+1}$ have the same sign (the opposite sign, resp.), then the interval $(t_\nu, t_{\nu+1})$ contains an odd (even, resp.) number of characteristic numbers of v.*

The proposition is an immediate consequence of the fact that there are exactly $n + k(u) - 1$ sign changes.

C. Positive Sums

In many applications, only γ-polynomials with nonnegative factors α_ν are admitted. We have encountered positive γ-polynomials in Chapter V in connection with rational approximation of Stieltjes functions. The theory for the approximation in the set of *positive* (more precisely: nonnegative) *sums*

$$G_n^+ := \left\{u \in C(X): u = \sum_{\nu=1}^{n} \alpha_\nu \gamma(t_\nu, .), \alpha_\nu \geq 0, t_\nu \in T\right\}$$

is more complete than for G_n^0 and easier than that for G_n. Usually, e.g., when $\gamma(t,x) = e^{tx}$, the set G_n^+ is an existence set while G_n^0 is not (cf. Exercise VI.2.12).

2.7 Uniqueness Theorem. *Assume that γ is SSR_{2n}. Then, to any $f \in C(X)$ there is at most one best approximation from G_n^+.*

Proof. (1) Let u^* be a best approximation from G_n^+ of order k^*. Firstly, we shall prove that u^* is also a best approximation to f from $G_{k^*}^0$. Indeed, u^* is optimal to f in the subset

$$\left\{u = \sum_{\nu=1}^{k^*} \alpha_\nu \gamma(t_\nu, .): \alpha_\nu > 0, t_1 < t_2 < \cdots < t_{k^*}\right\}.$$

This set is open in $G_{k^*}^0$ and varisolvent of constant degree $2k^*$. Hence, there is an alternant of length $2k^* + 1$ and by the characterization theorem 2.2, u^* is a best approximation from $G_{k^*}^0$.

(2) Suppose that u_1 and u_2 are two best approximations from G_n^+ and that $k(u_1) \geq k(u_2)$. Since u_1 is the unique solution for $G_{k(u_1)}^0 \supset G_{k(u_2)}^+$, we have $u_1 = u_2$. □

Finally, a best approximation can be characterized by alternants and their signs.

2.8 Characterization Theorem. *Assume that γ is SSR_{2n}. Then u is the best approximation to $f \in C(X)$ from G_n^+ if and only if one of the following conditions holds for $f - u$.*

1°. *There is an alternant of length $2n + 1$.*

2°. *There is an alternant of length $2k(u) + 1$ with sign $-\tilde{\varepsilon}_{2k(u)+1}$.*

Proof. (1) Assume that $k(u) = n$. Then from Theorem 2.2 it follows that 1° is sufficient for optimality and from the proof of Theorem 2.7 we conclude that 1° is also necessary. Condition 2° may be ignored here, because it is more restrictive than 1°.

(2) Assume that $k = k(u) < n$. If there is an alternant of length $2k + 1$ with sign $-\tilde{\varepsilon}_{k(u)+1}$, then by Proposition 2.5 each better approximating γ-polynomial contains a negative factor. Hence, u is the solution in G_n^+.

On the other hand, if there is no alternant of length $2k + 1$ with sign $-\tilde{\varepsilon}_{k(u)+1}$, then there is none of length $2k + 2$. By Theorem 2.2 there is a better approximating γ-polynomial v in G_{k+1}^0, which due to Proposition 2.5 must have one positive factor more than u. Hence, $v \in G_{k+1}^+ \subset G_n^+$, and u is not a best approximation. □

Since condition 2° is independent of n and refers only to the Descartes' rule for SSR_{2k+1}-kernels, we have the following

2.9 Corollary. *Let γ be SSR_{2n}. If the best approximation to $f \in C(X)$ from G_n^+ exists and is of order $k < n$, it is also the best approximation to f from G_m^+ for all $m > n$.*

As a specialization of Proposition 2.6, an interlacing property for the characteristic numbers is obtained.

2.10 Theorem. *Let γ be SSR_{2n}. If the best approximations to $f \in C(X)$ from G_{n-1}^+ and G_n^+ exist, then either they coincide or their characteristic numbers separate each other.*

2.11 Example. Consider the approximation of $f(x) = (1 + x)^{-1}$ on the interval $[0, 1]$ by sums of exponentials. From the representation

$$f(x) = \int_0^\infty e^{-tx} e^{-t}\, dt$$

it follows that f can be arbitrarily well approximated by positive sums. From Theorem 2.3 or Corollary 2.9, it follows that each best approximation is a positive sum (cf. Exercise 2.15).

Exercises

2.12. Characterize best approximations by functions of the form (1.18). Remark: This family may be treated similarly as the case, where one or more characteristic numbers are prescribed and fixed (Brink-Spalink 1975).

2.13. Show that the best rational approximation from $R_{n-1,n}$ to a Stieltjes function always belongs to the subset $R_{n-1,n}^p$. Use the results on γ-polynomials, in particular those on G_n^+ (and from Chapter V only the theorems on existence and on alternants).

2.14. Suppose that G_n is an existence set, and that both f and $g - f$ possess an integral representation of the form $\int_T \gamma(t, x)\, d\mu(t)$ with $d\mu$ being a nonnegative measure. Compare $d(f, G_n)$ and $d(g, G_n)$.

2.15. Let u be an exponential sum. Show that $1/(1 + x) - u$ has at most $k(u) + l(u)$ zeros. If moreover $u \in E_k^+$ and $1/(1 + x)$ has exactly $2k$ zeros, then spect(u) is contained in $(-\infty, 0)$. Prove that the best approximation to $1/(1 + x)$ from E_n is always a positive sum by using these observations and Proposition 2.5.

2.16. (1) Let $0 < r < n$ and $0 \le i_1 < i_2 < \cdots < i_r < i_{r+1} = n$. Show that the best approximation to $f(x) = x^n$ on $[0, 1]$ by *incomplete polynomials*

$$\sum_{\nu=1}^{r} \alpha_\nu x^{i_\nu}$$

is characterized by an alternant of length $r + 1$ and that the coefficients of the solution satisfy $\operatorname{sgn} \alpha_\alpha = (-1)^{r+1-\nu}$ for $\nu = 1, 2, \ldots, r$.
(2) Assume that $i_{j+1} \ge i_j + 2$ for some j. Show that the approximation to x^n by incomplete polynomials with the exponents $(i_1, i_2, \ldots, i_r)$ is improved, if i_j is replaced by $i_j + 1$. {From this observation, Borosh, Chui and Smith (1977) deduce that G.G. Lorentz' conjecture is true: The best incomplete polynomial with r terms has the form $\sum_{\nu=1}^{r} \alpha_\nu x^{n-\nu}$}.

§ 3. Approximation by Extended γ-Polynomials: Elementary Theory

The approximation problem is studied generally in the extended families G_n in order to guarantee the existence of a best approximation. Although G_n^0 is dense in G_n, it happens not only in exceptional cases that the best approximation is an extended γ-polynomial (see Section 5B).

On the other hand, for $n \geq 2$, the extended families are not varisolvent although Rice (1962) had claimed this for exponential sums. They are also neither asymptotically convex in the sense of Meinardus and Schwedt (1964) nor are they suns. Consequently, we cannot expect conditions in terms of alternants for the characterization of best approximations which are both necessary and sufficient. This gap will be closed later by turning to the consideration of local solutions instead of global ones.

A. Non-Uniqueness, Characterization of Best Approximations

Examples with more than one solution are easily constructed. As usual this is done by applying an invariance principle.

3.1 Example. Let $f \in C[-1, +1]$ be a positive function which is strictly decreasing in $[0, 1]$ and which has the symmetry property $f(-x) = f(x)$. Then f has at least two solutions in the family of exponentials E_2.

Suppose, to the contrary, that u^* is the unique best approximation. Note that $u \in E_2$, whenever $u(x) := u_0(-x)$, $u_0 \in E_2$. By the invariance principle, we conclude that $u^*(x) = u^*(-x)$ and u^* has the form

$$u^*(x) = \alpha \cosh tx \qquad t \geq 0.$$

Obviously, $\alpha > 0$ and $k(u^*) \geq 1$. Since $u^* \in E_2^0$, from the characterization theorem 2.2, it follows that $f - u^*$ must have an alternant of length $n + k + 1 \geq 2 + 1 + 1 = 4$. But $f - u^*$ is a strictly decreasing function in $[0, 1]$ and has at most two zeros in $[-1, +1]$. □

By similar arguments, one can see that E_2 is not varisolvent at $u_0(x) = 2x$. The interpolation problem

$$u(-1) = -2 + \delta, \qquad u(-\tfrac{1}{2}) = -1,$$

$$u(+1) = 2 - \delta, \qquad u(+\tfrac{1}{2}) = +1,$$

has no solution in E_2, whenever $\delta > 0$. Hence, E_2 is not locally solvent of degree 4 at u_0. On the other hand $u_0 - \sinh x$ has 3 zeros, and Property Z could be true only for a degree ≥ 4.

The violation of varisolvency also implies a gap in the characterization of best approximations.

3.2 Theorem. *Let $f \in C(X)$ and let G_n be an extended Descartes family.*

(1) *If there is an alternant of length $n + k(u) + 1$ for $f - u$, then u is the unique best approximation to f from G_n.*

(2) *If u is a (local) best approximation to f from G_n, then there is an alternant of length $n + l(u) + 1$ for $f - u$.*

Proof. (1) Suppose that there is an alternant of length $n + k(u) + 1$ to $f - u$ and that $v \in G_n$ is at least as good as an approximation as u. It follows from a de la Vallée-Poussin-argument (cf. the discussion of (2.1)) and the Descartes' rule, that sign $(v - u)$ is a vector with at least $n + k(u)$ sign changes. Since $k(v - u) \leq n + k(u)$, this is impossible.

(2) We write the best approximation u in the form

$$F(a) = \sum_{\nu=1}^{l} \sum_{\mu=0}^{m_\nu} \alpha_{\nu\mu} \gamma^{(\mu)}(t_\nu, .) + \sum_{\nu=k+1}^{n} \alpha_\nu \gamma(t_\nu, .)$$

with $\alpha_{k+1} = \cdots = \alpha_n = 0$, the numbers t_ν for $\nu \geq k + 1$, being distinct and not belonging to the spectrum of u. The derivatives

$$\begin{aligned} \partial F/\partial\alpha_{\nu\mu} &= \gamma^{(\mu)}(t_\nu, .), && \nu = 1, 2, \ldots, l, \quad \mu = 0, 1, \ldots, m_\nu, \\ \partial F/\partial t_\nu &= \sum_{\mu=0}^{m_\nu} \alpha_{\nu\mu} \gamma^{(\mu+1)}(t_\nu, .), && \nu = 1, 2, \ldots, l, \\ \partial F/\partial\alpha_\nu &= \gamma(t_\nu, .), && \nu = k + 1, k + 2, \ldots, n, \end{aligned} \tag{3.1}$$

form a basis for the linear space of γ-polynomials having the characteristic numbers $t_1, t_2, \ldots, t_l, t_{k+1}, \ldots, t_n$ with the multiplicities $m_1 + 2, m_2 + 2, \ldots, m_l + 2, 1, 1, \ldots, 1$. Since γ is $ESR_{2n}(t)$, this is a Haar subspace of dimension $k + l + (n - k) = n + l$. Moreover this Haar space is contained in $C_u G_n$. From Remark III.4.3 we know that there must be an alternant of the stated length. □

From Remark III.4.5, it also follows that G_n is locally solvent of degree $n + l(u)$ at u. For proper γ-polynomials, $k(u)$ and $l(u)$ coincide. Therefore, we have an alternative proof for the first part of Theorem 2.1 whenever γ is extended sign-regular. Moreover, we have the following result under the assumptions of Theorem 3.2.

3.3 Corollary. *Every best approximation in G_n^0 is the unique best approximation in G_n.*

B. γ-Polynomials of Order 2

The study of the family G_2 leads to the simplest case with nontrivial extended γ-polynomials. Here, the natural parametrization is not suitable. We obtain a complete theory by using the following parametrization F from a subset of $\mathbb{R}^2 \times T \times \mathbb{R}_+$ onto G_2:

$$
\begin{aligned}
F(a) = a_1 &\frac{\gamma(a_3+\sqrt{a_4},.)+\gamma(a_3-\sqrt{a_4},.)}{2} \\
&+ a_2 \frac{\gamma(a_3+\sqrt{a_4},.)-\gamma(a_3-\sqrt{a_4},.)}{2\sqrt{a_4}} \quad \text{with } a_4 \geq 0
\end{aligned}
\tag{3.2}
$$

and $a_3 \pm \sqrt{a_4} \in T$. The last quotient in (3.2) is just the divided difference $\gamma(a_3+\sqrt{a_4}, a_3-\sqrt{a_4}; .)$, and is to be read as $\gamma^{(1)}(a_3, .)$ when $a_4 = 0$. The order of the γ-polynomial $F(a)$ is not as easily seen from the parameter a as from the representation (1.4). However, this is no drawback in the investigations of this section.

Clearly, $\partial F/\partial a_j$, $j = 1, 2, 3$ are continuous. The derivative with respect to a_4 is evaluated by the rule I.1.2(3), that the derivative of a divided difference is obtained by repeating the variable:

$$
\begin{aligned}
&\frac{\partial}{\partial a_4}\gamma(a_3+\sqrt{a_4}, a_3-\sqrt{a_4}; .) \\
&= \frac{1}{2\sqrt{a_4}}\gamma(a_3+\sqrt{a_4}, a_3-\sqrt{a_4}, a_3+\sqrt{a_4}; .) \\
&\quad - \frac{1}{2\sqrt{a_4}}\gamma(a_3+\sqrt{a_4}, a_3-\sqrt{a_4}, a_3-\sqrt{a_4}; .) \\
&= \gamma(a_3+\sqrt{a_4}, a_3-\sqrt{a_4}, a_3+\sqrt{a_4}, a_3-\sqrt{a_4}; .).
\end{aligned}
$$

Consequently, all derivatives of first order are continuous. Moreover, if $a_2 \neq 0$, $a_4 = 0$, we have $F(a) \in G_2 \backslash G_2^0$ and

$$
\begin{aligned}
\partial F/\partial a_1 &= \gamma(a_3, .), \\
\partial F/\partial a_2 &= \gamma^{(1)}(a_3, .) \\
\partial F/\partial a_3 &= a_1\gamma^{(1)}(a_3, .) + a_2\gamma^{(2)}(a_3, .) \\
\partial F/\partial a_4 &= \tfrac{1}{2}a_1\gamma^{(2)}(a_3, .) + \tfrac{1}{6}a_2\gamma^{(3)}(a_3, .)
\end{aligned}
\tag{3.3}
$$

span a 4-dimensional Haar subspace. We recall that each γ-polynomial in G_2, which is not a proper γ-polynomial, either belongs to $G_2(-,+)$ or to $G_2(+,-)$. Now we can verify that these sign classes are uniqueness sets.

3.4 Theorem. *Let G_2 be an extended Descartes family. Then the γ-polynomial*

$$u = a_1\gamma(t,.) + a_2\gamma^{(1)}(t,.), \qquad \operatorname{sgn} a_2 =: s \neq 0$$

is the unique (local) best approximation to $f \in C(X)$ in the subfamily $G_2(-s,+s)$ if and only if $f-u$ has an alternant of length 4 whose sign equals $-\tilde{\varepsilon}_4 \cdot s$.

Proof. (1) Suppose that there is an alternant as stated and that $u_1 \in G_2(-s,+s)$ is at least as good an approximation as u. Without loss of generality, let $s = +1$. From the de la Vallée-Poussin argument and Descartes' rule of signs we get

$$\operatorname{sign}(u_1 - u) = (+, -, +, -).$$

We distinguish three cases.

(a) Both characteristic numbers of u_1 are larger than that of u. Then we have sign $(u_1 - u) = (+, -, -, +)$.

(b) Both characteristic numbers of u_1 are smaller than that of u. Then we have sign $(u_1 - u) = (-, +, +, -)$.

(c) The characteristic number of u lies between those of u_1. Then we have sign $(u_1 - u) = (-, +, -, +)$.

In each case we reach a contradiction.

(2) Suppose that there is no alternant as stated. Then there is no alternant of length 5. From the linear theory, we know that 0 is not a best approximation to $f - u$ from any 4-dimensional Haar space, and we have $\|f - u - h\| < \|f - u\|$ for a function of the form

$$h = \sum_{\mu=0}^{3} \beta_\mu \gamma^{(\mu)}(a_3, .).$$

Next, we may assume that $f - u$ has an alternant of exact length 4 because otherwise by Theorem 3.2 u is not an *LBA* and a better approximation in the same subfamily $G_2(-s, +s)$ may be constructed. Therefore, we are left with the case that the alternant has the exact length 4 and the sign $+\tilde{\varepsilon}_4 \cdot s$. By the de la Vallée-Poussin argument, h has 3 zeros $\xi_1 < \xi_2 < \xi_3$ and $sh(x) > 0$ for $x > \xi_3$. Hence, sign $\beta_3 = s$ and

$$h = \sum_{k=1}^{4} \delta_k \frac{\partial F}{\partial a_k}$$

with $\delta_4 > 0$. Therefore, $h \in C_u G_2 = C_u G_2(-s, +s)$, and u is not a critical point in $G_2(-s, +s)$. □

From Corollary 3.3 and Theorem 3.4, we have the following

3.5 Corollary. *Let G_2 be a Descartes family. Then to each $f \in C(X)$ at most two best approximations exist. If two best approximations exist, they have the form*

$$\beta_0^{(i)} \gamma(t^{(i)}, .) + \beta_1^{(i)} \gamma^{(1)}(t^{(i)}, .), \qquad i = 1, 2,$$

where $\beta_1^{(1)}$ and $\beta_1^{(2)}$ have opposite signs.

For $n = 2$, the theory is now quite complete. In view of Example 3.1, one cannot expect sharper results.

For $n = 2$, each sign class $G_2(s_1, s_2)$ contains at most one local best approximation. We will see that this is no longer true when $n \geq 3$.

Exercises

3.6. Assume that a best approximation to f from G_2^+ exists whose order equals 2. Show that there is no local best approximation to f of the form $\beta_0 \gamma(t, x) + \beta_1 \gamma^{(1)}(t, x)$.

3.7. If $f \in G_2$, then obviously the best approximation to f from G_2 is unique. However, can there exist another **local** best approximation to f from G_2?

3.8. Consider the approximation to $f(x) = \cos x$ on $[\alpha, \beta]$, where $\beta - \alpha < \pi$, by sums of exponentials. Show that each local best approximation to f from E_n is an exponential sum with only one characteristic number of multiplicity n. (Therefore, the results for E_n^c from Chapter IV may be used). Hint: If $u \in E_n$, then $A\cos x + B\sin x - u(x)$ has at most $n + 1$ zeros in $[\alpha, \beta]$ (Kammler 1973b).

3.9. Let $n \geq 1$. Consider the approximation of $f(x) = \sin x$ on $[-1, +1]$ by sums of exponentials. Show that there are an even number of local best approximations from E_{2n+1}. (Use the hint for the preceding exercise).

3.10. Assume that γ is $ETP_6(t)$ and

$$u = \alpha_{10}\gamma(t_1, .) + \alpha_{11}\gamma^{(1)}(t_1, .) + \alpha_2\gamma(t_2, .) \quad \text{where } \alpha_{11} \neq 0, \alpha_2 \neq 0, t_1 \neq t_2,$$

is a local best approximation to f from G_3. Show that $f - u$ has an alternant of length 6 whose sign is $-\operatorname{sgn}\alpha_{11}$. If, moreover sign $u = s = (-, -, +)$, then u is the unique best approximation from $G_3(s)$. If, on the other hand, sign $u = s = (+, -, +)$ and $t_1 > t_2$ then there is no better approximating γ-polynomial in $G_3(s)$ with spectrum in $T \cap (t_2, \infty)$.

3.11. Are G_n^0, G_n^+, G_n, $G_2(-+)$, $G_3(--+)$, $G_3(+-+)$ suns?
Hint: An answer for $G_3(+-+)$ may be obtained from Example IV.3.6.

§4. The Haar Manifold $G_n \backslash G_{n-1}$

A. Simple Parametrizations

It is the aim of this section to show that $G_n \backslash G_{n-1}$ is a Haar manifold. Then, from the general theory, we may obtain a characterization of LBA's in terms of alternants with signs. Moreover, this property can be used to achieve upper bounds for the number of best approximations.

The natural representation (1.4) for γ-polynomials has a serious drawback: one cannot see that in every neighborhood of $u = \gamma^{(1)}(t, .) \in G_2$ there are γ-polynomials of the form

$$\frac{1}{\delta}[\gamma(t + \delta, .) - \gamma(t, .)].$$

Therefore, one needs parametrizations for Descartes families which describe their topological structure.

Firstly we note that each γ-polynomial of order $\leq n$ can be written in the form

$$F_0(a, x) = \sum_{\mu=1}^{n} \beta_\mu \gamma_\mu(t_1, t_2, \ldots, t_\mu; x), \tag{4.1}$$

where

$$t_1 \leq t_2 \leq \cdots \leq t_n. \tag{4.2}$$

Here and in the following, the number of t-arguments in a divided difference is written as an index (Thus $\gamma_\mu(\ldots)$ is a divided difference of order $\mu - 1$). The characteristic number t_i appears $(m+1)$ times in (4.1), if a term with $\gamma^{(m)}(t_i, .)$ is present in (1.4). The characteristic numbers are labelled differently in (1.4) and (4.1); multiplicity is treated differently. We will understand (4.1) as a parametrization with the domain $A_0 \subset \mathbb{R}^{2n}$:

$$A_0 = \{a = (\beta_1, \beta_2, \ldots, \beta_n, t_1, \ldots, t_n): \beta_\mu \in \mathbb{R}, t_\mu \in T, t_1 \leq t_2 \leq \cdots \leq t_n\}.$$

To verify that F_0 is a map *onto* G_n, we claim that each $u \in G_n$ may be given in the form

$$u(x) = \sum_{\mu=1}^{k} \beta_\mu \gamma_\mu(t_1, t_2, \ldots, t_\mu; x)$$

+ terms with divided differences of order $k, k+1, \ldots, n-1$, and arguments which are subsets of $\{t_1, t_2, \ldots, t_n\}$.

For $k = 0$ this is obvious from $\gamma^{(m)}(t, x) = m! \gamma_{m+1}(t, t, \ldots, t; x)$ and the convention for counting multiplicities. Assume that we have already produced the representation for some $k-1$. Consider a divided difference with k arguments $\gamma(t_{i_1}, t_{i_2}, \ldots, t_{i_k}; .)$. Because of the symmetry of divided differences, we may reorder $t_1, t_2, \ldots, t_k$ and $t_{i_1}, t_{i_2}, \ldots, t_{i_k}$, resp. such that $t_\mu = t_{i_\mu}$, $\mu = 1, 2, \ldots, v$, and that v is maximal. Recalling that

$$\gamma_k(t_1, \ldots, t_v, t_{i_{v+1}}, \ldots, t_{i_k}; .) = \gamma_k(t_1, \ldots, t_v, t_{v+1}, t_{i_{v+2}}, \ldots, t_{i_k}; .) + (t_{i_{v+1}} - t_{v+1})\gamma_{k+1}(\ldots),$$

we get equality for $v+1$ arguments (modulo terms of higher order). By induction, equality is gained for all arguments, and by induction on k the representation (4.1) is established. □

The representation (4.1) may be used with any labelling of the t_μ's and the ordering (4.2) is not necessary for the induction above. Therefore, one may derive a more symmetric representation:

$$F_{\text{sym}}(a, x) = \sum_{\mu=1}^{n} \alpha_\mu \Gamma_\mu(t_1, t_2, \ldots, t_n; x) \tag{4.3}$$

where

$$\Gamma_\mu(t_1, t_2, \ldots, t_n; .) := \sum_{1 \leq i_1 < i_2 < \cdots < i_\mu \leq n} \gamma_\mu(t_{i_1}, t_{i_2}, \ldots, t_{i_\mu}; .).$$

Indeed, Γ_μ equals $\binom{n}{\mu}\gamma_\mu(t_1, \ldots, t_\mu; .)$ modulo terms of higher order and (4.3) is obtained by a repetition of the argumentation above.

The mappings F_0 and F_{sym} are continuous and injective on those subsets of A which are mapped onto $G_n \backslash G_{n-1}$. On the other hand, to guarantee continuity of the inverse mappings, a condition is postulated which is satisfied by the interesting kernels. The condition means that nothing severe (like the back-bending in Fig. 13) can happen when one or more characteristic numbers tend to infinity.

4.1 Definition. Let the families G_n be endowed with a topology $\mathcal{T}$. If, to each $u_0 \in G_n\backslash G_{n-1}$, there is a neighborhood $U(u_0)$ such that the characteristic numbers for all $u \in U(u_0)$ belong to a compact subset of T, then G_n is called *normal* to $\mathcal{T}$. If G_n is normal to the norm topology, then G_n is a *normal family*.

4.2 Lemma. *Let the Descartes family G_n be endowed with a topology $\mathcal{T}$ having the following properties*:

1°. *$\mathcal{T}$ is the norm-topology or a weaker one.*

2°. *The convergence of a filter implies the convergence of the functions at $2n$ points $x_i \in X$.*

3°. *G_n is normal relative to $\mathcal{T}$.*

Then the mappings defined by (4.1) *and* (4.3), *resp.*:

$$F_0^{-1}: G_n\backslash G_{n-1} \to A_0, \qquad F_{\text{sym}}^{-1}: G_n\backslash G_{n-1} \to A_0$$

are homeomorphisms.

Proof. We restrict ourselves to F_0. We already know that F_0 is continuous. Let $\mathcal{G}$ be a filter which converges to $u_0 \in G_n\backslash G_{n-1}$. From $\mathcal{G}$ we can select a sequence (u_j) such that

$$\lim_{j\to\infty} u_j(x_i) = u_0(x_i), \qquad i = 1, 2, \ldots, 2n,$$

holds for $2n$ points $x_i \in X$. By virtue of 3°, the spectra of the u_j's are contained in a compact subset of T. Thus, after passing to a subsequence if necessary, we may assume that the characteristic numbers converge to some $(t_1^0, t_2^0, \ldots, t_n^0)$ with all characteristic numbers written with their multiplicities. Let $\beta_1, \beta_2, \ldots, \beta_n$ be the solution of the n linear equations

$$\sum_{\mu=1}^{n} \beta_\mu \gamma_\mu(t_1^0, t_2^0, \ldots, t_\mu^0, x_i) = u_0(x_i), \qquad i = 1, 2, \ldots, n. \tag{4.4}$$

Since γ is $ESR(t)$, the coefficient matrix of the linear system is not singular. By perturbation theory for linear matrix equations, it follows that the coefficient vectors of the u_j's converge to $(\beta_1, \beta_2, \ldots, \beta_n)$. Since $u_0 \notin G_{n-1}$, we have uniqueness of the representation. This in turn implies uniform convergence of the sequence in $C(X)$ to the γ-polynomial specified by the left hand side of (4.4). Now 1° implies that $(\beta_1, \beta_2, \ldots, \beta_n)$ is the coefficient vector of u_0. Since the arguments hold for any subsequence, F_0^{-1} is continuous. □

By definition, if G_n is normal to some topology, then it is normal to any weaker topology. Therefore, from Lemma 4.2 we have the

4.3 Corollary. *Let G_n be a normal Descartes family. Then all topologies with the properties stated in Lemma* 4.2 *are equivalent to the norm-topology in $G_n\backslash G_{n-1}$.*

We recall that each bounded sequence in the family E_n of exponential sums contains a subsequence which either converges uniformly or converges pointwise to some element in E_{n-1}. Therefore, the exponential sums constitute normal Descartes families.

The representations F_0 and F_{sym} have a disadvantage, which is that one cannot immediately see the order k of an element. For this reason, the mappings are mostly used in a local fashion. Let u_0 be given in the form (1.4). Then each inner sum $\sum_\mu \alpha_{\nu\mu}\gamma^{(\mu)}(t_\nu, .)$ is replaced by an expression of the form (4.1) or (4.3). Such a representation is possible for a neighborhood of u_0; specifically, it is possible as long as the characteristic numbers remain in separated clusters. The change of representations means that we really need an atlas which does not consist of only one chart.

The features of this atlas are clear once we know the corresponding properties of the mapping F_0 and F_{sym}. Therefore, for ease of notation we will usually refer to (4.1) and (4.3), resp., when we study the features of the representations.

B. The Differentiable Structure

In the preceding section, we have seen that $G_n \backslash G_{n-1}$ is a *topological manifold* (with boundary). The first proof that this set is a C^1-*manifold*, used an involved analysis of the second order terms of the Taylor's series for the parametrization F_0. The following analysis due to Cromme (1982), generalizes the way in which the case $n = 2$ has been treated above. A *mean characteristic number* τ and $n-1$ *deviations* $\delta_1, \delta_2, \ldots, \delta_{n-1}$ become parameters which replace $t_1, t_2, \ldots, t_n$:

$$\begin{aligned}
t_1 &= \tau - \sqrt{\delta_1} - \sqrt{\delta_2} - \sqrt{\delta_3} - \cdots - \sqrt{\delta_{n-1}},\\
t_2 &= \tau + \sqrt{\delta_1} - \sqrt{\delta_2} - \sqrt{\delta_3} - \cdots - \sqrt{\delta_{n-1}},\\
t_3 &= \tau \qquad\quad\; + 2\sqrt{\delta_2} - \sqrt{\delta_3} - \cdots - \sqrt{\delta_{n-1}},\\
t_4 &= \tau \qquad\qquad\qquad\quad + 3\sqrt{\delta_3} - \cdots - \sqrt{\delta_{n-1}},\\
&\;\;\vdots\\
t_n &= \tau \qquad\qquad\qquad\qquad\qquad\quad + (n-1)\sqrt{\delta_{n-1}},
\end{aligned} \tag{4.5}$$

with $0 \le \delta_i \le \left(\dfrac{i+2}{i}\right)^2 \delta_{i+1}$, $i = 1, 2, \ldots, n-2$.

There are additional restrictions on the parameters, if T is a proper subset of $\mathbb{R}$. However, we may ignore them in the following study of local properties.

4.4 Theorem (Braess 1978). *Let G_n be a normal Descartes family. Then $G_n \backslash G_{n-1}$ is a Haar submanifold (with boundary) of $C(X)$. Specifically, if $u \in G_n \backslash G_{n-1}$ is given in the form* (1.4), *then*

$$C_u G_n = \left\{ h = \sum_{\nu=1}^{l} \sum_{\mu=0}^{m_\nu^*} \beta_{\nu\mu}\gamma^{(\mu)}(t_\nu, x) \colon \beta_{\nu\mu} \in \mathbb{R},\ \sigma_\nu \beta_{\nu m_\nu^*} \ge 0 \textit{ if } m_\nu \ge 2 \right\}. \tag{4.6}$$

where $\sigma_\nu := \operatorname{sgn} \alpha_{\nu m_\nu}$ and

$$m_\nu^* := \begin{cases} 1, & \textit{if } m_\nu = 0,\\ m_\nu + 2, & \textit{if } m_\nu > 0. \end{cases}$$

Proof. The main part of the proof will be concerned with the case where $u = u_0$ has only one characteristic number t_0 of multiplicity $n \geq 2$, i.e. when $l(u_0) = 1$. To obtain a suitable chart, choose the parameter set

$$A = \{a = (\beta_1, \ldots, \beta_n, \tau, \delta_1, \ldots, \delta_{n-1})\}$$

where the δ_j's are restricted as in (4.5), and define $F: A \to G_n$ by (4.3) and (4.5):

$$a \mapsto b = (\beta_1, \ldots, \beta_n, t_1, \ldots, t_n) \mapsto F_{\text{sym}}(b) = u. \tag{4.7}$$

When considering the differentiability of F, we anticipate a formula which will be deduced at the end of the section:

$$\frac{\partial}{\partial \delta_j} \Gamma_m(t_1, t_2, \ldots, t_n; x) = c_{mj} \gamma_{m+2}(\tau, \tau, \ldots, \tau; x) + o(1) \tag{4.8}$$

as $\delta_{n-1} \to 0$ $(1 \leq j \leq n-1)$, where $c_{mj} \geq 0$ and

$$c_{m,n-1} = (n + m - 2)\binom{n-1}{m-1} > 0. \tag{4.9}$$

Therefore, the first derivatives are continuous, and at $a = a_0 := F^{-1}(u_0)$, we have

$$\frac{\partial F}{\partial \beta_\mu} = \Gamma_\mu(t_0, \ldots, t_0; x) = \binom{n}{\mu} \gamma_\mu(t_0, t_0, \ldots, t_0; x), \tag{4.10}$$

$$\frac{\partial F}{\partial \tau} = \sum_{\mu=1}^{n} \beta_\mu \binom{n}{\mu} \gamma_{\mu+1}(t_0, \ldots, t_0; x), \tag{4.11}$$

$$\frac{\partial F}{\partial \delta_j} = c_{nj} \beta_n \gamma_{n+2}(t_0, \ldots, t_0; x) + \text{a } \gamma\text{-polynomial of order} \leq n + 1. \tag{4.12}$$

Hence,

$$d_{a_0} F \tilde{a} = \beta_n \gamma_{n+2}(t_0, \ldots, t_0; x) \sum_{j=1}^{n-1} c_{nj} \tilde{\delta}_j + \text{a } \gamma\text{-polynomial of order} \leq n + 1. \tag{4.13}$$

Note that the parameter set A is a (convex) subset of $\mathbb{R}^{2n}$ which is specified by finitely many linear restrictions. Therefore, in order to prove that the conditions for a C^1-manifold are satisfied at u_0, we only need to verify property III.1.7 (ii) which means that

$$d_{a_0} F \tilde{a} \neq 0 \quad \text{for all } \tilde{a} \in C, \tag{4.14}$$

where C is the tangent cone $C_{a_0} A$:

$$C = \Big\{ \tilde{a} = (\tilde{\beta}_1, \ldots, \tilde{\beta}_n, \tilde{\tau}, \tilde{\delta}_1, \ldots, \tilde{\delta}_{n-1}):$$

$$0 \leq \delta_j \leq \left(\frac{j+2}{j}\right)^2 \delta_{j+1} \text{ for } j = 1, 2, \ldots, n-2 \Big\}.$$

We consider three cases.

Suppose that $\tilde{\delta}_{n-1} > 0$. Then (4.8), (4.9), and $\tilde{\delta}_j \geq 0$ imply that $\sum_j c_{nj}\tilde{\delta}_j > 0$, and from (4.13) we conclude that $d_{a_0}F\tilde{a} \neq 0$.

Suppose that $\tilde{\delta}_{n-1} = 0$ and $\tilde{\tau} \neq 0$. Since also $\tilde{\delta}_1 = \tilde{\delta}_2 = \cdots = \tilde{\delta}_{n-2} = 0$, by inserting the derivatives, we obtain

$$d_{a_0}F\tilde{a} = \beta_n \tilde{\tau}\gamma_{n+1}(t_0,\ldots,t_0;x) + \text{a } \gamma\text{-polynomial of order} \leq n \neq 0.$$

Suppose that $\tilde{\delta}_{n-1} = \tilde{\tau} = 0$. Then (4.10) yields

$$d_{a_0}F\tilde{a} = \sum_{\mu=1}^{n-1} \tilde{\beta}_\mu \binom{n}{\mu} \gamma_\mu(t_0,\ldots,t_0;x)$$

which is not the zero function unless $\tilde{a} = 0$. Thus (4.14) is true.

With this, we have also shown that the tangent vectors may be represented in the form

$$h = \sum_{\mu=0}^{n+1} \theta_\mu \gamma^{(\mu)}(t_0, x) \quad \text{where } \theta_{n+1}\beta_n \geq 0. \tag{4.15}$$

On the other hand, given h as in (4.15), we can determine a vector $\tilde{a} = (\tilde{\beta}_1,\ldots,\tilde{\beta}_n,\tilde{\tau},0,\ldots,0,\tilde{\delta}_{n-1})$ with $\tilde{\delta}_{n-1} \geq 0$ such that $h = d_{a_0}F\tilde{a}$. Indeed, there is a $\tilde{\delta}_{n-1} \geq 0$ such that $h - (\partial F/\partial\delta_{n-1})\tilde{\delta}_{n-1}$ is a γ-polynomial of order $\leq n+1$. By choosing $\tilde{\tau}$, the order may be reduced to be $\leq n$. Finally, $\beta_1, \beta_2, \ldots, \beta_n$ are chosen such that the vector from the given cone is obtained.

Thus, the proof for the case $l(u_0) = 1$ and $n \geq 2$ is complete.

If t_0 has multiplicity 1, the natural chart $F\colon (\alpha, t) \to \alpha\gamma(t, .)$ provides a C^1-parametrization for which the non-singular derivative $d_a F$ yields the two-dimensional tangent space as stated.

If $l(u_0) \geq 2$, then for each partial sum a parametrization as above is taken (cf. the comment at the end of Section A). By putting different parameter sets and the mappings together in an obvious way, we get a chart for u_0 as required.

Since the cones of the form (4.9) satisfy the Haar condition, $G_n \backslash G_{n-1}$ is a Haar manifold. □

Proof of (4.8) and (4.9). For the calculation of the derivatives, we recall Rule I.1.2(3),

$$\frac{\partial}{\partial\delta_j}\gamma_m(t_1,\ldots,t_m;.) = \sum_{\nu=1}^{m} \gamma_{m+1}(t_1,\ldots,t_m,t_\nu)\frac{\partial t_\nu}{\partial\delta_j}$$

for $1 \leq j \leq m$. Since $2\sqrt{\delta_{n-1}}\,\partial t_j/\partial\delta_{n-1}$ equals $(n-1)$ for $j = n$ and -1 otherwise, we have

$$\begin{aligned}
&2\sqrt{\delta_{n-1}}\,\frac{\partial}{\partial\delta_{n-1}} \sum_{1\leq j_1<j_2<\cdots<j_m\leq n} \gamma_m(t_{j_1}, t_{j_2}, \ldots, t_{j_m}; x) \\
&= -\sum_{1\leq j_1<\cdots<j_m\leq n}\ \sum_{k=1}^{m} \gamma_{m+1}(t_{j_1},\ldots,t_{j_m},t_{j_k};x) \\
&\quad + (n-1) \sum_{1\leq j_1<\cdots<j_{m-1}<n} \gamma_{m+1}(t_{j_1},\ldots,t_{j_{m-1}},t_n,t_n;x).
\end{aligned} \tag{4.16}$$

Now the first sum on the right hand side is reorganized such that a sum over $r = j_k$ can be extracted. Writing only the summation indices and using symmetries, we get

$$\sum_{1 \le j_1 < \cdots < j_{m-1} < n} \left\{ \sum_{\substack{r=1 \\ (j_m = r) \\ j_i \ne r, (i \le m-1)}}^{n-1} + \sum_{\substack{k=1 \\ j_m = n \\ r = j_k}}^{m-1} \right\} \tag{4.17}$$

for this double sum. Note that the sum within the braces contains $[n - 1 - (m - 1)] + m - 1 = n - 1$ terms. Since the last term in (4.16) contains a factor $(n - 1)$, it may also be rewritten as a sum of the form (4.17).

With this reorganization, we obtain from (4.16)

$$\begin{aligned}
&\sum_{1 < j_1 < \cdots < j_{m-1} < n} \left\{ \sum_{\substack{r=1 \\ j_i \ne r (1 m-1)}}^{n-1} \right. \\
&[\gamma_{m+1}(t_{j_1}, \ldots, t_{j_{m-1}}, t_n, t_n; x) - \gamma_{m+1}(t_{j_1}, \ldots, t_{j_{m-1}}, t_r, t_r; x)] \\
&\left. + \sum_{k=1}^{m-1} [\gamma_{m+1}(t_{j_1}, \ldots, t_{j_{m-1}}, t_n, t_n; x) - \gamma_{m+1}(t_{j_1}, \ldots, t_{j_{m-1}}, t_n, t_{j_k}; x)] \right\} \\
&= \sum \left\{ \sum_r (t_n - t_r) [\gamma_{m+2}(t_{j_1}, \ldots, t_{j_{m-1}}, t_n, t_n, t_r; x) \right. \\
&\quad + \gamma_{m+2}(t_{j_1}, \ldots, t_{j_{m-1}}, t_n, t_r, t_r)] \\
&\quad \left. + \sum_k (t_n - t_{j_k}) \gamma_{m+2}(t_{j_1}, \ldots, t_{j_{m-1}}, t_n, t_n, t_{j_k}] \right\}.
\end{aligned}$$

Next, we observe that $t_n - t_i$ may be bounded by a multiple of $\sqrt{\delta_{n-1}}$. Since the divided differences are continuous functions of the parameters, the last expression equals

$$\begin{aligned}
&\sum \left\{ \sum_r 2(t_n - t_r) + \sum_k (t_n - t_{j_k}) \right\} \gamma_{m+2}(\tau, \ldots, \tau; x) + o(\sqrt{\delta_{n-1}}) \\
&= 2\sqrt{\delta_{n-1}} \binom{n-1}{m-1} (2n + m - 3) \gamma_{m+2}(\tau, \ldots, \tau; x) + o(\sqrt{\delta_{n-1}}).
\end{aligned}$$

Here we have made use of the fact that the sum over the permutations is symmetrical. Specifically, it is a multiple of $\sum_{j=1}^{n-1} (t_n - t_j) = 2(n-1)\sqrt{\delta_{n-1}}$. Summing, we have

$$\frac{\partial}{\partial \delta_{m-1}} \Gamma_m(t_1, t_2, \ldots, t_n; x) = (n + m - 2) \binom{n-1}{m-1} \gamma_{m+2}(\tau, \ldots, \tau; x) + o(1). \tag{4.18}$$

The derivatives with respect to the other δ_j's can be determined by a reduction to the arguments above. Given $j < n - 1$, the sum over the permutations of t_1, $t_2, \ldots, t_n$ may be split into outer and inner sums such that the inner sums refer to permutations of $t_1, t_2, \ldots, t_{j+1}$. The numbers $t_{j+2}, \ldots, t_n$ may be considered as fixed, and in view of the structure of (4.5), the inner sums can be treated as above. We finally obtain (4.8) with $c_{mj} \geq 0$. □

C. Families with Bounded Spectrum

Henceforth, we will sometimes restrict ourselves to the subset of those γ-polynomials whose spectrum is restricted to a given compact interval $T_0 = [t_-, t_+] \subset T$. By choosing T_0 appropriately, we may eliminate some complications in the proofs, even if we do not know in advance a compact set which contains the spectra of the solutions.

The subset

$$\{u \in G_n \backslash G_{n-1} \colon \operatorname{spect}(u) \subset [t_-, t_+]\} \tag{4.19}$$

is again a Haar manifold. Referring to the representation (1.4), the tangent cone at u in this set is given by

$$\begin{aligned} C_u G_n = \Bigg\{ h = \sum_{\nu=1}^{l} \sum_{\mu=0}^{m_\nu^*} \beta_{\nu\mu} \gamma^{(\mu)}(t_\nu, x)&: \\ \beta_{\nu\mu} \in \mathbb{R},\ \sigma_\nu \beta_{\nu m_\nu^*} \geq 0 \text{ if } m_\nu \geq 0, \text{ or } t_\nu = t_-, \text{ or } t_\nu = t_0 \Bigg\}&, \end{aligned} \tag{4.20}$$

where

$$m_\nu^* = \begin{cases} m_\nu + 2, & \text{if } m_\nu > 0,\ t_\nu \neq t_-, t_+, \\ m_\nu + 1, & \text{otherwise,} \end{cases}$$

and

$$\sigma_\nu = \begin{cases} -\operatorname{sgn} \alpha_{\nu m_\nu}, & \text{if } t_\nu = t_+, \\ \operatorname{sgn} \alpha_{\nu m_\nu}, & \text{otherwise.} \end{cases}$$

To verify this, we consider a term of order $k \geq 1$ with the characteristic number $t = t_-$. Then the parametrization F_0 from (4.1) is differentiable and we have at $F_0^{-1}(u)$:

$$\begin{aligned} &\sum_{\nu=1}^{k} \left(\delta_\nu \frac{\partial F_0}{\partial \beta_\nu} + \delta_{k+\nu} \frac{\partial F_0}{\partial t_\nu} \right) \\ &\quad = \left(\sum_{\nu=1}^{k} \delta_{k+\nu} \right) \beta_k \gamma_{k+1}(t_-, t_-, \ldots, t_-; x) + \text{a } \gamma\text{-polynomial of order } \leq k. \end{aligned} \tag{4.21}$$

Since only $\delta_{k+\nu} \geq 0$ for $\nu = 1, 2, \ldots, k$ is admitted, the relevant combinations are contained in the cone (4.20). Moreover, the function in (4.21) is not the zero γ-polynomial, unless $\delta_1 = \delta_2 = \cdots = \delta_{2k} = 0$. Therefore, Condition III.1.7(ii)′ is true and F_0 defines a chart as required.

The case $t_\nu = t_+$ is analogous. □

Exercise

4.5. Let u_0 be a *proper* γ-polynomial in a Descartes family G_n and $k(u_0) = n$. Show that there exists a neighborhood $U(u_0)$ in G_n such that spect(u) is contained in a compact subset of T whenever $u \in U(u_0)$.
{This means that normality needs only be postulated for the extended γ-polynomials.}

§ 5. Local Best Approximations

A. Characterization of Local Best Approximations

Since the tangent cones to a Descartes family satisfy the Haar condition, critical points can be characterized in terms of signed alternants. In view of Theorem I.3.15, the maximal number of the zeros of the γ-polynomials in those cones has to be determined. For this, another integer $L = L(u)$ with

$$l(u) \le L(u) \le k(u)$$

will be defined.

Let $u \in G_n \backslash (G_{n-1} \cup G_n^0)$ be given in the form (1.4) and let u have l_1 characteristic numbers with a multiplicity greater than one. We relabel the t_v's such that these characteristic numbers are

$$t_1 < t_2 < \cdots < t_{l_1}.$$

Let

$$\begin{aligned} \sigma_v &:= \operatorname{sgn} \alpha_{v m_v}, \qquad v = 1, 2, \ldots, l_1, \\ r_v &:= \begin{cases} 1, & \text{if } \sigma_v \sigma_{v+1} (-1)^{m_{v+1}} < 0,\ v \le l_1 - 1, \\ 0, & \text{otherwise}, \end{cases} \\ L &:= L(u) = l + \sum_{v=1}^{l_1} r_v. \end{aligned} \tag{5.1}$$

5.1 Lemma. *Let $u \in G_n \backslash (G_n^0 \cup G_{n-1})$. Then, for each $h \in C_u G_n$ there are at most $n + L(u)$ sign changes in the sequences $s_1, s_2, \ldots, s_r$ of generalized signs of h. If moreover, the number of sign changes equals $n + L(u)$, then $s_r = \sigma_{l_1}$ holds.*

The inductive proof is left to the reader. (Observe that the signs assigned to the terms with index $v > l_1$ appear in pairs and have no influence on the deficiency of sign changes).

Since $G_n \backslash G_{n-1}$ is a Haar manifold, from the alternation theorem for Haar cones (Theorem I.3.15) and the lemma above, the following is immediate.

5.2 Theorem. *Let G_n be a normal Descartes family. Then $u \in G_n \backslash (G_n^0 \cup G_{n-1})$ is a local best approximation to f from G_n, if and only if there is an alternant of length*

$n + L(u) + 1$ to $f - u$ whose sign equals $-\tilde{\varepsilon}_{n+L(u)}\sigma_{l_1}(u)$. Moreover, each critical point in $G_n \backslash G_{n-1}$ is a strongly unique local best approximation.

For ease of reference, we rewrite the continuity result III.1.13 for approximation by γ-polynomials.

5.3 Lemma (Perturbation Lemma). *Let u_0 be a local best approximation to $f_0 \in C(X)$ from the normal Descartes family G_n and suppose that $u_0 \neq G_{n-1}$. Then there is an ε-neighborhood U_ε in G_n of u_0, and a δ-neighborhood V_δ of f_0 in $C(X)$ such that, to each $f \in V_\delta$, there is a unique local best approximation u from G_n in U_ε and the mapping $f \mapsto u$ is continuous.*

B. The Generic Viewpoint

By Corollary 3.3, all *LBA*'s which are not global solutions, must be extended γ-polynomials and only the best one may be a proper γ-polynomial. The extended γ-polynomials are understood to be those γ-polynomials having characteristic numbers with a multiplicity greater than one. At first glance, one may expect that two of the characteristic numbers will almost never coalesce and that small perturbations of a given function f will remove multiplicities greater than 1 from a solution to f.

Lemma 5.5 below will show, however, that in most cases small perturbations do not change the multiplicities of the characteristic numbers (cf. Exercise III.1.17 for a similar but elementary phenomenon). In fact, for the solutions to most of the functions in $C(X)$, the parameter L equals l. (This means that there is even the tendency for $k - L$ to become maximal. We note that $k - L$ is just the gap between the length of an alternant which is sufficient for a global solution and the length of an alternant which is necessary for a local solution.)

The case $L > l$ and some other degeneracies occur very rarely. It is crucial for the analysis in the next section that we can eliminate the troublesome exceptional cases by perturbation arguments.

5.4 Definition. The function $f \in C(X)$ is said to be an *nonexceptional point* in $C(X)$ (with respect to G_n) if the following properties hold for the approximation from G_k for $k \leq n$:

(i) For each local best approximation u to f from G_k, there is an alternant of exact length $k + l(u) + 1$. Consequently, $l(u) = L(u)$ holds.

(ii) Each local best approximation to f from G_k has the maximal order k.

(iii) There is a neighborhood U of f in $C(X)$ such that the number of local best approximations in G_k is the same for all $f_1 \in U$, and the local best approximations are continuous functions of f_1.

For the moment, we consider the term "*exceptional point*" as a purely formal definition. It will turn out later that (i), (ii) and (iii) are generic properties in $C(X)$, (cf. Section II.1B).

It is the aim of this chapter to get bounds for

$$c_n := \sup_{f \in C(X)} \{\text{number of } LBA\text{'s to } f \text{ from } G_n\}, \qquad n = 1, 2, \ldots. \tag{5.2}$$

The justification of the definition above will now be established under the hypothesis that the c_k's are finite.

5.5 Lemma. *Assume that $c_k < \infty$, $k = 1, 2, \ldots, n$. Then the elements of $C(X)$ which are not exceptional with respect to the Descartes family G_n form an open dense subset of $C(X)$. Moreover, there is an unexceptional f having exactly c_n local best approximations from G_n.*

Proof. We start the inductive proof at $n = 0$. Recall that $G_0 = \{0\}$. The set of functions f such that there is only an alternant of length 1 to $f - 0$ is obviously open and dense in $C(X)$.

Let $n \geq 1$ and assume that the elements of $C(X)$ which are not exceptional with respect to G_{n-1} form an open dense subset $C_1 \subset C(X)$. The property (i) for $n - 1$ and Theorem 3.2 imply that no LBA from G_n to $f \in C_1$ has an order less than n. Hence, (ii) (from Definition 5.4) is a generic property.

Given a positive integer j, by Lemma 5.3 the subset of elements in C_1 having at least j local solutions in $G_n \backslash G_{n-1}$ is open. This and $c_n < \infty$ imply that the subset $C_2 \subset C(X)$ with functions satisfying condition (iii) is open and dense.

Assume that u_0 *is an* LBA to $f_0 \in C_2$ and that the length of the alternant does not exceed $n + l(u_0) + 1 =: r$. For each u in a sufficiently small neighborhood U of u_0 in G_n, we have $l(u) \geq l(u_0)$. Moreover, for each f in a sufficiently small neighborhood of $f_0 \in C(X)$, the length of the alternant of $f - u$, $u \in U$ cannot exceed r. From Theorem 3.2(2), it follows that $l(u) = l(u_0)$, if $u \in U$ is an LBA to f. Consequently, (i) holds for an open set.

To prove the density, let u_0 be an arbitrary LBA to f_0 from $G_n \backslash G_{n-1}$. Suppose that $L(u_0) > l(u_0)$. Write u_0 in the form (1.4) with the t_ν's being ordered as for the evaluation of (5.1). Given $\delta > 0$, let

$$u_1 = \sum_{\substack{\nu=1 \\ (r_\nu=1)}}^{l_1-1} \left\{ \sum_{\mu=0}^{m_\nu-1} \alpha_{\nu\mu}\gamma_{\mu+1}(t_\nu, t_\nu, \ldots, t_\nu; x) + \alpha_{\nu m_\nu}\gamma_{m_\nu+1}(t_\nu, t_\nu, \ldots, t_\nu, t_\nu + \delta; x) \right\} \tag{5.3}$$

$$+ \text{ all other terms of } u_0 \text{ unchanged.}$$

The new γ-polynomial u_1 has been constructed from u_0 by separating a single characteristic number from each term, which contributes to the difference $L(u_0) - l(u_0)$. The splitting process changes the parity check in (5.1), such that all parameters r_ν for u_1 are zero. Hence, $L(u_1) = l(u_1) = L(u_0)$ and by Theorem 5.2, u_1 is an LBA to $f_1 = f_0 + (u_1 - u_0)$. Moreover, if there is an alternant of length greater than $n + l(u_1) + 1$, by slightly modifying f_1 near superfluous points of the alternant, the length of the alternant may be reduced to the minimal one, which is $n + l(u_1) + 1$.

What happens with the other LBA's during the perturbation process above? From the discussion at the beginning of the proof we know that the LBA's already having the required properties will keep them as long as the perturbation parameter δ in (5.3) is small. Since $c_n < \infty$, by a finite number of repetitions we gain a function with property (i). This proves the density.

Finally, the last statement of the lemma is a consequence of (iii). □

§6. Maximal Components

It is very easy to construct a function $f \in C(X)$ with at least n local best approximations from G_n. We describe this construction, since a generalization will provide all local solutions in the generic case.

Let $\hat{u}$ be a γ-polynomial of order $n-1$ in G_{n-1}^+. Let $f \in C(X)$ be such that there is an alternant of exact length $2n-1$ to $f - \hat{u}$. Moreover we may choose f such that the alternant has a given sign. Specifically, due to Proposition 2.5, one choice implies that each each better approximating γ-polynomial has a term with a negative factor.

By construction, $\hat{u}$ is not a best approximation to f from G_n. Given $\tau \in T \backslash \operatorname{spect}(\hat{u})$, the function $\hat{u} + 0 \cdot \gamma(\tau, .)$ is not a critical point in the subset of γ-polynomials having the form $\sum_{\nu=1}^{n-1} \alpha_\nu \gamma(t_\nu, .) + \alpha_n \gamma(\tau, .)$. There is a better approximating γ-polynomial u_1 in that subset, close to $\hat{u} + 0 \cdot \gamma(\tau, .)$. Then $\alpha_\nu > 0$ for $\nu \le n-1$, and $\alpha_n < 0$. It follows that $s := \operatorname{sign} u_1 = (+, +, \ldots, +, -, +, \ldots, +)$ with $s_j = -1$, $s_i = +1$ for $i \ne j$ if τ is larger than exactly $j-1$ characteristic numbers of $\hat{u}$. Usually not only G_n, but also $G_n(s) \cup G_{n-1}$ is an existence set for any $s \in \{+1, -1\}^n$. Then a nearest point from $G_n(s) \cup G_{n-1}$ with $s = \operatorname{sign} u_1$, is contained in the subset $G_n(s)$, since $\|f - u_1\| < d(f, G_{n-1})$. Moreover, a solution in $G_n(s)$ is an *LBA* in G_n. Obviously there are n choices of τ which lead to n distinct sign-classes $G_n(s)$, each of which contains an *LBA*. □

The aim of this chapter will be to show that there are at most $n!$ *LBA*'s in G_n to each $f \in C(X)$ whenever T is an interval. By (5.2) this may be expressed as:

$$c_n \le n! \tag{6.1}$$

To this end we extend the construction above such that all local solutions are found.

Roughly speaking, we proceed as follows. Take any *LBA* $\hat{u}$ to f from G_{n-1} and any $\tau \in T \backslash \operatorname{spect}(\hat{u})$. Then

$$\hat{u}(x) + 0 \cdot \gamma(\tau, x) \tag{6.2}$$

formally is in G_n. Determine a better approximation than $\hat{u}$ of the form $v + \alpha_n \gamma(\tau, .)$ with $v \in G_{n-1}$ close to $\hat{u}$ and α_n small. Then apply a Newton-type iteration for successive improvement until a local solution is established. It is crucial that, given $\hat{u}$ (as in the special case above), the result only depends on the location of τ in relation to the ($\le n-1$) characteristic numbers of $\hat{u}$ and that there are at most n inequivalent choices.

By means of this algorithm, for each solution in G_{n-1} and each extra characteristic number, we may construct an *LBA* from G_n. We want to show that in this way, each *LBA* in G_n is obtained. For this purpose, we have to proceed in the reverse direction. Given an arbitrary *LBA* u_0 from G_n, we have to associate a local solution from G_{n-1} to it and an additional characteristic number τ, such that the construction above yields u_0.

The appropriate tool is the investigation of the level sets and their components.

A. Introduction of Maximal Components

For the most interesting Descartes families, the domain of the t-variable is an open interval. For convenience, we will restrict ourselves to compact sets until the end of the next section (without mentioning the fact). The restriction can eventually be abandoned by an exhaustion argument, and it simplifies the discussion because of the following:

6.1 Remark. Each bounded and closed subset of a Descartes family G_n is compact whenever T is a compact subset of $\mathbb{R}$. □

Given $f \in C(X)$, the *level sets* of the distance function $\rho: G_n \to \mathbb{R}$, $\rho(u) = \|f - u\|$ are denoted by

$$\rho^\alpha := \{u \in G_n: \|f - u\| \leq \alpha\},$$

where $\alpha \in \mathbb{R}_+$. A direct application of the mountain pass theorem is impossible because only the subset $G_n \backslash G_{n-1}$ is a manifold. It may be applied, however, to get a local result.

6.2 Lemma. *Let G_n be a Descartes family and let $f \in C(X)$. Assume that $C\rho^\alpha$ is a (connected) component of ρ^α for some $\alpha > 0$, which does not intersect G_{n-1}. Then $C\rho^\alpha$ contains exactly one local best approximation.*

Proof. By Remark 6.1, $C\rho^\alpha$ is compact and contains at least one *LBA*. Since $C\rho^\alpha$ is a connected and compact level set in some subset of the Haar embedded manifold $G_n \backslash G_{n-1}$, by the mountain pass theorem we have uniqueness of the *LBA*. □

Let $u^* \in G_n \backslash G_{n-1}$ be an *LBA* to f. Denote the component of the level set ρ^α which contains u^* by $C\rho^\alpha$. Furthermore, put

$$\beta := \sup\{\alpha: C\rho^\alpha \cap G_{n-1} = \phi\}. \tag{6.3}$$

By the lemma above, u^* is the unique *LBA* to f in

$$C\rho := \bigcup_{\alpha < \beta} C\rho^\alpha. \tag{6.4}$$

For this reason, $C\rho$ is called the *maximal component* associated to u^*.

First we observe that β as defined by (6.3) satisfies $\beta \leq \|f\| = \|f - 0\|$, because the level set Cp^α, $\alpha = \|f\|$, contains the arc $\{\lambda u^*\}_{0 \le \lambda \le 1}$. Furthermore, we claim that $\beta > \|f - u^*\|$. Indeed, since u^* is a strongly unique *LBA*, there are $c > 0$, $r > 0$ such that

$$\|f - u\| \ge \|f - u^*\| + c\|u - u^*\|, \quad \text{whenever } u \in G_n,\ \|u - u^*\| \le r.$$

We may assume that $r < d(u^*, G_{n-1})$. Let $\alpha_1 = \|f - u\| + cr/3$. Then ρ^{α_1} contains no elements of G_n whose distance from u^* lies between $r/2$ and r. Hence, the component Cp^{α_1} is disjoint from G_{n-1} and $\beta \ge \alpha_1$. The next step is the following

6.3 Assertion. *The closure $\overline{Cp}$ of the maximal component with respect to the uniform norm intersects G_{n-1}.*

Proof. Suppose to the contrary that the assertion is not true. Then we can construct an extension of Cp (like a collar to the boundary ∂Cp), which is also disjoint from G_{n-1}.

Since $\overline{Cp}$ is bounded and closed, it is compact by Remark 6.1. Obviously, we have $\|f - u\| = \beta$ for each $u \in \partial Cp$. Since $\overline{Cp}$ is connected, ∂Cp contains no strict *LBA* and therefore no critical point. Given $u_0 \in \partial Cp$, by Lemma IV.2.2 there is an open neighborhood $U = U(u_0)$ and a continuous mapping ψ: $[0,1] \times \bar{U} \to G_n$ such that

$$\psi(0, u) = u,$$
$$\|f - \psi(t,u)\| < \|f - \psi(s,u)\|, \qquad u \in \bar{U},\ 0 \le s < t \le 1. \tag{6.5}$$

After reducing U, if necessary, the following properties hold:

(i) $\bar{U}$ does not intersect G_{n-1},

(ii) $\bar{U}$ is compact and path connected.

(iii) $\|f - \psi(1,u)\| < \beta$ for each $u \in \bar{U}$.

If $u_1 \in Cp \cap U$, the orbit $\{\psi(t,u_1)\colon 0 \le t \le 1\}$ belongs to Cp. Therefore, $\psi(1,\bar{U})$ intersects Cp and it follows from (iii) that the path-connected set $\psi(1,\bar{U})$ is contained in Cp. Moreover, given $u \in \bar{U}$, by (6.5) the points from the orbit $\{\psi(t,u)\colon 0 \le t \le 1\}$ whose distance from f is $\le \beta$, form a (connected) arc. Thus $\psi(t,u) \in \overline{Cp}$ whenever $0 \le t \le 1$, $u \in \bar{U}$ and $\|f - \psi(t,u)\| \le \beta$. Using this observation for $t = 0$, we see that

$$\|f - u\| > \beta \quad \text{whenever } u \in \bar{U} \backslash \overline{Cp}. \tag{6.6}$$

A finite number of such open sets, say $U_1, U_2, \ldots, U_m$, cover ∂Cp.

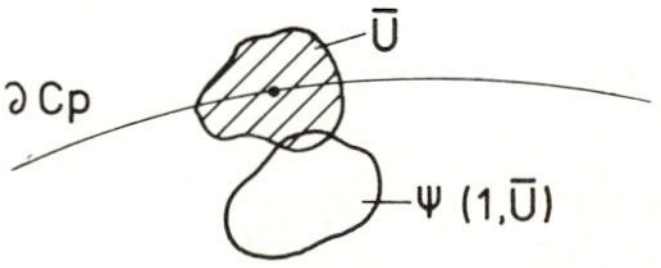

Fig. 29

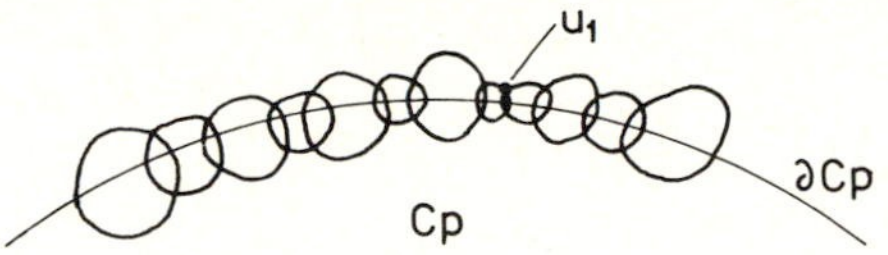

Fig. 30. Construction of a collar for the extension of Cp

$$U = \bigcup_{j=1}^{m} U_j \supset \partial Cp. \tag{6.7}$$

The set

$$M = \bar{U} \backslash (U \cup Cp)$$

is compact and the distance function $\|f - .\|$ attains its minimum at some $u_1 \in M$. We conclude from $\overline{Cp} \subset Cp \cup U$ that $u_1 \notin \overline{Cp}$. By (6.6) we have $\beta_1 := \|f - u_1\| > \beta$. Since $Cp \cap U$ is connected, it contains the level set $\rho^{(\beta+\beta_1)/2}$ which contradicts the maximality of β. Therefore, ∂Cp intersects G_{n-1}. □

B. The Boundary of Maximal Components

From the proof of the preceding lemma, it is clear that the critical points of mountain pass type are only expected in the intersection of the boundary of maximal components with G_{n-1}. The study of the intersection requires the handling of sequences $(u_r) \subset G_n \backslash G_{n-1}$ which converge to a γ-polynomial $\hat{u}$ of degree $k = k(\hat{u}) \leq n - 1$. We will separate some $v_r \in G_k$ from u_r for $r = 1, 2, \ldots$, such that $v_r \to \hat{u}$. Unfortunately, it cannot be done by a simple splitting with the complement satisfying $k(u_r - v_r) = n - k$. This is elucidated by the sequence in G_2 having the elements

$$u_r = 5\gamma(t, .) - 2\gamma(t + 2^{-r}, .), \qquad r = 1, 2, \ldots \tag{6.8}$$

which converges to $3\gamma(t, .) \in G_1$.

For an appropriate splitting, let $\{t_\mu^{(r)}\}_{\mu=1}^n = \text{spect}(u_r)$. We relabel the characteristic numbers such that

$$\lim_{r\to\infty} t_\mu^{(r)} = \hat{t}_\mu, \qquad \mu = 1, 2, \ldots, k,$$

and $\text{spect}(\hat{u}) = \{\hat{t}_\mu\}_{\mu=1}^k$. Here characteristic numbers are written according to their multiplicity. If we put

$$u_r = \sum_{\mu=1}^{n} \beta_\mu^{(r)} \gamma_\mu(t_1^{(r)}, t_2^{(r)}, \ldots, t_\mu^{(r)}; .),$$

then we have $\lim_{r\to\infty} \beta_\mu^{(r)} = 0$, $\mu \geq k + 1$. Moreover, we put

$$v_r = \sum_{\mu=1}^{k} \beta_\mu^{(r)} \gamma_\mu(t_1^{(r)}, t_2^{(r)}, \ldots, t_\mu^{(r)}; .),$$

and $w_r = u_r - v_r$.

6.4 Lemma. *Let G_n be a Descartes family and u^* be a local best approximation to f from $G_n \backslash G_{n-1}$. Then each element $\hat{u}$ from the intersection of the boundary of the maximal component and G_{n-1}, is a local best approximation to f from $G_{k(\hat{u})}$.*

Proof. Suppose to the contrary, that $\hat{u} \in \overline{Cp} \cap G_{n-1}$ is not a critical point in G_k, where $k = k(\hat{u})$. Consider a sequence $(u_r) \subset Cp$ which converges to $\hat{u}$. Let $u_r = v_r + w_r$ according to the splitting process described above.

Apply Lemma IV.2.2 on flows for noncritical points to $\hat{u} \in G_k$ and denote the resulting flow by ψ. Its domain $U \subset G_k$ contains v_r for r being sufficiently large. Put

$$v_r(\lambda) = \psi(\lambda, v_r).$$

In particular, the parameters $\xi^r(\lambda) = F^{-1}(v_r(\lambda)) \in \mathbb{R}^{2k}$ have the form

$$\xi^r(\lambda) = \xi^r + \lambda b, \tag{6.9}$$

when the parametrization F from (4.7) is used for a neighborhood of $\hat{u} \in G_k$. The elements of the w_r-sequence may be written as follows

$$w_r = \sum_{\mu=k+1}^{n} \beta_\mu^{(r)} \gamma_\mu(t_1(\xi^r), \ldots, t_k(\xi^r), t_{k+1}^{(r)}, \ldots, t_\mu^{(r)}). \tag{6.10}$$

Moreover, let $w_r(\lambda)$ be the γ-polynomial which results from (6.10), when ξ^r is replaced by $\xi^r(\lambda)$. From the differentiability of F, it follows that $t_1, t_2, \ldots, t_k$ are continuously differentiable functions of ξ. Hence,

$$\left| \frac{\partial}{\partial \lambda} \gamma_\mu(t_1(\xi(\lambda)), \ldots, t_k(\xi(\lambda)), t_{k+1}, \ldots, t_\mu; x) \right| \leq M_1 < \infty$$

for all ξ's in a neighborhood of $F^{-1}(\hat{u})$ and $t_{k+1}, \ldots, t_n \in T$. Set

$$u_r(\lambda) := v_r(\lambda) + (1 - \lambda) w_r(\lambda), \qquad 0 \leq \lambda \leq 1.$$

By applying Lemma IV.2.2 (on the flow for noncritical points) to the approximation of the perturbed function $f - w_r$:

$$\|f - w_r - \psi(\lambda, v)\| \leq \|f - w_r - v\| - c\lambda, \qquad 0 \leq \lambda \leq 1, \quad v \in U,$$

we obtain

$$\begin{aligned} \|f - u_r(\lambda)\| &\leq \|f - w_r - v_r(\lambda)\| + \lambda \|w_r\| + (1 - \lambda)\| w_r - w_r(\lambda)\| \\ &\leq \|f - w_r - v_r\| - c\lambda + \lambda \|w_r\| + (1 - \lambda) M_1 \lambda \sum_{\mu > k} |\beta_\mu^{(r)}| \\ &\leq \|f - u_r\|. \end{aligned}$$

whenever

$$\|w_r\| < \frac{c}{2} \quad \text{and} \quad \sum_{\mu>k} |\beta_\mu^{(r)}| < \frac{c}{2M_1}.$$

These conditions hold for sufficiently large r. Since $u_r \in Cp$, the arc $\{u_r(\lambda): 0 \le \lambda \le 1\}$ belongs to Cp. But $u_r(1) = v_r(1) \in G_{n-1}$ contradicts $Cp \cap G_{n-1} = \emptyset$. □

Let (u_r) be a sequence in G_n which converges to $\hat{u} \in G_{n-1}$. If the sequence of the spectra: $(\text{spect}(u_r))$ has an accumulation point $\tau \notin \text{spect}(\hat{u})$, then obviously we can perform a splitting $u_r = v_r + w_r$, where $w_r \to 0$ with $k(w_r) \ge 1$ and $k(v_r) + k(w_r) = n$. This means that a situation like (6.8) is excluded. We will show that given $\hat{u} \in \partial Cp \cap G_{n-1}$, such a sequence can indeed be always constructed.

Let $t_1, t_2, \ldots, t_p$ be not necessarily disjoint numbers in T, $p \le n$. Then

$$G_n(t_1, t_2, \ldots, t_p\} := \text{closure}\{u \in G_n \colon \{t_1, \ldots, t_p\} \subset \text{spect}(u)\} \qquad (6.11)$$

denotes the subset of G_n with p characteristic numbers fixed. We emphasize that some t_j with $j \le p$ may be missing in the spectrum of $u \in G_n(t_1, \ldots, t_p)$ if $k(u) < n$. There is no serious confusion with the notation for the sign classes $G_n(s_1, s_2, \ldots, s_n)$ because the letters s and t never occur at the same time for a specification of a subset of G_n. Further, if $t_1 = t_2 = \cdots = t_m$ with $m \le p$, the abbreviation

$$G_n(m \times t_1, t_{m+1}, \ldots, t_p) := G_n(t_1, t_1, \ldots, t_1, t_{m+1}, \ldots, t_p)$$

is used.

The difficulties in the analysis arise from the fact that a γ-polynomial $\hat{u}$ of order $n-1$ is a singular point in the set G_n. However, this is changed when we consider $\hat{u}$ in $G_n(\tau)$ provided that $\tau \notin \text{spect}(\hat{u})$. Then some neighborhood of $\hat{u}$ in $G_n(\tau)$ is again a Haar manifold. Thus we will analyse the singular points by considering $G_n(\tau)$ instead of G_n, and (6.2) is understood in this setting.

Similarly, an analysis with m fixed characteristic numbers will be performed, if m characteristic numbers from $(\text{spect}(u_r))$ converge to some $\tau \in T$ while τ is found only with multiplicity $m-1$ in the spectrum of the limit.

6.5 Assertion. *Assume that $\hat{u}$ is a local best approximation to f from $G_{n-1} \backslash G_{n-2}$ and τ is a characteristic number of $\hat{u}$ with multiplicity $m-1$, $m \ge 2$. Given $\varepsilon_0 > 0$, there are a $\delta > 0$ and $0 < \varepsilon < \varepsilon_0$ such that there is a unique local best approximation u_1 to f from $G_n(t_1, t_2, \ldots, t_m)$ with $\|u_1 - \hat{u}\| < \varepsilon$ whenever*

$$|t_i - \tau| \le \delta \qquad i = 1, 2, \ldots, m. \qquad (6.12)$$

Proof. First, we note that $C_{\hat{u}} G_n(m \times \tau) \subset C_{\hat{u}} G_{n-1}$. *Hence,* $\hat{u}$ is a critical point in $G_n(m \times \tau)$. Since a neighborhood of $\hat{u}$ in $G_n(m \times \tau)$ is a Haar manifold, $\hat{u}$ is the strongly unique best approximation from some neighborhood $U := G_n(m \times \tau) \cap \mathring{B}_r(\hat{u})$.

Next, given $t_1, t_2, \ldots, t_m \in T$, we define a one-one mapping $\Phi\colon U \to G_n(t_1, t_2, \ldots, t_m)$ by the following procedure: The partial sum $\sum_{\mu=1}^m \beta_\mu \gamma_\mu(\tau, \ldots, \tau; x)$ in the representation of a γ-polynomial in $U \subset G_n(m \times \tau)$ is replaced by $\sum_{\mu=1}^m \beta_\mu \gamma_\mu(t_1, t_2, \ldots, t_\mu; x)$. Thereby, we may keep the factors β_μ. Let $\varepsilon > 0$. If δ is sufficiently small, then (6.12) implies that

$$\|\Phi(v) - v\| < \tfrac{1}{3}\varepsilon, \quad \text{for } v \in \bar{U}. \tag{6.13}$$

Now the assertion follows from the perturbation lemma IV.2.8. □

Moreover, from uniqueness and local compactness it follows that the *LBA* is a continuous function of $t_1, t_2, \ldots, t_m$.

Recalling that T is assumed to be compact, we are now in a position to prove

6.6 Lemma. *Let G_n be a Descartes family and let $f \in C(X)$ not be exceptional with respect to G_{n-1}. Let Cp be a maximal component to a local best approximation u^* to f from G_n. Then, to each $\hat{u} \in \partial Cp \cap G_{n-1}$ there is a sequence (u_r) in C_p which converges to $\hat{u}$ and which allows a splitting*

$$u_r = v_r + w_r, \qquad k(v_r) = n - 1, \qquad k(w_r) = 1$$

such that $w_r \to 0$ and spect(w_r) *has an accumulation point τ which is disjoint from* spect$(\hat{u})$. *Moreover, $\partial Cp \cap G_{n-2} = \varnothing$.*

Proof. 1. Assume that $\hat{u} \in \overline{Cp} \cap G_{n-1}$ and $k(\hat{u}) = n - 1$. Let (u_r) be a sequence in Cp which converges to $\hat{u}$. If (spect(u_r)) contains an accumulation point disjoint from spect$(\hat{u})$, then the splitting is obvious.

Otherwise (since T is bounded) we may assume (after going to a subsequence if necessary), that m characteristic numbers of u_r converge to some $\tau \in T$, with τ being a characteristic number *uf* $\hat{u}$ with multiplicity only $m - 1$.

Since there is an alternant only of length $k(\hat{u}) + l(\hat{u}) + 1$ to $f - \hat{u}$, there is an ε-neighborhood of $\hat{u}$ in $C(X)$ such that $f - v$ has no longer alternant whenever $\|v - \hat{u}\| < \varepsilon$. Now let $\delta > 0$ be as in Assertion 6.5 and let Φ and U be as in the proof. Moreover, let $\delta_2 \in (0, \delta)$ be a number to be fixed later. Then there is a best aproximation $u(\delta_2)$ to f from $Cp \cap W$ where

$$W = \bigcup \Phi(U) \subset \bigcup G_n(t_1, t_2, \ldots, t_m)$$

with the union being taken over

$$\tau - \delta \le t_1 \le \tau + \delta,$$

$$\tau - \delta_2 \le t_j \le \tau + \delta_2, \qquad j = 2, 3, \ldots, m.$$

By assumption, $u_r \in W$ holds for sufficiently large r. Clearly, $u(\delta_2)$ is an *LBA* to f from W. The tangent cone $C_{u(\delta_2)}W$ contains an $(n + l(\hat{u}))$-dimensional subspace, if t_1 lies in the open interval $(\tau - \delta, \tau + \delta)$ or if $\tau - \delta_2 < t_j < \tau + \delta_2$ for some $j \in \{2, 3, \ldots, m\}$. Since the length of the alternant of $f - u(\delta_2)$ does not exceed $n + l(\hat{u})$, it follows that $t_1 = \tau + \delta$ or $\tau - \delta$ and $t_j = \tau + \delta_2$ or $\tau - \delta_2$ $(j = 2, 3, \ldots, m)$ belong to spect$(u(\delta_2))$.

Next, we note that $\hat{u}$ is an *LBA* in $G_n(t_1, (\mathrm{m} - 1) \times \tau)$ for any t_1. By continuity, we have $u(\delta_2) \to \hat{u}$, as $\delta_2 \to 0$. Specifically, let $\delta_2 = \delta/N$ for $N = 1, 2, \ldots$. Then $u(\delta/N) \in Cp$ for $N = 1, 2, 3, \ldots$ and $u(\delta/N) \to \hat{u}$. Since $\tau + \delta$ or $\tau - \delta$ is contained in spect$(u(\delta/N))$, but not in spect$(\hat{u})$, we get a sequence as desired.

2. Assume that $\hat{u} \in \partial Cp \cap G_{n-2}$ and let (u_r) be a sequence in Cp converging to $\hat{u}$.

If $\operatorname{spect}(u_r)$ has an accumulation point $\tau \notin \operatorname{spect}(\hat{u})$, then $\hat{u} + 0 \cdot \gamma(\tau, .)$ is not a critical point in $G_{k(\hat{u})+1}(\tau)$. Therefore a flow in a neighborhood of $\hat{u}$ as in Lemma 6.4 may be constructed. By a perturbation argument, one gets a flow on $G_{k(\hat{u})+1}(t)$ for t close to τ. Now arguments like those in the proof of Lemma 6.4 yield a contradiction.

If more than m characteristic numbers of u_r converge to some $\tau \in T$, with τ being a characteristic number of $\hat{u}$ with multiplicity $m - 1$, then $\hat{u}$ is not a critical point in $G_{k(\hat{u})+2}((m+1) \times \tau)$. A contradiction is obtained as above.

In any other case, all characteristic numbers of (u_r) converge to spectral values of $\hat{u}$, but for each $\tau \in \operatorname{spect}(\hat{u})$, the multiplicity is increased by at most one.

Then by applying the arguments of the first part of the proof to each cluster, one gets another sequence with an accumulation point outside of $\operatorname{spect}(\hat{u})$. Once more, one obtains a contradiction and $\hat{u} \in G_{n-2}$ is excluded. □

§7. The Number of Local Best Approximations

A. The Construction of Local Best Approximations

A method will be considered which enables us to construct *LBA*'s in G_n using the knowledge of the critical points in G_{n-1}. From the results of the last section, we will deduce that, at least when the given f is not exceptional, all critical points may be obtained in this way. Consequently, if f is not exceptional, there are at most $n!$ critical points. From the discussion of the generic viewpoint, it is clear that the number of critical points for any $f \in C(X)$ may not exceed $n!$.

7.1 Construction (Standard Construction). Let $\hat{u}$ be a local best approximation to f from G_{n-1}. Choose an interval $\hat{T} \subset T \backslash \operatorname{spect}(\hat{u})$ and $\tau \in \hat{T}$. Let $h \in C_{\hat{u}} G_n(t)$ satisfy

$$\|f - \hat{u} - h\| < \|f - \hat{u}\|. \tag{7.1}$$

By the definition of tangent rays there is a curve $\{u_\lambda : 0 \leq \lambda \leq 1\} \subset G_n(\tau)$ such that $\|u_\lambda - \hat{u} - \lambda h\| = o(\lambda)$ as $\lambda \to 0$. Then for sufficiently small $\lambda > 0$:

$$\|f - u_\lambda\| < \|f - \hat{u}\|. \tag{7.2}$$

There are two possibilities: The component of the level set $\{u \in G_n : \|f - u\| < \|f - \hat{u}\|\}$ containing the curve $\{u_\lambda : \lambda > 0, \lambda \text{ sufficiently small}\}$ is disjoint from G_{n-1}. Then a maximal component and a local solution to f has been constructed. If, on the other hand, the component intersects G_{n-1}, it may be discarded. (The local solutions in this component are obtained from starting points at lower levels).

Repeat the procedure for some τ from another interval $\hat{T}$ of $T \backslash \operatorname{spect}(\hat{u})$ until all (the $\leq n$) intervals have been scanned.

Repeat the procedure for another local best approximation from G_{n-1} until all of them (there are at most c_{n-1}), have been scanned. □

To verify that the procedure is reasonable, let $\hat{u}$ and τ be as above. We may choose an open neighborhood V of $\hat{u}$ in G_{n-1} such that τ is disjoint from spect(v) for each $v \in V$. Put

$$W = V + \text{span}[\gamma(\tau, .)] = \{u = v + \alpha\gamma(\tau, .): v \in V, \alpha \in \mathbb{R}\} \subset G_n(\tau).$$

If $v + \alpha\gamma(\tau, .) \in W$, then

$$C_{v+\alpha\gamma(\tau,.)}G_n(\tau) = C_v G_{n-1} + \text{span}[\gamma(\tau, .)],$$

and W is a Haar embedded manifold. In particular, $C_{\hat{u}}G_n(\tau)$ contains an $n + l(\hat{u})$-dimensional Haar subspace. Since f is assumed to be not exceptional with respect to G_{n-1}, there is no alternant of length $n + l(\hat{u}) + 1$ to $f - \hat{u}$ and $\hat{u}$ is indeed not a critical point in $G_n(\tau)$.

We remark that the generic assumption cannot be abandoned here. To understand this, assume that $L(\hat{u}) > l(\hat{u})$. We have $r_\nu \neq 0$ for at least one $\nu < l_1$. If

$$t_\nu < \tau < t_{\nu+1}$$

then the parity is changed because each element in the tangent cone has the spectral value τ at most with the odd multiplicity 1. Hence $\hat{u}$ remains a critical point in $G_n(\tau)$.

7.2 Lemma. *Let f be a nonexceptional point with respect to G_{n-1}. Then the component constructed depends only on the choice of $\hat{u}$ and on the interval of* $\mathbb{R}\backslash$spect($\hat{u}$) *containing τ.*

Proof. Given $\tau \in$ spect($\hat{u}$), the standard construction yields a curve in $G_n(\tau)$ such that its elements

$$u_\lambda = \alpha_\lambda\gamma(\tau, .) + v_\lambda, \qquad v_\lambda \in G_{n-1} \tag{7.3}$$

satisfy

$$\|f - u_\lambda\| < \|f - \hat{u}\| \tag{7.4}$$

for sufficiently small λ, say for $0 < \lambda < 1$. By the local uniqueness theorem, there is a neighborhood U of $\hat{u}$ in G_{n-1}, and a $\delta > 0$ such that there is a unique best approximation to g in U provided that $\|f - g\| < \delta$. Hence, the level set

$$\{v \in U: \|g - v\| < \|f - \hat{u}\|\} \tag{7.5}$$

is either connected or empty.

When we compare the construction with two different curves, then we can fix on each one an element such that the coefficients α_λ in (7.3) coincide. By setting $g = f - \alpha_\lambda\gamma(\tau, .)$ we conclude from the connectedness of the set in (7.5) that the curve (u_λ) intersects a component which is independent of the tangent vector h satisfying (7.1) and of the special choice of the curve.

Let τ_1 and τ_2, $\tau_1 < \tau_2$, be contained in the same interval of $T\backslash\text{spect}(\hat{u})$. Assume that U is so small that the spectral values are contained in $T\backslash[\tau_1, \tau_2]$ whenever $u \in U$. We claim, if we choose α appropriately, that

$$\|f - \alpha\gamma(t, .) - v_{\alpha,t}\| < \|f - \hat{u}\| \tag{7.6}$$

where $v_{\alpha,t}$ is the best approximation to $f - \alpha\gamma(t, .)$ from U, and $\tau_1 < t < \tau_2$.

Indeed, since $\hat{u}$ is not a critical point to f from $G_n(\tau)$, it follows that

$$\|f - \alpha\gamma(\tau, .) - v_{\alpha,\tau}\| \leq \|f - \hat{u}\| - c \cdot s \cdot \alpha, \tag{7.7}$$

for $0 \leq s\alpha \leq \delta_\tau$, where $c > 0$, $\delta_\tau > 0$ and $s = \text{sgn}\,\alpha_\lambda$, with α_λ from (7.3). Obviously, (7.6) implies that

$$\|f - \alpha\gamma(t, .) - v_{\alpha,\tau}\| \leq \|f - \hat{u}\| - \frac{c}{2} \cdot s \cdot \alpha$$

for $0 \leq s\alpha \leq \delta_\tau$ and t sufficiently close to τ. Therefore, we have that (7.5) holds for some α and the t's in some open neighborhood of $\tau \in [\tau_1, \tau_2]$. It follows from a compactness argument, that we may choose an α which is suitable for the interval $[\tau_1, \tau_2]$. Therefore, we get the same component for the construction with τ_1 as with τ_2. □

From the lemma above, it is clear why we have to restrict ourselves to nonexceptional points. The number of *LBA*'s in G_n which can be constructed from an *LBA* $\hat{u} \in G_{n-1}$ depends on the number of intervals into which T is divided by spect($\hat{u}$). Therefore, the number of such intervals is $l(\hat{u}) + 1$. Now, in view of the perturbation lemma, the statement of the lemma is ony reasonable if this number is insensitive to small perturbations of the given function f. This is precisely the case when f is a nonexceptional point.

B. Completeness of the Standard Construction

Now we are ready to prove the main theorem

7.3 Theorem (Braess 1978). *Let G_n be a Descartes family. Moreover, assume that T is an interval. Then to each $f \in C(X)$ there are at most $n!$ local best approximations from G_n.*

Proof. Since G_1 is a uniqueness set, the theorem is obviously true for $n = 1$. Assume that the theorem has been already proved for $1, 2, \ldots, n - 1$.

From the hypothesis and Lemma 5.5, it follows that the functions which are not exceptional with respect to G_{n-1} are dense in $C(X)$. In view of the perturbation lemma, it is sufficient to consider the number of solutions for an f which is not exceptional with respect to G_{n-1}.

Suppose that there are more than $n!$ *LBA*'s to f from G_n. Then select a subset of $1 + n!$ of them. Next, choose a compact interval $T_0 \subset T$ which contains the spectra of those *LBA*'s from G_n and of all *LBA*'s in G_k for $k \leq n - 1$. Now consider the restriction to the kernel, whose t-domain is only T_0. By Lemma 6.6 all LBA's of order n are obtained by the standard construction. From Lemma 7.2, we conclude that we get at most $n \cdot c_{n-1} \leq n \cdot (n-1)!$ solutions. Therefore, it is impossible that there are more than $n!$ LBA's, and the proof that $c_n \leq n!$ is complete. □

The bound in Theorem 7.3 is not optimal. There is only one case in which exactly n LBA's are obtained from one solution $\hat{u}$ in G_{n-1}. This happens only if $\hat{u} \in G_{n-1}^+$ or $(-\hat{u}) \in G_{n-1}^+$, i.e., if $\hat{u}$ is the only critical point. Hence, $c_n \leq (n-1)c_{n-1}$ for $n \geq 3$. Since one obtains $c_3 = 3$ by some technical considerations, it follows that $c_n \leq (3/2)(n-1)!$.

On the other hand, Verfürth (1982) has shown that c_n grows faster than any polynomial:

$$c_n \geq e^{(2/3)n}/\sqrt{40n}$$

and that for small n better lower bounds may established. The best known results for some n's are

$$c_1 = 1, \quad c_2 = 2, \quad c_3 = 3, \quad 5 \leq c_4 \leq 9, \quad 8 \leq c_5 \leq 36, \quad 12 \leq c_6 \leq 180.$$

Chapter VIII. Approximation by Spline Functions with Free Nodes

In the last two decades, approximation by spline functions has become an important tool in applied mathematics, since the accuracy of the approximation depends only on local properties of the given function. As a consequence, spline functions with fixed nodes lead only to weak Haar spaces and the interpolation is governed by an interlacing property between the nodes and the points of interpolation. Nevertheless, a best approximation can be characterized in terms of alternants (with a possible degenerate length).

The approximation by spline functions with free nodes shows many features which are familiar from γ-polynomials. But the similarities are often hidden because of the degeneracies which are caused by the interlacing properties mentioned above.

When the approximation of functions which are convex in a generalized sense is considered, all degeneracies are excluded. In this theory of monosplines, there are nice results for interpolation and approximation, as for Stieltjes functions in rational approximation and for completely monotone functions in exponential approximation. Here, the theory is even more developed. Specifically, the investigation of nodes with prescribed multiplicities and the L_p-case have been done for monosplines. In this case topological methods are an essential tool again. However, in order to avoid some technical complications, the results have often been shown in a framework which afterwards turned out to be that of γ-polynomials.

In the study of families of spline functions with free nodes, we encounter tangent spaces which consist of spline functions with fixed nodes. Therefore, we start by recalling some basic facts on approximation by spline functions with fixed nodes.

§ 1. Spline Functions with Fixed Nodes

A. Chebyshevian Spline Functions

The starting point for the investigation of spline functions is the set of *polynomial splines* with simple nodes:

$$s(x) = \sum_{j=0}^{n} b_j x^j + \sum_{i=1}^{k} a_i (x - t_i)_+^n, \qquad \alpha < t_1 < t_2 < \cdots < t_k < \beta, \tag{1.1}$$

where the truncated powers are defined by

$$x_+^n = \begin{cases} x^n & \text{for } x \geq 0, \\ 0 & \text{otherwise.} \end{cases}$$

The restriction of the spline function (1.1) to every interval $(\alpha, t_1), (t_1, t_2), \ldots, (t_k, \beta)$ is a polynomial from Π_n, and s is $n-1$ times continuously differentiable at the nodes. More generally, let $\{w_j\}_{j=0}^n$ be $n+1$ positive functions on $[\alpha, \beta]$ with $w_j \in C^{n-j}$ for $j = 0, 1, \ldots, n$. Define

$$\Phi_j(x,t) = \begin{cases} w_0(x) \displaystyle\int_t^x w_1(\xi_1) \int_t^{\xi_1} w_2(\xi_2) \ldots \int_t^{\xi_{j-1}} w_j(\xi_j)\, d\xi_j \ldots d\xi_1 & \text{for } x \geq t, \\ 0 & \text{for } x < t, \end{cases} \tag{1.2}$$

and

$$v_j(x) = \Phi_j(\alpha, x), \qquad j = 0, 1, \ldots, n. \tag{1.3}$$

The *Chebyshevian spline functions* with the (simple) *nodes* $t_1, t_2, \ldots, t_k$, are expressions of the form

$$s(x) = \sum_{j=0}^{n} b_j v_j(x) + \sum_{i=1}^{k} a_i \Phi_n(t_i, x). \tag{1.4}$$

Obviously, the polynomial splines are obtained by setting $w_i \equiv 1, i = 0, 1, \ldots, n$.

In the theory of spline functions, it is reasonable to consider nodes with multiplicities up to $n+1$. Specifically, s has the nodes $\alpha < t_1 < \cdots < t_l < \beta$, and t_i is a node of multiplicity m_i for $i = 1, 2, \ldots, l$, if

$$s(x) = \sum_{i=0}^{l} \sum_{j=1}^{m_i} a_{ij} \Phi_{n+1-j}(t_i, x), \tag{1.5}$$

where $t_0 = \alpha$ and $m_0 = n+1$. Note that a spline function has jump discontinuities at nodes of multiplicity $n+1$. To be definite, we set $s(t_i) = s(t_i + 0)$.

Another definition of spline functions refers to their local properties. Consider the differential opertors defined by

$$L_j f = \frac{1}{dx} \frac{1}{w_j(x)} \frac{d}{dx} \frac{1}{w_{j-1}(x)} \cdots \frac{d}{dx} \frac{1}{w_0(x)} f(x), \qquad j = 0, 1, \ldots, n. \tag{1.6}$$

It follows from (1.2) that $L_n \Phi_j(t, x) = L_j \Phi_j(t, x) = 0$ for $x \neq t$ and $L_n s(x) = 0$ for $x \neq t_i$, $i = 1, 2, \ldots, l$. Therefore, the restriction of s to any subinterval $[t_i, t_{i+1}]$, $0 \leq i \leq l$, belongs to (the restriction of) the $(n+1)$-dimensional linear space $V = \operatorname{span}\{v_0, v_1, \ldots, v_n\}$. Moreover, s has a node of multiplicity at most m at $t \in (\alpha, \beta)$ if it has a continuous $(n-m)$-th derivative in a neighborhood of t. This definition is equivalent to that of (1.5).

Since $L_n u = 0$ for $u \in V$, it follows that V is a Haar subspace of $C[\alpha, \beta]$. In the case $w_i = 1$ for $i = 0, 1, \ldots, n$, one obviously has that $V = \Pi_n$. (Therefore, the functions in V are sometimes called V-polynomials.)

There is a strong connection with the theory of γ-polynomials. Chebyshevian spline functions are *K-polynomials*, i.e. they may be written in the form

$$s(x) = \sum_{i=0}^{l} \sum_{j=0}^{m_i-1} b_{ij} \frac{\partial^j}{\partial t^j} K(t_i, x), \tag{1.7}$$

with the kernel $K(t, x) = \Phi_n(t, x)$. In the theory of splines, the notation γ-polynomial is avoided, since the kernel K is not strictly totally positive, but only totally positive in the sense of Theorem 1.1 of this chapter. The equivalence of (1.5) and (1.7) results from

$$\frac{\partial}{\partial t} \Phi_j(t, x) = -w_j(t) \Phi_{j-1}(t, x), j = 1, 2, \ldots, n. \tag{1.8}$$

Indeed, Φ_j may be written in the form $\Phi_j(t, x) = w_0(x) \int_t^x \Psi(t, \xi_1)\, d\xi_1$, where Ψ is given by a $(j-1)$-fold integral. By taking derivatives and noting that $\Psi(t, t) = 0$ we get

$$\begin{aligned} \frac{\partial}{\partial t} \Phi_j(t, x) &= -w_0(x) \Psi(t, t) + w_0(x) \int_t^x \frac{\partial}{\partial t} \Psi(t, \xi_1)\, d\xi_1 \\ &= w_0 \int_t^x \frac{\partial}{\partial t} \Psi(t, \xi_1)\, d\xi_1, \quad \text{if } j \geq 2. \end{aligned} \tag{1.9}$$

By repeating the argument, we may shift the differentiation under the innermost integral. Then $\partial/\partial t \int_t^{\xi_{j-1}} w_j(\xi_j)\, d\xi_j = -w_j(t)$ yields (1.8).

From (1.8), it follows that $\Phi_n(t, .), \Phi_{n-1}(t, .), \ldots, \Phi_{n+1-m}(t, .)$ may be written as linear combinations of $\Phi_n(t, .), (\partial/\partial t)\Phi_n(t, .), \ldots, (\partial/\partial t)^{m-1}\Phi_n(t, .)$ and vice versa. Thus (1.5) and (1.7) are equivalent. □

Henceforth, we shall write $s \in S_{n,k} := S_{n,k}[\alpha, \beta]$ if s has k nodes in (α, β) counting multiplicities. For convenience, $s \in S_{n,r}[\alpha_1, \beta_1]$, with $\alpha \leq \alpha_1 < \beta_1 \leq \beta$, will mean that the restriction of s to the subinterval $[\alpha_1, \beta_1]$ belongs to $S_{n,r}[\alpha_1, \beta_1]$.

Mostly, we employ the spline functions in the form given by (1.5) with $t_i \neq t_j$ for $i \neq j$. Sometimes it is more appropriate to enumerate the nodes with their multiplicities without changing the symbol. This cannot cause confusion, since in cases of ambiguity, the correct interpretation will be indicated by the restriction $t_1 < t_2 < \cdots < t_l$ for distinct nodes and $t_1 \leq t_2 \leq \cdots \leq t_k$ for nodes labelled with their multiplicities. The same convention holds for zeros of spline functions: x_0 is a zero of multiplicity m if $f(x_0) = f'(x_0) = f'(x_0) = \cdots = f^{(m-1)}(x_0) = 0$.

The interpolation by spline functions is governed by the fact that the kernel $\Phi_n(t, x)$ is totally positive. The determinants which correspond to the associated linear system were already introduced in (VII.1.8).

1.1 Fundamental Theorem of Algebra for Spline Functions (Karlin and Ziegler 1966). *Let $n \geq 1$, $k \geq 0$, and Φ_n be defined by* (1.2). *Let $\{t_i\}_{i=1}^k$, $\{x_i\}_{i=1}^{n+k+1}$ satisfy* *the following conditions:*

(a) $\alpha \le t_1 \le t_2 \le \cdots \le t_k \le \beta$ *and* $\alpha \le x_1 \le x_2 \le \cdots \le x_{n+k+1} \le \beta$.

(b) *Whenever m of the* t_i*'s and j of the* x_i*'s* coincide, *then* $m + j \le n + 1$.

Then

$$\Phi_n^* \begin{pmatrix} \alpha, \ldots, & \alpha, & t_1, \ldots, & t_k \\ x_1, \ldots, & x_{n+1}, & x_{n+2}, \ldots, & x_{n+k+1} \end{pmatrix} \ge 0 \qquad (1.10)$$

with strict inequality if and only if the interlacing conditions

$$x_i < t_i < x_{i+n+1}, \qquad i = 1, 2, \ldots, k, \qquad (1.11)$$

hold.

B. Zeros of Spline Functions

Theorem 1.1 was originally proved by the repeated use of composition rules for determinants. Here, we establish the equivalent results on the zeros of spline functions, since the latter are more often used in the treatment of approximation problems.

It is typical that a non-zero spline function may vanish on a non-degenerate interval. Therefore, we shall adopt the convention that an interval on which a spline function vanishes will be henceforth counted as $n + 1$ zeros.

1.2 Lemma *Let* $n \ge 1$, $k \ge 0$. *Then each non-zero* $s \in S_{n,k}$ *has at most* $n + k$ *zeros counting multiplicities.*

Proof. Let $n = 1$. If s has only simple nodes, i.e. if s does not vanish on an interval, the lemma is obvious (see Fig. 31). If s also has double nodes then the same splitting argument as for $n > 1$ below applies.

Let $n > 1$. We distinguish several cases.

If $k = 0$, then s has at most n zeros since V is a Haar space.

If t_1 is a node of multiplicity $n + 1$ and $s = 0$ for $\alpha \le x < t_1$, then by induction on k, s has at most $n + (k - n - 1)$ zeros in $[t_1, \beta]$. Together with $n + 1$ zeros from $[\alpha, t_1)$, we have at most $n + k$ zeros in $[\alpha, \beta]$.

If t_1 is a node of multiplicity n and $s = 0$ for $\alpha \le x \le t_1$, then by induction s has at most $n + (k - n)$ zeros in $[t_1, \beta]$, one of them being t_1. Therefore, one of the $n + 1$ zeros from $[\alpha, t_1]$ has been already counted and the statement of the lemma is true.

The same arguments hold if the last node t_l has multiplicity n or $n + 1$ and s vanishes on $[t_l, \beta]$.

Fig. 31. Polynomial spline function with $n = 1$

Let s vanish on $[t_p, t_{p+1})$ where $1 \leq p < l$. Then we have $s \in S_{n,r}[\alpha, t_{p+1}]$ and $s \in S_{n,k-r}[t_p, \beta]$ for some $r > 0$. By applying the lemma to the specified subintervals and noting that the $n + 1$-fold zero from $[t_p, t_{p+1})$ has been counted twice, we obtain that the number of zeros is at most $(n + r) + (n + k - r) - (n + 1) = n + k - 1$.

Let t_p be a node of multiplicity n or $n + 1$. We may restrict ourselves to the case where $s \neq 0$ for at least one point in $[\alpha, t_p)$ and for one in $(t_p, \beta]$. For some $r \geq 0$ we have $s \in S_{n,r}[\alpha, t_p]$ and $s \in S_{n,k-r-n}[t_p, \beta]$. From what we know for the subintervals, we conclude that s has at most $(n + r) + (n + k - r - n) = n + k$ zeros.

We are left with the case that s has only nodes of multiplicity $\leq n - 1$ and s does not vanish on a subinterval. Then $s \in C^1$, and $\tilde{s} := d/dx(s/w_0)$ is a spline function in $S_{n-1,k}$ with the same nodes and of the same multiplicities as s. For completeness, we note that $S_{n-1,k}$ refers to the modified kernel $\Psi(t, x)$, (see (1.9)). By induction on n, we know that $\tilde{s}$ has at most $n - 1 + k$ zeros counting multiplicities. By Rolle's theorem, s has at most $n + k$ zeros. □

As an immediate consequence, we obtain

1.3 Corollary. *Let $n \geq 1$, $k \geq 0$. If $s \in S_{n,k}$ has $n + k + 1$ distinct zeros, then s vanishes between two of them.*

The corollary shows that the $(n + k + 1)$-dimensional linear spaces of Chebyshevian spline functions with k fixed nodes are *weak Haar subspaces.*

1.4 Definition. An m-dimensional subspace $V \subset C[\alpha, \beta]$ is called a *weak Haar subspace* if, for each set of $m + 1$ points $\alpha \leq x_0 < x_1 < \cdots < x_m \leq \beta$, there is no $u \in V$ such that

$$(-1)^i u(x_i) > 0, \qquad \text{for } i = 0, 1, \ldots, m.$$

We are now in a position to improve Lemma 1.2.

1.5 Theorem. *Let $s \in S_{n,k}$ have the nodes $\{t_i\}_{i=1}^{k}$ and the zeros $\{x_i\}_{i=1}^{n+k+1}$, where in each set the members are written with their multiplicities. Moreover, assume that the conditions* (a) *and* (b) *of Theorem* 1.1 *hold. If the interlacing conditions*

$$x_i < t_i < x_{n+i+1}, \; i = 1, 2, \ldots, k,$$

are also satisfied, then $s = 0$.

Proof. For $k = 0$ the theorem is obvious, so suppose that $k > 0$. From the assumptions and Lemma 1.2, we conclude that s vanishes on some interval, say on $[t_p, t_{p+1}]$ with $t_p < t_{p+1}$. Let t_p be a knot of multiplicity m, i.e., $t_{p-m} < t_{p-m+1} = t_{p-m+2} = \cdots = t_p$. Set

$$z_i = \begin{cases} x_i, & i = 1, 2, \ldots, p, \\ t_p, & i = p + 1, \ldots, p + n - m + 1. \end{cases}$$

Notice that $s \in C^{n-m}$ in some neighborhood of t_p, and $s^{(j)}(t_p - 0) = 0$ for $j = 0, 1, \ldots, n - m$. The restriction of s to $[\alpha, t_p]$ is a spline function with the nodes $\{t_i\}_{i=1}^{p-m}$ and the zeros $\{z_i\}_{i=1}^{n+p-m+1}$. The interlacing conditions are again satisfied. By induction, it follows that $s(x) = 0$ for $\alpha \le x \le t_p$. Similarly, s vanishes on $[t_{p+1}, \beta]$, and we have $s = 0$. □

Proof of Theorem 1.1. (1) Assume that the interlacing conditions are violated, say $x_j \ge t_j$. The restriction of $S_{n,k}$ to $[t_j, \beta]$ is an $n + 1 + k - j$ dimensional space. On the other hand there are $n + 2 + k - j$ interpolation conditions concerning $\{x_i\}_{i=j}^{n+k+1}$. Therefore, the interpolation is not possible for arbitrary data, and the determinant is zero.

(2) Assume that the interlacing conditions hold. By Theorem 1.5 the homogeneous interpolation problem has only the trivial solution $s = 0$. Hence the determinant is not zero.

Next observe that the set of t's and x's which satisfy the interlacing conditions (1.11) is convex. Since the determinant (1.10) does not vanish on this connected set, it does not change its sign.

We do not give the proof of the fact that the sign is positive if the interlacing conditions are satisfied, since it will not be used in our context. □

C. Characterization of Best Uniform Approximations

The $(n + k + 1)$-dimensional spaces of spline functions with k fixed nodes are only weak Haar subspaces of $C[\alpha, \beta]$. Therefore the best uniform approximation is not always unique. Nevertheless, a best Chebyshevian spline function may be characterized in terms of alternants. Typically, the length of the alternant depends on the subinterval on which its points are located.

1.6 Characterization Theorem (Rice 1967, Schumaker 1968a). *Let $n \ge 1, k \ge 0$ and $\alpha < t_1 \le t_2 \le \cdots \le t_k < \beta$ be fixed. A spline s is a best approximation to $f \in C[\alpha, \beta]$ from the set of spline functions with nodes $\{t_i\}_{i=1}^{k}$ if and only if $f - s$ has an alternant of length $n + r + 2$ in some interval $[t_p, t_q]$ such that $s \in S_{n,r}[t_p, t_q]$.*

Proof. (1) Suppose that there is an alternant as stated and that s_1 is a better approximation. By a de la Vallée-Poussin type of argument, $s - s_1$ has $n + r + 1$ distinct zeros. Since $s - s_1 \in S_{n,r}[t_p, t_q]$ by Corollary 1.3, $s - s_1$ is zero between two of the zeros. But there is a point of the alternant between the two zeros as specified, and so s_1 cannot be a better approximation.

(2) Assume that there is no alternant as stated. Since the set of approximating functions contains the $(n + 1)$-dimensional Haar subspace V, we need only consider the case where $f - s$ has an alternant of length $\ge n + 2$ on $[\alpha, \beta]$. Let $\alpha \le x_1 < x_2 < \cdots < x_{n+r+1} \le \beta$ with $r \ge 1$ be the points of a maximal alternant. Then $\sigma(-1)^i(f - s)(x_i) = \|f - s\|$, $i = 1, 2, \ldots, n + r + 1$, holds for $\sigma = +1$ or for $\sigma = -1$. Moreover, for $i = 1, 2, \ldots, n + r + 1$, define $[\underline{x}_i, \overline{x}_i]$

to be the maximal interval containing x_i such that $(f-s)(\underline{x}_i) = (f-s)(\overline{x}_i) = \sigma(-1)^i \|f-s\|$ but $(f-s)(x) \neq -\sigma(-1)^i \|f-s\|$ for all $x \in [\underline{x}_i, \overline{x}_i]$.

We now claim that there are $\{\tau_i\}_{i=1}^r \subset \{t_i\}_{i=1}^k$ such that

$$\overline{x}_i < \tau_i < \underline{x}_{n+i+1}, \qquad i = 1, 2, \ldots, r. \tag{1.12}$$

To prove this, we will repeatedly make use of the argument that by assumption there are at least j nodes out of the set $\{t_i\}_{i=1}^k$ in $(\overline{x}_p, \underline{x}_{p+n+j+1})$, whenever $j \geq 1$, $1 \leq p \leq k - j$.

In particular, at least one of the given nodes is in $(\overline{x}_1, \underline{x}_{n+2})$. Choose τ_1 as the smallest of them.

For an inductive construction, assume that $\tau_1, \tau_2, \ldots, \tau_{j-1}$ satisfy (1.12) and that the smallest possible index v_i for which $\tau_i = t_{v_i}$ has been chosen. Set

$$\tau_j = \min\{t_p \colon \overline{x}_j < t_p < \underline{x}_{n+j+1} \text{ and } p > v_{j-1}\}. \tag{1.13}$$

We claim that this set is not empty.

By the assumption on the alternant, some t_q lies in $(\overline{x}_j, \underline{x}_{n+j+1})$. Let q be the largest possible such index. Since there are at least j nodes in $(\overline{x}_1, \underline{x}_{n+j+1})$, we have $q \geq j$. If $v_{j-1} < q$, then t_q is in the set (1.13). Otherwise, for some $1 \leq r \leq j-2$, the nodes $t_q, t_{q-1}, \ldots, t_{q-r+1}$ have been selected for $\{\tau_i\}_{i=1}^{j-1}$, while t_{q-r} has not. We have $t_{q-r} \leq \overline{x}_{j-r}$, for otherwise t_{q-r} would have been taken as τ_{j-r-1}. Hence, there are only r nodes in $(\overline{x}_{j-r-1}, \underline{x}_{n+j+1})$, which is a contradiction. Therefore, the set (1.13) is indeed non-empty and $\tau_1, \tau_2, \ldots, \tau_r$ satisfying (1.12) can be constructed.

Next we choose $z_1, z_2, \ldots, z_{n+r}$ such that

$$\max\{\tau_{i-n}, \overline{x}_i\} < z_i < \min\{\tau_i, \underline{x}_{i+1}\},$$

where the τ's are ignored when they carry a subscript less than 1 or greater than r. By (1.12), this is possible and we get

$$z_i < \tau_i < z_{n+i}, \qquad i = 1, 2, \ldots, r. \tag{1.14}$$

By Theorem 1.1 there is a spline function $\hat{s} \in S_{n,r}$ with nodes from $\{\tau_i\}_{i=1}^r$ such that

$$\hat{s}(z_i) = 0, \qquad i = 1, 2, \ldots, n+m,$$

$$\hat{s}(x_1) = (f-s)(x_1).$$

Then $\hat{s}$ has no more zeros than specified above. Indeed, if there were an additional zero $\xi \in [\alpha, \beta]$, then the augmented set $\{\xi, z_i, i = 1, 2, \ldots, n+m\}$ would still satisfy the interlacing condition and by Theorem 1.5 we would have $\hat{s} = 0$, contradicting $\hat{s}(x_1) \neq 0$. Since all zeros are simple zeros, we know that $\hat{s}$ changes its sign at z_i, for $i = 1, 2, \ldots, n+m$. Hence,

$$(f-s)(x) \cdot \hat{s}(x) > 0, \text{ for } \underline{x}_i \leq x \leq \overline{x}_i,\ i = 1, 2, \ldots, n+r+1.$$

Now it follows from the Kolomogorov criterion that $s + \delta\hat{s}$ is a better approximation for some $\delta > 0$. □

Finally it follows from a general result on weak Haar spaces (cf. Exercise 1.9) that there is always a best approximation with an alternant of maximal length.

1.7 Theorem (Jones and Karlovitz 1970). *Let $n \geq 1$, $k \geq 0$ and $\alpha < t_1 \leq t_2 \leq \cdots \leq t_k < \beta$ be fixed. Then to each $f \in C[\alpha, \beta]$ there is a best approximation to f from the set of spline functions with nodes $\{t_i\}_{i=1}^k$ such that $f - s$ has an alternant of length $n + k + 1$.*

Exercises

1.8. Show that no nonzero spline function from $S_{n,k}$ vanishes on two distinct non-degenerate intervals whenever $1 < k \leq n + 1$.

1.9. Let M be an n-dimensional weak Haar subspace of $C[\alpha, \beta]$. Show that to each $f \in C[\alpha, \beta]$, $f \notin M$, there is a (best approximation) $u \in M$ such that $f - u$ has an alternant of length $n + 1$.
Hint: Consider the $(n + 1)$-dimensional subspace of $C[\alpha, \beta]$ which contains M and f. Recall Exercise II.3.21.

1.10. Assume that $s^* \in S_{nk}[t_1, t_2, \ldots, t_k]$ is a best uniform approximation to f. Let $[t_p, t_q]$ be a minimal critical interval in the sense of Theorem 1.6. Prove the following version of strong uniqueness:

$$\|f - s\|_{[a,b]} \geq d(f, S_{n,k}[t_1, \ldots, t_k]) + c\|s - s^*\|_{[t_p, t_q]}$$

for all $s \in S_{n,k}[t_1, \ldots, t_k]$ with some $c > 0$.

§2. Chebyshev Approximation by Spline Functions with Free Nodes

A. Existence

When the approximation by spline functions with free nodes is considered, functions with multiple nodes are naturally included in the theory. If, for example, the given function f is a spline function with a node of multiplicity $m \leq n$, then it may be arbitrarily well approximated by functions with m simple nodes. Indeed, by (1.7) f may be written in terms of the derivatives of the kernel Φ_n, and thus it is the limit of expressions of the form

$$\sum_{i=1}^{m} a_i \Phi_n(t_i, x).$$

Therefore, spline functions with multiple nodes are needed to obtain an existence set. Moreover, the tangent spaces always contain spline functions whose nodes have a multiplicity which is augmented by one or two. Thus, functions with multiple nodes are used for the characterization of best approximations.

In the theory with free nodes, we first establish the existence of a *weak solution* of the approximation problem, namely splines which may not be continuous. Regularity properties, such as continuity or differentiability, are shown afterwards. Similarly, when later situations are treated in which the nodes are separated, it will usually not be the first step in the proofs to show that two (or more) nodes cannot coalesce.

2.1 Remark. Let $n \geq m \geq 1$. Assume that (s_r) is a bounded sequence in $S_{n,m} \cap C[\alpha, \beta]$ and that all nodes converge to some $z \in (\alpha, \beta)$ as $r \to \infty$. Then a subsequence of (s_r) converges uniformly to some $s \in S_{n,m} \cap C^{n-m}[\alpha, \beta]$.

Proof. Let $0 < \delta < \min\{\beta - z, z - \alpha\}$. Then all nodes of s_r are in $[z - \delta, z + \delta]$ for r sufficiently large. Since the restrictions of s_r to $[\alpha, z - \delta]$ and $[z + \delta, \beta]$, resp., belong to finite dimensional subspaces, after passing to subsequences, if necessary, we have uniform convergence on $[\alpha, z - \delta] \cup [z + \delta, \beta]$.

Next choose $n + m + 1$ points

$$\alpha < x_0 < x_1 < \cdots < x_n < z - \delta < z + \delta < x_{n+1} < \cdots < x_{n+m} < \beta.$$

The spline functions are represented in the form

$$s_r(x) = \sum_{j=0}^{n} b_j^r \Phi_{n-j}(\alpha, x) + \sum_{i=1}^{m} c_i^r \Phi_n(t_1^r, t_2^r, \ldots, t_i^r, x), \tag{2.1}$$

where the repeated arguments in the second sum refer to divided differences. By the fundamental theorem (Theorem 1.1), the determinant of the matrix for the interpolation at $x_0, x_1, \ldots, x_{n+m}$ by linear combinations of $\Phi_n(\alpha, .)$, $\Phi_{n-1}(\alpha, .), \ldots,$ $\Phi_0(\alpha, .)$, $\phi_n(z, .)$, $\ldots$, $\phi_n(z, z, \ldots, z, .)$ does not vanish. By a continuity argument, the determinants of the matrices which correspond to (2.1) are bounded away from zero. Therefore, the sequences (a_j^r), (b_j^r) are bounded. They converge, after passing to subsequences if necessary. This implies uniform convergence for the corresponding spline functions.

Since the limit function possesses the node z with multiplicity at most m, it belongs to the class $C^{n-m}[\alpha, \beta]$. □

2.2 Proposition (de Boor 1969). *To each $f \in C[\alpha, \beta]$ there is a best uniform approximation from $S_{n,k}$.*

Proof. Let (s_r) be a bounded sequence in $S_{n,k}$. After going over to a subsequence, we may assume that the sequences of the nodes converge, say to $\alpha = z_0 < z_1 < \cdots < z_l < z_{l+1} = \beta$. Similarly, we may assume that the restrictions of s_r to any subinterval (z_i, z_{i+1}) converge to V-polynomials. The limits define a function u on $[\alpha, \beta] \backslash \{z_1, z_2, \ldots, z_l\}$. We complete the definition by setting $u(z_i) = u(z_i + 0)$. If $m \leq n$ nodes in the sequences converge to z_i $(1 \leq i \leq l)$, we make use of the preceding remark for $[z_{i-1} + \delta, z_{i+1} - \delta]$. We obtain $u \in C^{n-m}(z_{i-1}, z_{i+1})$, and z_i is a node of multiplicity at most m. If, on the other hand, more than n nodes from the sequence converge to some z_i, then z_i may be counted with multiplicity $n + 1$. Hence $u \in S_{n,k}$, and a subsequence of (s_r) converges to u uniformly on each compact subset of $[\alpha, \beta] \backslash \{z_1, z_2, \ldots, z_l\}$.

With this, we obtain the existence of a best approximation from Lemma II.1.4. □

B. Continuity and Differentiability Properties

In the preceding proposition, we did not exclude the possibility that the constructed best spline function has jump discontinuities. Nevertheless, in that case there also exists a best continuous spline function.

2.3 Theorem (Schumaker 1968b). *Let $n \geq 1$, $k \geq 0$. Then $S_{n,k} \cap C[\alpha, \beta]$ is an existence set in $C[\alpha, \beta]$.*

Proof. Let s_0 be a nearest point to f from $S_{n,k}$. Assume that s_0 is discontinuous at the node t_j for some $1 \leq j \leq l$, i.e., t_j is an $(n+1)$-fold node. We will split this node into two without deteriorating the approximation.

Set $\varepsilon(x) = f(x) - s_0(x)$ and denote by u the V-polynomial which coincides with s_0 on (t_{j-1}, t_j). Without loss of generality, assume that the jump $c_1 := \varepsilon(t_j + 0) - \varepsilon(t_j - 0)$ is positive. By continuity, there is a $\delta > 0$ such that

$$\left.\begin{aligned} &|\{f(x) - u(x)\} - \varepsilon(t_j - 0)| < \tfrac{1}{4}c_1 \qquad (2.2)\\ &\tfrac{1}{2}w_0(t_j) < w_0(x) < \tfrac{3}{2}w_0(t_j) \qquad (2.3)\end{aligned}\right\} \quad \text{for } |x - t_j| < \delta.$$

Moreover, there are constants $0 < c_2 \leq c_3$ such that

$$c_2(x-t)^n \leq \Phi_n(t,x) \leq c_3(x-t)^n \quad \text{for } \alpha \leq t < x \leq \beta. \tag{2.4}$$

Now we set $\delta_1 = c_1 c_2 \delta / 6 c_3 \|\varepsilon\|$ and redefine s_0 by

$$\tilde{s}(x) = u(x) + A\Phi_n(t_j - \delta_1, x),$$

where A is determined by $A\Phi_n(t_j - \delta_1, t_j) = c_1/2$. From (2.4) we get $A \geq \frac{1}{2}(c_1/c_3)\delta_1^{-n} \geq 3\|\varepsilon\|/c_2 \cdot \delta^n$. Hence,

$$A\Phi_n(t_j - \delta_1, t_j + \delta) \geq 3\|\varepsilon\|.$$

Consequently, we have $\tilde{s}(x_1) = s(x_1)$ for some $x_1 \in (t_j, t_j + \delta)$. The spline function

$$s_1(x) = \begin{cases} \tilde{s}(x) & \text{for } x \in [t_j - \delta_1, x_1], \\ s(x) & \text{otherwise,} \end{cases} \tag{2.5}$$

has a simple node at $t_j - \delta_1$ and an n-fold one at x_1.

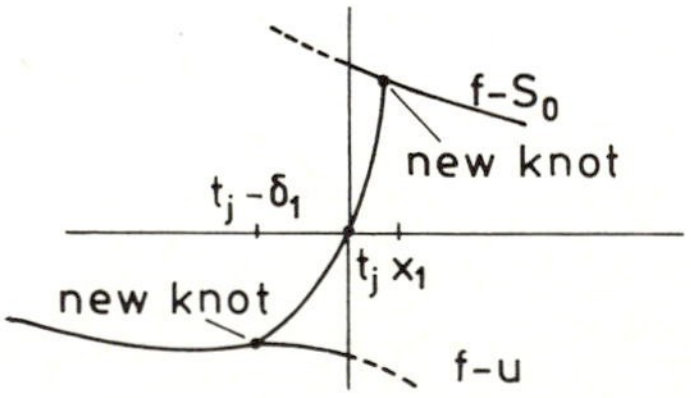

Fig. 32. Modification of the error curve in the neighborhood of a jump discontinuity by splitting the node

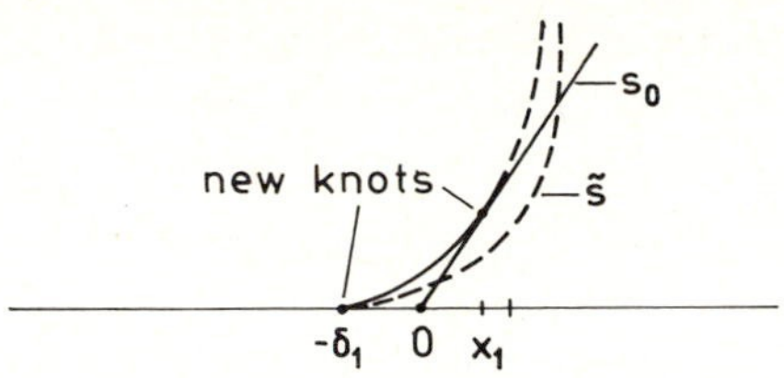

Fig. 33. Construction of a differentiable spline function

Notice that $\Phi_n(t, .)/w_0$ is strictly monotone. By (2.3), $A\Phi_n(t,x)$ is smaller than $(3/4)c_1$ on $(t_j - \delta, t_j)$ and larger than $(1/4)c_1$ on $(t_j, t_j + \delta)$. With this and (2.2), s_1 is also a best approximation.

Obviously, the splitting process can be performed for each node of multiplicity $n + 1$. □

For differentiable functions there are even best approximations with continuous derivatives.

2.4 Proposition (Schumaker 1968b). *Let $n \geq 2$, $k \geq 0$, and $f \in C^1[\alpha, \beta]$. Then there is a best uniform approximation to f from $S_{n,k}$ which is contained in $C^1[\alpha, \beta]$.*

Proof. Assume that a best approximation s_0 to f has a node t_j of multplicity n. Let $u \in V$ be the V-polynomial which coincides with f on $[t_{j-1}, t_j]$. After replacing f by $f - u$, we have $s_0(x) = 0$ on $[t_{j-1}, t_j]$. Next, by a change of the variable we may assume that $t_j = 0$. Finally, let $c := s_0'(+0)$ be positive. Then

$$\tfrac{1}{2}cx \leq s_0(x) \leq 2cx \quad \text{for } 0 \leq x \leq \delta,$$

for some $\delta > 0$. Let c_2 and c_3 be as in (2.5). Set $\delta_1 = (c_2/8c_3)$, $A_0 = 2(c/c_2)\delta^{-n+1}$ and

$$\tilde{s}(x) = A_0 \Phi_n(-\delta_1, x).$$

By a trivial calculation, one obtains $\tilde{s}(\delta_1) \leq s_0(\delta_1)$ and $\tilde{s}(\delta) > s_0(\delta)$. Let

$$A_1 = \max_{0 \leq x \leq \delta} \frac{s_0(x)}{\Phi_n(-\delta, x)}.$$

The maximum is attained at some $x_1 \in (0, \delta)$. Then

$$s_1(x) = \begin{cases} A_1 \Phi_n(-\delta_1, x) & \text{for } -\delta_1 \leq x \leq x_1, \\ s_0(x) & \text{otherwise.} \end{cases}$$

is differentiable in $(-\delta, \delta)$.

Since $f \in C^1[\alpha, \beta]$ implies that $(f - s_0)(0) \neq +\|f - s_0\|$, by choosing δ sufficiently small, s_1 is also a best approximation to f. □

On the other hand, if $p > n/2$ and $k \geq n - p$, one cannot exclude the possibility that there is no best spline function with a continuous p-th derivative. This may happen even if $f \in C^\infty[\alpha, \beta]$ (see Remark 2.9).

C. Characterization of Best Approximations

The necessary and the sufficient conditions for best spline functions with free nodes refer to alternants on subintervals between the nodes. This is not surprising, in view of the theory with fixed nodes. The gap between the necessary and the sufficient conditions for global solutions, which one finds in the study of γ-polynomials, becomes even larger here. In the special situation of monosplines, which will be investigated in the next section, this difference disappears.

The first result provides a sufficient condition.

2.5 Theorem (Braess 1971). *Let $f \in C[\alpha, \beta]$ and $s \in S_{n,k} \cap C$ possess the nodes $\alpha = t_0 < t_1 < \cdots < t_{l+1} = \beta$. If $f - s$ has an alternant of length $n + k + r + 2$ in some interval $[t_p, t_q]$ such that $s \in S_{n,r}[t_p, t_q]$, then s is a best approximation to f from $S_{n,k}$. Moreover, s coincides with any other best approximation to f on some interval.*

Proof. Let there be an alternant as stated. Suppose that $s_1 \in S_{n,k}$ is a better approximation. By standard arguments $s - s_1$ has $n + k + r + 1$ zeros in $[t_p, t_q]$. Since $s - s_1 \in S_{n,k+r}[t_p, t_q]$, by Corollary 1.3, $s - s_1$ vanishes between two of the zeros. But between the two zeros lies a point of the alternant, which is a contradiction.

The second statement follows with similar arguments. □

By considering the special case when $t_p = \alpha$, $t_q = \beta$, we obtain

2.6 Corollary (Schumaker 1968b). *Let $f \in C[\alpha, \beta]$. If $f - s$ has an alternant of length $n + 2k + 2$, then s is a best approximation to f from $S_{n,k}$.*

The sufficient conditions above do not guarantee uniqueness of the solution. In connection with uniqueness problems, we typically encounter interlacing conditions, in which each given node is formally counted twice.

2.7 Lemma. *Assume that $s \in S_{n,k}[\alpha, \beta]$ possesses the nodes $t_1 \le t_2 \le \cdots \le t_k$, $s_1 \in S_{n,k}$, and that $s_1 - s$ has $n + 2k + 1$ zeros at $x_1 \le x_2 \le \cdots \le x_{n+2k+1}$. If the interlacing conditions*

$$x_{2i} < t_i < x_{n+2i} \quad \textit{for } i = 1, 2, \ldots, k. \tag{2.6}$$

hold, then $s - s_1$ vanishes on $[\alpha, \beta]$.

Sketch of proof. Since $s - s_1 \in S_{n,2k}$, by Corollary 1.3 $s - s_1$ vanishes on some subinterval, say on (t_p, t_{p+1}). Then the restriction to $[\alpha, t_p]$ or to $[t_{p+1}, \beta]$ again satisfies interlacing conditions which are analogous to (2.6), and by induction it follows that $s = s_1$ on $[\alpha, \beta]$. For the details of the proof, the reader is referred to Arndt (1974). □

With this lemma, Arndt (1974) proved the uniqueness result announced by Schumaker (1968b).

2.8 Uniqueness Theorem. *Let $n \geq 2, k \geq 1$, and let $s \in S_{n,k} \backslash S_{n,k-1}$ have the nodes $\alpha = t_0 < t_1 < \cdots < t_{l+1} = \beta$. Assume that $f - s$ has an alternant of length $n + 2k + 2$ on $[\alpha, \beta]$ but does not have an alternant of length $n + 2r + 2$ on any subinterval $[t_p, t_q]$, where $s \in S_{n,r}[t_p, t_q]$ and $r < k$. Then s is the unique best approximation to f from $S_{n,k}$.*

Proof. Let $x_0 < x_1 < \cdots < x_{n+2k+1}$ be points of an alternant. Then by assumption

$$x_{2i} < t_i < x_{n+2i-1}, \qquad i = 1, 2, \ldots, k. \tag{2.7}$$

Therefore, s has nodes of multiplicity at most $n/2$ and $s \in C^1[\alpha, \beta]$. Assume that $\|f - s_1\| \leq \|f - s\|$. Because of Proposition 2.4, we may restrict ourselves to the case $s_1 \in C^1[\alpha, \beta]$. By the de la Vallée-Poussin argument, $s - s_1$ has $n + 2k + 1$ zeros $z_1, z_2, \ldots, z_{n+2k+1}$ with

$$x_{i-1} \leq z_i \leq x_i, \qquad i = 1, 2, \ldots, n + 2k + 1. \tag{2.8}$$

Here, we have counted nonnodal zeros twice. With (2.7) and (2.8), we conclude from the preceding lemma that $s - s_1 = 0$. □

2.9 Remark. Let $n \geq 3$ and $p \geq n/2 + 1$. By means of the uniqueness theorem, one can easily see that there may be no best spline to a given $f \in C^\infty[\alpha, \beta]$ with a continuous p-th derivative. Indeed, let $k \geq n - p + 1$, and choose a spline function s_0 with one node t_1 of multiplicity $n - p + 1$ and $k - (n - p + 1)$ additional simple nodes. Obviously, one may construct a C^∞-function f such that $f - s_0$ has an alternant $x_0 < x_1 < \cdots < x_{n+2k+1}$, where $x_{n+2j-2} < t_j < x_{n+2j-1}$ for $j = 1, 2, \ldots$. The spline function s_0 is the unique best approximation to f and $s_0 \notin C^p[\alpha, \beta]$.

With some special arguments Schumaker (1969) showed that $(n + 1)/2$ knots may coalesce and that one still cannot avoid $s \notin C^p$, $p \geq (n + 1)/2$ for a unique best spline function. For another way to obtain the improvement, see Exercise 2.14.

For deriving necessary conditions using critical point theory, we consider the parametric representation

$$F(b, t, x) = \sum_{i=0}^{l} \sum_{j=0}^{m_i - 1} b_{ij} \frac{\partial^j}{\partial t^j} \Phi_n(t_i, x). \tag{2.9}$$

The derivatives

$$\frac{\partial F}{\partial b_{ij}} = \frac{\partial^j}{\partial t^j} \Phi_n(t_i, x), \qquad j = 0, 1, \ldots, m_i - 1, \quad i = 0, 1, \ldots, l, \tag{2.10}$$

$$\frac{\partial F}{\partial t_i} = \frac{\partial^{m_i}}{\partial t^{m_i}} \Phi_n(t_i, x), \qquad i = 1, 2, \ldots, l\,(m_i \leq n - 1), \tag{2.11}$$

span a linear space of spline functions with fixed nodes, where the multiplicity of the interior nodes is increased by 1 as long as the resulting multiplicity does not exceed n. From (III.1.2), we know that this linear space is contained in the tangent cone. Recalling Theorem 1.6 for splines with fixed nodes and the lemma of first variation, we obtain a necessary condition.

2.10 Theorem (Braess 1971). *Let s^* be a best approximation to f from $S_{n,k} \cap C$ and let s^* possess the nodes $\alpha = t_0 < t_1 < \cdots < t_{l+1} = \beta$. Then $f - s^*$ has an alternant of length $n + r + l_1 + 2$ in some interval $[t_p, t_q]$, where $s \in S_{n,r}[t_p, t_q]$ and l_1 nodes out of $\{t_i\}_{i=p+1}^{q-1}$ have a multiplicity $m_i \leq n - 1$.*

The cases where the spline function in question is differentiable or has only simple nodes admit simpler formulations.

2.11 Corollary. *Let $n \geq 2$, and let s^* with nodes $\{t_i\}_{i=1}^{l}$ be a best approximation to f from $S_{n,k}$.*

(1) *If moreover, $s^* \in C^1[\alpha, \beta]$, then $f - s^*$ has an alternant of length $n + q + r + 2$ in some interval $[t_p, t_{p+q+1}]$, where $q \geq 0$ and $s^* \in S_{n,r}[t_p, t_{p+q+1}]$.*

(2) *If moreover, s^* has only simple nodes, then $f - s^*$ has an alternant of length $n + 2q + 2$ in some interval $[t_p, t_{p+q+1}]$.*

There is a gap between the necessary condition and the sufficient one, even if spline functions with only one node are considered. This is quite different from the theory of γ-polynomials, since in that theory no gap appeared in G_1, i.e. in the family with one nonlinear parameter. The following example shows that the gap cannot be bridged.

2.12 Example. Consider approximation in $S_{n,1}[0, 3]$ with polynomial splines, i.e., with $\Phi_n(t, x) = (x - t)_+^n$. Let $\alpha > 0$ and

$$f(x) = \begin{cases} (-1)^{n+1} \sin(n+2)\pi x, & 0 \leq x \leq 1, \\ 0, & 1 \leq x \leq 2, \\ \alpha(x-2)^n, & 2 \leq x \leq 3 \end{cases} \tag{2.12}$$

and

$$s(x) = \alpha(x - 2)_+^n.$$

Obviously, there is an alternant of length $n + 2$. If $\alpha \geq 6^n + 1$, s is a best approximation. Suppose to the contrary that $\|f - s_1\| < \|f - s\| = 1$ for some $s_1 \in S_{n,1}$. Then s_1 has a node in $[0, 1]$ and the restriction of s_1 to the interval $[1, 3]$ is a polynomial.

Moreover $|s(x)| \leq 1$ for $1 \leq x \leq 2$. Applying the well-known theorem of Bernstein on the growth of polynomials, we get $|s_1(3)| \leq (3 + \sqrt{8})^n < 6^n$. This contradicts $|f(3) - s_1(3)| \leq \|f - s_1\| < 1$, and s is optimal.

If, on the other hand $|\alpha| < \frac{1}{2}$, then s is not optimal in $S_{n,1}$. Indeed, observe that the function $s_1 - 6^{-n}\left(x - \dfrac{n+1}{n+2}\right)_+^n$ satisfies $\|f - s_1\| = \|f - s\|$, but s_1 is

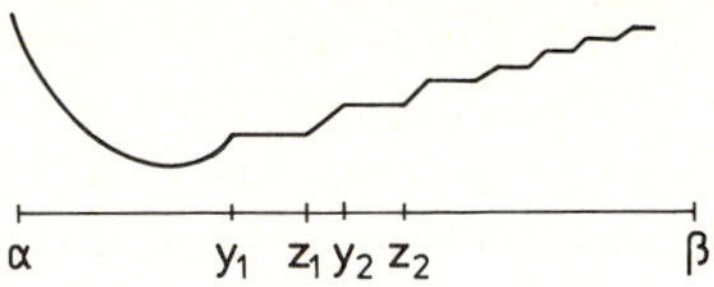

Fig. 34. Distance of f from $S_{n,1}(t)$

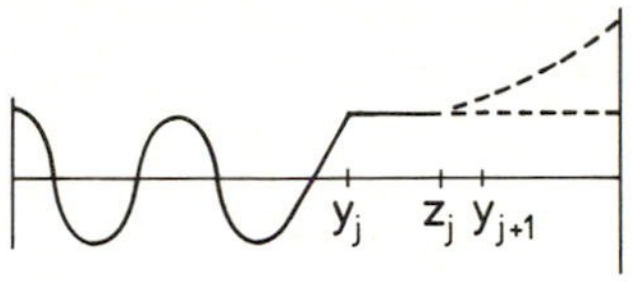

Fig. 35. Error curve when f is defined on $[y_j, y_{j+1}]$

not a best approximation, since $f - s_1$ has only an alternant of length $n + 1$. Hence, for $|\alpha| < \frac{1}{2}$, neither s nor s_1 is optimal. □

There is another difference to the results for γ-polynomials. It is possible to find a function f with an arbitrarily large number of connected components of local best approximations in $S_{n,1}$. To this end, let $S_{n,1}(t)$ denote the linear subspace of polynomial spline functions with one node at the point t. Moreover, let

$$\alpha < y_1 < z_1 < y_2 < z_2 < \cdots < \beta.$$

Following Brink-Spalink (1975), we will construct a function f such that $t \mapsto d(f, S_{n,1}(t))$ is constant for $y_j < t < z_j$, for $j = 1, 2, \ldots$. Then the nearest points from $S_{n,1}(t)$ for $y_j < t < z_j$ are local best approximations in the nonlinear set $S_{n,1}$.

First we define f in $[\alpha, y_1]$ such that f has an alternant of exact length $n + 2$ and such that $f(y_1) = \|f\| = 1$.

More generally, assume that f has already been defined on $[\alpha, y_j]$. Let u_j be the unique best approximating polynomial to f on that interval and let $(f - u_j)(y_j) = c_j := \|f - u_j\|_{[\alpha, y_j]}$ (see Fig. 35).

Define f on $[y_j, z_j]$ such that $f - u_j$ is constant in that region. If moreover,

$$0 < (f - u_j)(x) \leq c_j\left[1 + \left(\frac{x - z_j}{\beta - y_j}\right)^n\right] \text{ for } z_j \leq x \leq \beta, \tag{2.13}$$

then $d(f, S_{n,1}(t)) = c_j$ for $y_j \leq t \leq z_j$. Next we set

$$f(x) = u_j(x) + c_j + \varepsilon_j(x - z_j)^n \text{ for } z_j \leq x \leq y_{j+1}$$

where $\varepsilon_j > 0$ is chosen so small that the conditions (2.13) hold for $1, 2, \ldots, j$. This is possible since $\varepsilon_j = 0$ would imply that $u_{j+1} = u_j$. Obviously, $\|f - u_{j+1}\|_{[\alpha, y_{j+1}]} \geq \|f - u_{j+1}\|_{[\alpha, y_j]} > \|f - u_j\|_{[\alpha, y_j]}$, and the levels for $[y_j, z_j]$ and $[y_{j+1}, z_{j+1}]$ in Fig. 34 are indeed different. □

Exercises

2.13. Show that $S_{n,k}$ is an existence set in $L_p[\alpha, \beta]$ for $1 \le p < \infty$.

2.14. In order to improve the statement of Remark 2.9 show the following: Let $t_1 \le t_2 \le \cdots \le t_k$, and p_i be the number of distinct nodes in (α, t_i). Assume that there is an alternant $x_0 < x_1 < \cdots < x_{n+2k+l+1}$ such that

$$x_{2i+p_i} < t_i < x_{n+2i+p_i}, \qquad i = 1, 2, \ldots, k.$$

Show that s is the unique best approximation in $S_{n,k}$. Why can this criterion by applied to a node of multiplicity $(n+1)/2$ to get a unique best spline function $s_0 \notin C^p$ with $p \ge (n+1)/2$?

2.15. Let $r, s \ge 1$, $n + 1 = r + s$, and $p \in \Pi_n$ satisfy the interpolation conditions

$$\begin{aligned} p^{(j)}(0) &= 0 \quad \text{for } j = 0, 1, \ldots, r-1,\\ p(1) &= 1,\\ p^{(j)}(1) &= 0 \quad \text{for } j = 1, 2, \ldots, s-1. \end{aligned}$$

Show that p is strictly monotone in $(0, 1)$ and that $p(x)/x^r$ is strictly positive on $[0, 1]$. If this polynomial is extended to a spline function by setting

$$s(x) = 0 \text{ for } x \le 0 \text{ and } s(x) = 1 \text{ for } x \ge 1,$$

what are the multiplicities of the nodes?

§ 3. Monosplines of Least L_∞-Norm

A. The Family $S_{n,k}^+$

In rational approximation the Stieltjes functions are distinguished by nice properties, as are the completely monotone functions in approximation by exponential sums. The monosplines play the same role in the theory of spline functions. They lead to the consideration of the family $S_{n,k}^+$:

$$S_{n,k}^+ = \left\{ s \in S_{n,k} \colon s(x) = \sum_{j=0}^{n} b_j \Phi_j(\alpha, x) + \sum_{i=1}^{k} a_i \Phi_n(t_i, x),\ a_i \ge 0 \text{ for } i = 1, 2, \ldots, k \right\}. \tag{3.1}$$

3.1 Proposition. *Let $n \ge 1$, $k \ge 1$. Then $S_{n,k}^+$ is an existence set in $C[\alpha, \beta]$.*

Proof. It is sufficient to prove that $S_{n,k}^+$ is boundedly compact in the topology of compact convergence on (α, β), i.e., each bounded sequence contains a subsequence which converges uniformly on each compact subset of (α, β).

Let (s_ν) be a bounded sequence. After passing to a subsequence if necessary, the number of nodes is constant and the sequences of nodes converge. Specifically, let

$$\lim t_i^{(\nu)} = z_i,$$

where, for some $\delta > 0$ and $1 \le p \le q \le k$:

$$z_i = \alpha \quad \text{for } i < p, \quad \alpha + 3\delta \le z_i \le \beta - 3\delta \quad \text{for } p \le i \le q,$$

$$\text{and} \quad z_i = \beta \quad \text{for } i > q.$$

For sufficiently large v, we have the representation

$$s_v(x) = \sum_{j=0}^{n} \tilde{b}_j^{(v)} \Phi_j(\alpha, x) + \sum_{i=p}^{q} a_j^{(v)} \Phi_n(t_i^{(v)}, x) \tag{3.2}$$

for $\alpha + \delta \le x \le \beta - \delta$. For large v, there are no nodes in $[\alpha + \delta, z_p - \delta]$ and from the boundedness of (s_v) it follows that the sequences $(\tilde{b}_j^{(v)})$ for $j = 0, 1, \dots, n$ are bounded. We may assume that they converge. Therefore the first sum converges to a V-polynomial as $v \to \infty$, and the second sum is also bounded. Notice that $\Phi_n(t_i^{(v)}, \beta - \delta) \ge \Phi_n(\beta - 2\delta, \beta - \delta)$. Hence, the sequences $(a_j^{(v)})$ for $j = p, p + 1, \dots, q$ are also bounded and converge, after passing to subsequences.

With this, we have shown that a subsequence converges uniformly on $[\alpha + \delta, \beta - \delta]$ to some $s^* \in S^+_{n,q-p+1}$. Since the representation (3.2) holds on $[t_{p-1}^{(v)}, t_{q+1}^{(v)}]$, and $t_{p-1}^{(v)} \to \alpha$, $t_{q+1}^{(v)} \to \beta$, we get convergence as stated.

By Lemma II.1.4, $S^+_{n,k}$ is an existence set. □

B. Monosplines

3.2 Definition. Let $s \in S_{n,k}$, $f \in C^{n-1}[\alpha, \beta]$ be piecewise C^n and

$$w_n^{-1} L_{n-1} f \text{ be strictly increasing on } (\alpha, \beta), \tag{3.3}$$

where the differential operator L_{n-1} is defined by (1.6). Then $M = f - s$ is called a *monospline* of type $(n + 1, k)$.

The differentiability requirements may be slightly relaxed. Let $\{v_j\}_{j=1}^{n+1}$ be a basis of the $(n + 1)$-dimensional subspace $V \subset C[\alpha, \beta]$. If

$$\det \begin{pmatrix} v_1(x_1) & v_1(x_2) & \dots & v_1(x_{n+2}) \\ \dots & & & \\ v_{n+1}(x_1) & v_{n+2}(x_2) & \dots & v_{n+1}(x_{n+2}) \\ f(x_1) & f(x_2) & \dots & f(x_{n+2}) \end{pmatrix} > 0$$

whenever $\alpha \le x_1 < x_2 < \dots < x_{n+2} \le \beta$, then f is said to be in the *convexity cone* of V. The convexity cone is independent of the chosen basis for V up to a sign.

Assume that $u \in V$ and that f satisfies (3.3). Then $w_n^{-1} L_{n-1}(f - u)$ is strictly monotone and has at most one zero. By Rolle's theorem, $f - u$ has at most $n + 1$ zeros. Therefore f (or $-f$) is in the convexity cone of V. (Instead of using (3.3) for the definition of monosplines, one could have required that f belongs to the convexity cone of V.)

We note that any $f \in C^{n-1}$ with $f^{(n-1)}$ convex belongs to the convexity cone of Π_n with the basis $\{1, x, x^2, \dots, x^n\}$. In particular, the most often investigated monosplines are

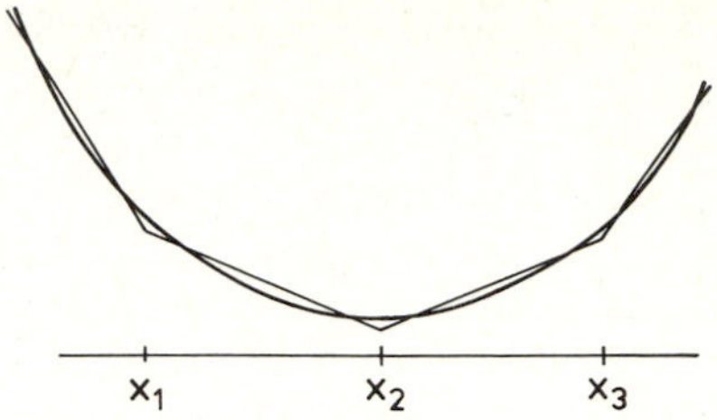

Fig. 36. A convex function f and a spline function $S \in S_{1,3}$ such that $f - s$ has 8 zeros

$$x^{n+1} - s_{n,k}(x) \tag{3.4}$$

where $s_{n,k}$ are polynomial spline functions in $S_{n,k}$.

In 1960, Johnson proved uniqueness of the monospline with least uniform norm by topological methods. We prefer a quick proof which makes use of the approximation in $S_{n,k}^{+}$. The advantage will become clear later, when the proofs for the study of monosplines with nodes of prescribed multiplicity will be carried out.

The following lemma describes the special case of the fundamental theorem of algebra for monosplines, where all nodes are simple:

3.3 Lemma. *Let $n \geq 1$, $k \geq 0$, and let the monospline M of class $(n+1, k)$ have k simple nodes $t_1 < t_2 < \cdots < t_k$.*
(1) *M has at most $n + 2k + 1$ zeros in $[\alpha, \beta]$.*
(2) *M' has at most $n + 2k$ zeros provided that $w_0(x) = 1$.*
(3) *If M has $n + 2k + 1$ zeros in (α, β), then $M(\beta) > 0$ holds.*

Proof. Let $M = f - s$. From (1.7) we know that $w_n^{-1} L_{n-1} s$ is constant in each interval $[t_i, t_{i+1}]$. Hence, $w_n^{-1} L_{n-1} M$ is strictly increasing and has at most one zero. By Rolle's theorem, $L_{n-2} M$ has at most two zeros in $[t_i, t_{i+1}]$ for $i = 0, 1, \ldots, k$. Now, $L_{n-2} M$ is continuous in $[\alpha, \beta]$ and there are at most $2(k+1)$ zeros. By applying Rolle's theorem $(n-1)$ times, we obtain the statement (1). As an intermediate result, we have obtained at most $n + 2k$ zeros for $L_0 M = (d/dx)(w_0^{-1} M) = M'$, provided that $w_0 = 1$. This yields (2). Finally the maximal number of zeros is only attained if for each $0 < j < n - 1$, the weighted derivative $L_j M$ is positive between β and the last zero. □

3.4 Theorem (Johnson 1960). *Let $n \geq 2$, $w_0 = 1$ and let f satisfy* (3.3). *Then s^* is the unique best approximation to f from $S_{n,k}$ if and only if $f - s^*$ has an alternant of length $n + 2k + 2$. Furthermore, s^* has k distinct nodes and belongs to $S_{n,k}^{+}$.*

Proof. First it is not clear whether the best approximation from $S_{n,k}$ has multiple nodes. Therefore we start the proof by considering a best approximation s^* to f from the subset $S_{n,k}^{+}$; this exists by Proposition 3.1.

Let s^* have exactly r nodes. By considering the derivative with respect to the parameters in (3.1) with k replaced by r, we conclude that the tangent cone $C_{s^*} S_{n,k}^{+}$

contains the space of spline functions in $S_{n,2r}$ having the double nodes $t_1, t_2, \dots, t_r$. By the lemma of first variation and the characterization theorem 1.6, there is an alternant of length $n + 2q + 2$ in some interval $[t_p, t_{p+q+1}]$. Assume that $t_p > \alpha$ or $t_{p+q+1} < \beta$. Then $(f - s^*)'$ has $n + 2q + 1$ zeros in $[t_p, t_{p+q+1}]$. Since the restriction of $f - s^*$ to $[t_p, t_{p+q+1}]$ is a monospline of class $(n + 1, r)$, we have a contradiction to Lemma 3.3(2).

Now we know that $f - s^*$ has an alternant of length exactly $n + 2r + 2$ and that the alternant includes the points α and β. Lemma 3.3(3) yields $f(\beta) - s^*(\beta) > 0$. Continuity implies that

$$f(x) - s^*(x) > 0 \quad \text{for } z \leq x \leq \beta$$

for some $z < \beta$. Then for sufficiently small $\varepsilon > 0$, the function

$$s_1(x) := s^*(x) + \varepsilon \Phi_n(z, x) \in S^+_{n,r+1}$$

is also a best approximation if $r < k$. Since $f - s_1$ has an alternant only of length $n + 2r + 1$, this contradicts the arguments above with r replaced by $r + 1$.

Therefore, $r = k$ and s^* has an alternant of length $n + 2k + 2$. The conditions of the uniqueness theorem 2.8 concerning subintervals are also satisfied. Hence, s^* is the unique best approximation from $S_{n,k}$. □

As a consequence we get the generalization of Bernstein's comparison theorem, which is analogous to the comparison theorem V.3.9 for rational approximation.

3.5 Comparison Theorem (Braess 1975c). *Let $n \geq 1$, $k \geq 0$. Assume that*

$$0 \leq f^{(n+1)}(x) \leq g^{(n+1)}(x) \quad \textit{for } \alpha \leq x \leq \beta. \tag{3.5}$$

Then for the approximation by polynomial spline functions one has

$$d(f, S_{n,k}) \leq d(g, S_{n,k}). \tag{3.6}$$

Proof. It suffices to prove (3.6) under the assumption that $0 < f^{(n+1)}(x) < g^{(n+1)}(x)$ for $x \in [\alpha, \beta]$. Indeed, we might replace f by $f_\varepsilon = f + \varepsilon x^{n+1}$ and g by $g_\varepsilon = g + 2\varepsilon x^{n+1}$ with ε arbitrarily small and note that $d(f_\varepsilon, S_{n,k})$ and $d(g_\varepsilon, S_{n,k})$ are continuous in ε.

Let s_f and s_g be the best approximations to f and g respectively. By the preceding theorem, there are $n + 2k + 2$ points $\alpha = x_1 < x_2 < \cdots < x_{n+2k+2} = \beta$ such that

$$(-1)^{n-i}(f - s_f)(x_i) = \|f - s_f\| > 0. \tag{3.7}$$

Moreover, s_f has a convex $(n - 1)$-st derivative. From this and the assumption, it follows that $h = (g - f) + s_f$ has a strictly convex $(n - 1)$-st derivative.

Suppose that (3.6) does not hold. Set $M = (g - s_g) - (f - s_f) = h - s_g$. By a de la Vallée-Poussin type argument we have

$$(-1)^{n-i} M(x_i) < 0 \quad \text{for } i = 1, 2, \dots, n + 2k + 2.$$

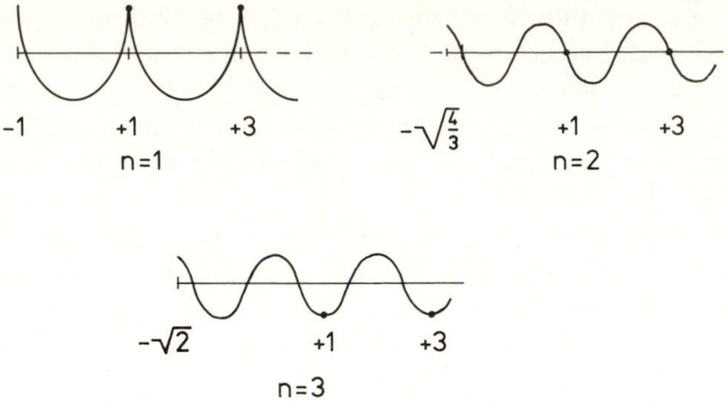

Fig. 37. Optimal polynomial monosplines

Consequently, M has $n + 2k + 1$ zeros and $M(\beta) < 0$. Since M is a monospline, this contradicts Lemma 3.3(3). □

There is a difference from Bernstein's classical comparison theorem for polynomials. Condition (3.5) cannot be replaced by the relaxed condition $|f^{(n+1)}(x)| \leq g^{(n+1)}(x)$. In this connection the following problem was solved by Pinkus (1981): Given a positive function $w(x)$, for which f is $d(f, S_{n,k})$ maximal under the restriction $|f^{(n+1)}(x)| \leq w(x)$ almost everywhere? The solution is a function with $|f^{(n+1)}(x)| = w(x)$ almost everywhere, for which $f^{(n+1)}(x)$ changes sign at exactly k points in (α, β).

3.6 Example. The optimal polynomial monosplines with simple nodes of the form (3.4) are explicitly known up to $n = 3$ (Johnson 1960).

For $n = 1$ set

$$M(x) = \begin{cases} x^2 - \frac{1}{2}, & -1 \leq x \leq +1, \\ M(x-2), & 1 \leq x \leq 2k+1. \end{cases}$$

M is continuous in $[-1, 2k+1]$ and has an alternant of length $2k + 3 = n + 2k + 2$ (see Fig. 37). Given an interval $[\alpha, \beta]$, $\alpha < \beta$, an affine transformation of the variable x yields

$$d_{[\alpha,\beta]}(x^2, S_{1,k}) = \frac{1}{8}\left(\frac{\beta - \alpha}{k+1}\right)^2.$$

For $n = 2$, a monospline is constructed which is periodic apart from the boundary intervals (see Fig. 37).

$$M(x) = \begin{cases} x^3 - x, & -2/3^{1/2} \leq x \leq +1, \\ M(x-2), & 1 \leq x \leq 2k+1, \\ \text{anal. cont.} & 2k+1 \leq x \leq 2k + 2/3^{1/2}. \end{cases}$$

We first obtain a periodic C^1-function from an antisymmetric polynomial which vanishes at the nodes and has 2 points of alternation between each pair of successive nodes. Then the domain is enlarged such that two points are added to the alternant. Therefore, there is an alternant of length $1 + 2(k + 1) + 1 = n + 2k + 2$. Again by an affine transformation of x one gets

$$d_{[\alpha,\beta]}(x^3, S_{2,k}) = \frac{1}{12 \cdot 3^{1/2}} \left(\frac{\beta - \alpha}{k + 2/3^{1/2}} \right)^3.$$

For $n = 3$, a similar construction yields

$$M(x) = \begin{cases} x^4 - 2x^2 + \frac{1}{2}, & -2^{1/2} \le x \le 1, \\ M(x-2), & 1 \le x \le 2k+1, \\ \text{by anal. cont. on} & 2k+1 \le x \le 2k + 2^{1/2}. \end{cases}$$

The construction starts with a symmetric polynomial whose derivative vanishes at the nodes. After enlarging the boundary intervals, the length of the alternant is $1 + (2k + 3) + 1 = n + 2k + 2$, and one obtains

$$d_{[\alpha,\beta]}(x^4, S_{3,k}) = \frac{1}{32} \left(\frac{\beta - \alpha}{k + 2^{1/2}} \right)^4.$$

C. The Fundamental Theorem of Algebra for Monosplines

For all investigations of monosplines in which the multiplicities of the nodes are prescribed, an estimate of the number of zeros is essential. The conventions on the counting of multiplicities and the restrictions are the same as in Theorem 1.1. Fortunately, the treatment of monosplines is much simpler because they do not vanish on intervals of positive length.

3.7 Fundamental Theorem of Algebra for Monosplines (Part 1) (Micchelli 1972). *Let $M = f - s$ be a monospline of class $(n + 1, k)$ with nodes of multiplicity $m_1, m_2, \ldots, m_l$. Let $\sigma_i = 1$, if m_i is odd, and zero otherwise. Then M has at most*

$$N := n + \sum_{i=1}^{l} (m_i + \sigma_i) + 1 \tag{3.8}$$

zeros. Moreover, if M has exactly N zeros in (α, β), then

$$M(\beta) > 0, \qquad (-1)^N M(\alpha) > 0, \tag{3.9}$$

and in the representation (1.5) *for the spline part s, one has*

$$a_{im_i} > 0, \quad \text{if } m_i \text{ is odd.} \tag{3.10}$$

Remark. There is a strong connection between this theorem and Descartes' rule of signs. In this context, the role of the parameters σ_i for the parity check of the multiplicities becomes apparent. The fundamental theorem yields exactly the same bound on the number of zeros which is given for the analogous γ-polynomials by Remark VII.1.11. An extension of (3.10) to non-leading terms which goes beyond the framework of the Descartes' rule was given by Micchelli (1972).

Proof of Theorem 3.7. The proof proceeds by a simultaneous induction on n and k. If $n = 1$, then all nodes are simple. If $k = 0$, there are no nodes. In both cases, the theorem follows from Lemma 3.3.

Let $n \geq 2$. If there is no node of multiplicity n, then $\tilde{M} = d/dx(w_0^{-1}M)$ is a monospline of class $(n - 1 + 1, k)$ and has at most $n - 1 + \sum(m_i + \sigma_i) + 1$ zeros. Moreover, if the number of zeros is maximal, then $\tilde{M}(\beta) \geq 0$. By Rolle's theorem, we get the statements on the zeros.

Let $n \geq 2$ and assume that t_j is a node of multiplicity $m_j = n$. Notice that $N - n - 1$ is an even number. By induction on k, we get preliminary bounds of $n + 1 + \sum_{i<j}(m_i + \sigma_i)$ for the number of zeros in $[\alpha, t_j]$, and of $n + 1 + \sum_{i>j}(m_i + \sigma_i)$ for the interval $[t_j, \beta]$. Moreover, the first bound can be reduced by one if $(-1)^{n+1}M(\alpha) < 0$, and the latter one if $M(\beta) < 0$. We distinguish three cases.

Case 1. n is odd. Noting that $m_j + \sigma_j = n + 1$, by adding the bounds for the number of zeros in $[\alpha, t_j]$ and $[t_j, \beta]$ we obtain the stated bound for the whole interval.

Case 2. n is even and $M(t_j) = 0$. Since the zero at t_j is counted in both subintervals, the total number is the sum reduced by one. As $m_j + \sigma_j = n$, we get the statement of the theorem.

Case 3. n is even and $M(t_j)$ is positive (resp. negative). From (3.9), we know that the preliminary bound for $[t_j, \beta]$ (resp. for $[\alpha, t_j]$) can be reduced by one. Again we get the statement of the theorem.

Finally, (3.9) and (3.10) are consequences of the statement for the subintervals. □

Since we may apply (3.8) to subintervals, we get immediately

3.8 Corollary. *Let M be a monospline of class $(n + 1, k)$ with nodes $t_1 < t_2 < \cdots < t_l$ of multiplicity $m_1, m_2, \ldots, m_l$. Put $k(i) = \sum_{j\leq i}(m_i + \sigma_i)$. If M has $N = n + k(l) + 1$ zeros $x_1 \leq x_2 \leq \cdots \leq x_N$, then*

$$x_{k(i)} < t_i < x_{n+k(i-1)+1} \quad \textit{for } i = 1, 2, \ldots, l. \tag{3.11}$$

From Theorem 3.4, we know that best monosplines with simple nodes have the form $f - s$ with $S \in S_{n,k}^{+}$, i.e., the coefficients a_i are positive. More generally, the leading signs of monosplines with nodes of odd multiplicity are given by (3.10), whenever the number of zeros is maximal. A compactness result for these monosplines will be important in the sequel.

3.9 Lemma. *Let $m_1, m_2, \ldots, m_l$ be positive odd numbers and assume that* (3.3) *holds. Let A be the subset of monosplines with l nodes of given multiplicities m_i and with $N := n + \sum_{i=1}^{l}(m_i + 1) + 1$ zeros. Then A is relatively compact in the set of monosplines of class $(n + 1, k)$, where $k = \sum_{i=1}^{l} m_i$.*

Proof. Let $0 \leq j \leq n - 1$. From the proof of the fundamental theorem, we see that each maximal interval $[t_p, t_q] \subset [\alpha, \beta]$ in which L_jM is continuous contains a zero of L_jM. Since $L_nM = L_nf$ is bounded in $[\alpha, \beta]$, we conclude by induction that $L_{n-1}M, L_{n-2}M, \ldots, L_0 M$ are bounded independently of the location of the nodes for all $M \in \Lambda$. In particular, the jumps at the nodes and therefore the coefficients in the representation (1.5) for the spline part are bounded. Hence, from each sequence one can choose a subsequence for which the parameters converge. This implies convergence of the monosplines to a limit function, which is again a monospline. □

3.10 Remark. Let $0 < \delta < (\beta - \alpha)/(N + 3)$ and

$$\Lambda_\delta = \{M \in \Lambda\colon M \text{ has } N \text{ distinct zeros } x_i \ (i = 1, 2, \ldots, N) \text{ such that}$$
$$x_{i+1} - x_i \geq \delta \quad \text{for } i = 0, 1, \ldots, N\}$$

where $x_0 := \alpha$, $x_{N+1} := \beta$. Then Λ_δ is compact.

Indeed, by Lemma 3.9, each sequence in Λ_δ contains a subsequence which converges to a monospline with N zeros such that between two of them there is always an interval of length δ.

On the other hand, if we do not restrict ourselves to Λ_δ for some $\delta > 0$, the limit of a sequence of monosplines from the set mentioned in Lemma 3.9 need not belong to that set. If, for example, $n + 2$ zeros tend to some $z \in (\alpha, \beta)$, then z will not be a zero of multiplicity $n + 2$.

With this we are ready to prove the

3.11 Fundamental Theorem of Algebra for Monosplines (Part 2) (Micchelli 1972). *Assume that f satisfies* (3.3) *and let $m_1, m_2, \ldots, m_l$ be odd integers with $m_i \leq n$, and let N be defined by* (3.7). *Then for any set of N nodes $\alpha < x_1 < \cdots < x_N < \beta$, there is exactly one monospline $M = f - s$ with nodes of multiplicity m_1, $m_2, \ldots, m_l$ which vanishes at x_i, $i = 1, 2, \ldots, N$.*

Proof. (1) Let

$$\Lambda_0 = \bigcup_{\delta > 0} \Lambda_\delta$$

denote the set of monosplines with N distinct zeros in (α, β). Referring to (1.5), the spline portion of a monospline is specified by a parameter vector $a \in \mathbb{R}^N$. The set A of those parameters which are sent to an $M \subset \Lambda_0$ is obviously open in n-space.

Let $M \in \Lambda_0$ have the zeros $x_1 < x_2 \cdots < x_N$ and the nodes $t_1, t_2, \ldots, t_l$. By taking derivatives we see that the tangent space $T_M \Lambda_0$ consists of the spline functions with nodes $t_1, t_2, \ldots, t_l$ of multiplicities $(1 + m_1), \ldots, (1 + m_l)$, respectively. The dimension of $T_M \Lambda_0$ is N. By (3.11), $\{x_i\}_1^N$ satisfies the interlacing properties. Therefore given $y_1, y_2, \ldots, y_N$, there is a unique $h \in T_M \Lambda_0$ with

$$h(x_i) = y_i, \qquad i = 1, 2, \ldots, N.$$

Hence, the Jacobian

$$\left(\frac{\partial M(x_i)}{\partial a_k}\right)_{i,k=1}^N \tag{3.12}$$

is not singular. Consider the mapping which sends the parameter to the set of zeros. Since all zeros are simple, we have

$$\frac{\partial x_i}{\partial a_k} = -\frac{1}{M'(x_i)}\frac{\partial M(x_i)}{\partial a_k}.$$

Therefore, the Jacobian $(\partial x_i/\partial a_k)_{i,k=1}^N$ is obtained by multiplying (3.12) by a diagonal matrix, and is thus non-singular. By the Inverse Function Theorem, the mapping $A \to \mathbb{R}^n$ which sends the parameter a to the set of zeros of the associated monosplines is a local homeomorphism.

(2) Let $\mathbf{x} = (x_1, \ldots, x_N)$ denote a collection of N distinct zeros. From the compactness of Λ_δ and part (1), it follows that there are only a finite number of monosplines in Λ_δ which vanish at $x_1, x_2, \ldots, x_N$. We claim that this number is independent of $\mathbf{x}$.

Let $\mathbf{z} = (z_1, \ldots, z_N)$ be a second collection of zeros. Set

$$\delta_0 = \min_{0 \le i \le N} \min\{x_{i+1} - x_i, z_{i+1} - z_i\}$$

where $x_0 := z_0 := \alpha$, $x_{N+1} := z_{N+1} := \beta$. Let Ω be the subset of those parameters $a \in A$ which correspond to monosplines in $\bigcup\{\Lambda_\delta: \delta > \delta_0/2\}$. Let $\mathbf{x}_t = t\mathbf{x} + (1-t)\mathbf{z} \in \mathbb{R}^N$ for $0 \le t \le 1$. Given a monospline M_0 which vanishes at z_1, z_2, ..., z_n, then by the path lifting lemma, (Lemma IV.1.7), there is a unique one-parameter family $(M_t)_{0 \le t \le 1}$ in Λ_0 such that $M_t(x_{t,i}) = 0$ for $1 \le i \le N$. With this, a $1-1$-correspondence between the monosplines with zeros at $\mathbf{x}$ and at $\mathbf{z}$ is established.

(3) By induction on l, we will show that Λ_0 is not empty. There is nothing to prove for $l = 0$ and so we assume that $\Lambda_0 \ne \emptyset$ holds for $l - 1$.

Let $p = N - (m_l + 1)$. Given $x_1 < x_2 < \cdots < x_N$, then by the induction hypothesis, there exists a monospline $\underline{M}$ with $l - 1$ nodes of the prescribed multiplicities which satisfies $\underline{M}(x_i) = 0$, $i = 1, 2, \ldots, p$. We add another node in (x_p, x_{p+1}), and adjust the parameters a_{lj} so that

$$M_0 = \underline{M} - \sum_{j=1}^{m_l} a_{lj}\Phi_{n+1-j}(t_l, .)$$

satisfies $M_0(x_i) = 0$, $i = p+1, p+2, \ldots, N-1$. This is possible because there is a V-polynomial u with $u(t_l) = u'(t_l) = \cdots = u^{(n-m_l+1)}(t_l) = 0$ which interpolates Lagrangian data at $x_{p+1}, x_{p+2}, \ldots, x_{N-1}$. If M_0 has a total of N zeros, then we are done. Therefore the interesting case is when M_0 has only $N - 1$ zeros, i.e. when $M_0(x_N) < 0$.

Here we consider the subset

$$\begin{aligned}\Lambda_\delta^- = \{&M \textit{ is a monospline with } l \text{ nodes of multiplicity } m_1, m_2, \ldots, m_l,\\ &0 \ge M(x_N) \ge \delta^{-1} \text{ and } M(x_i) = 0 \quad \text{for } i = 1, 2, \ldots, N-1\\ &\text{such that } x_{i+1} - x_i \ge \delta \quad \text{for } i = 0, 1, \ldots, N-1\}.\end{aligned}$$

For any $\delta > 0$, the compactness of Λ_δ^- and the interlacing property (3.11) can be shown as for Λ_δ. By applying the path lifting theorem, we get a curve M_λ such that

$$M_\lambda(x_i) = (1 - \lambda)M_0(x_i) \quad \text{for } i = 1, 2, \ldots, N \text{ and } 0 \le \lambda \le 1. \tag{3.13}$$

Hence, there is a monospline with the prescribed zeros.

(4) In view of (2), it is sufficient to prove uniqueness for one particular set of N zeros. Choose a set of N zeros and construct M_0 as above. After replacing one knot from $\{x_i\}_{i=p+1}^N$, we may assume that $M_0(x_N) \le 0$. By the induction hypothesis, M_λ $(0 \le \lambda \le 1)$, is unique whenever $t_l \ge x_p$. Since each monospline can be connected to one with $t_l \ge x_p$ by a path in the sense of (3.13), uniqueness is established. For details we refer to Micchelli (1972). □

D. Monosplines with Multiple Nodes of Least L_∞-Norm

As we saw in the last section, the set of monosplines with nodes of prescribed multiplicities contains a connected N-dimensional manifold, namely the subset of monosplines with N distinct zeros. Because of its nice properties, the use of global methods is possible. In particular, in this way the uniqueness theorem 3.4 can be extended to monosplines with multiple nodes.

3.12 Theorem (Barrar and Loeb 1974). *Assume that f satisfies* (3.3). *Let m_1, $m_2, \ldots, m_l$ be odd integers which do not exceed n. Then there is a unique spline function s among those with l nodes $t_1 < t_2 < \cdots < t_l$ of multiplicity $m_1, m_2, \ldots, m_l$ for which the monospline $f - s$ has the least uniform norm.*

The proof proceeds in several steps. An important step is the proof that only functions with the maximal number of zeros are candidates for the monosplines of least L_∞-norm. [Exercise IV.3.10 shows us that we must be careful.]

3.13 Proposition. *There is a best monospline in the sense of Theorem* 3.12 *and each best one has an alternant of length $N + 1$.*

Proof. First we admit that two or more nodes coalesce, with multiplicities counted appropriately. Then we get the existence of a best spline approximation u to f from this extended family W, as in the proof of Proposition 2.2. We distinguish several cases.

Case 1. Let u have exactly l distinct nodes $t_1, \ldots, t_l$ with exact multiplicities $m_1, m_2, \ldots, m_l$. Then $T_u W$ consists of the splines with nodes $t_1, \ldots, t_l$ of multiplicities $m_1 + 1, m_2 + 1, \ldots, m_l + 1$. From critical point theory and Theorem 1.6, we know that $f - u$ has an alternant of length $n + 2 + \sum_{i=p+1}^{q-1} (m_i + 1)$ in some interval $[t_p, t_q]$. If $t_p > \alpha$ or $t_q < \beta$, then at most one point of the alternant is not a boundary point of the domain. Therefore $(f - u)'$ has $n + 1 + \sum_{i=p+1}^{q-1} (m_i + 1)$ zeros in $[t_p, t_q]$. Since $(f - u)'$ is a monospline, this contradicts the fundamental theorem 3.7. Therefore $t_p = \alpha$, $t_q = \beta$ and M has an alternant of length $N + 1$. We note that M has the maximal number of zeros.

Case 2. Let u have l distinct nodes $t_1, t_2, \ldots, t_l$ and a node, say t_j, of multiplicity $\mu_j < m_j$. Then, when computing $T_u W$ we have to leave away the derivative $\partial u/\partial t_j$ and the dimension of $T_u W$ is reduced by 1. On the other hand, by the fundamental theorem the maximal number of zeros is reduced by two at least. Thus M cannot have the required alternant and u is not optimal. Therefore Case 2 is excluded.

Case 3. Let $t_i = t_{i-1} = \cdots = t_{i-r}$ for some $r > 0$, and let the accumulated multiplicity m of this node satisfy $m_i < m \leq n - 1$. Consider the corresponding partial sum in the representation of the spline function

$$\sum_{j=0}^{m-1} a_{ij} \Phi_{n+1-j}(t_i, .). \tag{3.14}$$

Choose a parametrization with m_i nodes of the form $t + \alpha_1 \sqrt{\delta}$, while the other $m - m_i$ ones are located at $t - \alpha_2 \sqrt{\delta}$ with $\delta \geq 0$. Here, the factors $\alpha_1 > 0$ and $\alpha_2 > 0$ are determined by specializing (VII.4.5) to the case of two nodes. When the tangent cone is computed, the partial sum (3.14) generates the terms

$$\sum_{j=0}^{m+1} \delta_{ij} \Phi_{n+1-j}(t_i, .) \tag{3.15}$$

with $\delta_{ij} \in \mathbb{R}$ for $j = 0, 1, \ldots, m$ and

$$\delta_{i,m+1} \cdot a_{i,m-1} \geq 0. \tag{3.16}$$

Assume that u is a critical point. From Case 2 we know that m must be odd in order to have an alternant of the appropriate length. Moreover, (3.10) implies that $a_{i,m-1} > 0$. The usual counting of signs shows that the restriction (3.16) does not prevent the construction of a better approximation in the tangent cone $C_u W$. Hence Case 3 is also excluded.

Case 4. If multiplicities of nodes greater than or equal to n occur, a splitting of the nodes as in Section 2 may be performed without a deterioration of the approximation, (see Exercise 2.15). Since this splitting enlarges the dimension of the tangent space but not the length of the alternant, we again have a contradiction.

Summing up, only Case 1 is possible. This is the statement of the proposition. □

Proof of Theorem 3.12. Given $M \in \Lambda_0$, let $z_1 < z_2 < \cdots < z_N$ be its zeros and

$$\lambda_i(M) := \max_{z_i \leq x \leq z_{i+1}} |M(x)| \quad \text{for } i = 0, 1, \ldots, N. \tag{3.17}$$

Here, the usual convention $z_0 = \alpha$ and $z_{N+1} = \beta$ is adopted.

We consider the mapping $G: \Lambda_0 \to \mathbb{R}^N$,

$$G(M) = \left(\log\frac{\lambda_0}{\lambda_1}, \log\frac{\lambda_1}{\lambda_2}, \ldots, \log\frac{\lambda_{N-1}}{\lambda_N} \right),$$

where $\lambda_i = \lambda_i(M)$ are defined as in (3.17). Critical points for the approximation problem under consideration are characterized by $G(M) = 0$. We will show that G is a homeomorphism. Then there is a unique M with $G(M) = 0$.

Let $F: A \to \Lambda_0$ be the parametrization as in the proof of the second part of the fundamental theorem. Let $x_0 < x_1 < \cdots < x_N$ be the points of M for which $\lambda_i(M) = (-1)^{N-i}M(x_i)$ holds. Since we have $x_0 = \alpha$, $x_N = \beta$, and $M'(x_i) = 0$ for $1 \le i \le N-1$, it follows that

$$\frac{\partial \lambda_i}{\partial a_k} = (-1)^{N-i}\frac{\partial M(x_i)}{\partial a_k}, \qquad i = 0, 1, \ldots, N.$$

Suppose that the Jacobian of $G \circ F$ is singular at $a = F^{-1}(M)$. Then there are real numbers $c_1, c_2, \ldots, c_N$ which are not all zero such that

$$\sum_{k=1}^{N} c_k \frac{\partial}{\partial a_k} \log \frac{\lambda_{i-1}}{\lambda_i} = 0 \quad \text{for } i = 1, 2, \ldots, N,$$

which in turn is

$$\sum_{k=1}^{N} c_k \left[\frac{1}{M(x_{i-1})}\frac{\partial M(x_{i-1})}{\partial a_k} - \frac{1}{M(x_i)}\frac{\partial M(x_i)}{\partial a_k}\right] = 0. \tag{3.18}$$

Note that $h := \sum_k c_k \frac{\partial F}{\partial a_k} \in S_{n,N-n}$ together with (3.18) implies that

$$\frac{h(x_{i-1})}{M(x_{i-1})} = \frac{h(x_i)}{M(x_i)} \quad \text{for } i = 1, 2, \ldots, N.$$

Thus h has N sign changes like M, which is impossible.

Hence the Jacobian of G is not singular, and G is a local homeomorphism.

Next, define

$$\Delta_N[\alpha, \beta] := \{z \in \mathbb{R}^N: \alpha < z_1 < z_2 < \cdots < z_N < \beta\}. \tag{3.19}$$

By the fundamental theorem, the mapping $H: \Lambda_0 \to \Delta_N[\alpha, \beta]$, which sends M to its zeros, is a homeomorphism.

The resulting map $G \circ H^{-1}: \Delta_N[\alpha, \beta] \to \mathbb{R}^N$ is a local homeomorphism. Moreover, $\mathbf{z} \to \partial\Delta_N[\alpha, \beta]$ corresponds to the fact that the distance between two zeros becomes close. Then since M' is bounded in the compact set Λ_0, it follows that $\min\{\lambda_i\} \to 0$ and $\max\{\lambda_i\} \ge \operatorname{dist}(f, S_{nk})$ implies that $\|G(M)\| \to \infty$. By Theorem IV.1.8, $G \circ H^{-1}$ is even a global homeomorphism and $G(M) = 0$ is true for exactly one monospline. □

E. Perfect Splines and Generalized Monosplines

A polynomial spline function f in $S_{n+1,k}$ for which $|f^{(n+1)}| = 1$ a.e. is called a *perfect* spline. The nodes of f are the points where $f^{(n+1)}$ changes its sign (Tikhomirov 1969, and Cavaretta 1975).

This is a special case of a Chebyshevian perfect spline. Let $w \in C[\alpha, \beta]$ be a positive weight function, and let L_n be the differential operator given by (1.6). A function $f \in C^n[\alpha, \beta]$ whose $(n+1)$-st derivative exists and is continuous except at k points is called a *perfect spline* if $|L_n f(x)| = w(x)$. The points at which $L_n f$ changes its sign are called the nodes.

Perfect splines and monosplines may be considered as special cases of generalized monosplines (Braess and Dyn 1982): Let $m_1, m_2, \ldots, m_l$ be nonnegative numbers. Then

$$f - s$$

is a *generalized monospline* if $L_n f$ is piecewise continuous,

$$\operatorname{sgn} L_n f(x) = \operatorname{sgn} \prod_{i=1}^{l} (t_i - x)^{m_i}, \tag{3.20}$$

and s is a spline function with nodes $t_1 < t_2 < \cdots < t_l$ of multiplicity $m_1 - 1$, $m_2 - 1, \ldots,$ and $m_l - 1$.

Note that $\operatorname{sgn} L_n f(x) > 0$ if all m_i are even and that on the other hand $f - s = f$ is a perfect spline if $m_i = 1$ for $i = 1, 2, \ldots, l$. Generally, the sign of $L_n f$ depends on the parity of the multiplicities of the nodes. It is not a coincidence that the sign function in (3.20) coincides with the sign function (I.4.5) from the theory of generalized Gaussian quadrature formulas.

We note that the generalized monosplines of least L_∞-norm are unique. The fundamental theorem of algebra may be shown for them by generalizing the proofs of the theorems 3.7 and 3.11. An alternative proof makes use of the theory of generalized Gaussian quadrature formulas: The interpolation problem for generalized monosplines is equivalent to an L_1-approximation problem if the multiplicities of the nodes and the multiplicities of the zeros are interchanged (Micchelli and Pinkus 1977, see also Barrar and Loeb 1980). This equivalence will be described for a slightly different concept in the next section.

We note that this relation between the fundamental theorem and quadrature formulas should not be confused with the correspondence between optimal quadrature formulas and monosplines of least norm (see Nikolskii 1974, Korneichuk 1984).

Exercise

3.14. Construct periodic monosplines of the form $M_n = x^n - s$ with nodes at $\pm 1, \pm 3, \pm 5, \ldots$ such that

$$M_n' = nM_{n-1} + a_n \quad \text{and} \quad M_n(-x) = (-1)^n M(x)$$

for $-1 \leq x \leq +1$, see Example 3.6.

§4. Monosplines of Least L_1-Norm

Monosplines of least L_1-norm are not always unique, so the situation is different from approximation in the supremum norm. Uniqueness was first shown by Jetter and Lange (1978) and Jetter (1978) for the one-sided and for the unrestricted

L_1-approximation of polynomial monosplines of the form $t^{n+1} - s(t)$. The counterexamples show that uniqueness is related to the uniformity of the generating measure.

For convenience, we will elaborate the theory in the framework of monosplines for γ-polynomials. In this theory, one avoids the technical complications which arise in the treatment of spline functions since the associated kernels are only totally positive and not extended totally positive.

The theory of generalized Gaussian quadrature formulas which was discussed in Chapter I, §4, leads to a duality of nodes and zeros of monosplines. The L_1-approximation problem for monosplines with a given kernel is equivalent to an interpolation problem for monosplines with the dual kernel.

In this framework, the kernels are usually denoted by $K(t, x)$ and not by $\gamma(t, x)$, though there is indeed no difference with γ-polynomials. We assume that the reader is familar with the notation and the results from Chapter VII, §1 and Chapter I, §4.

A. Examples of Nonuniqueness

Let $n, k \geq 1$ and $1 \leq p < \infty$. It is our aim to construct a function $f \in C^\infty[\alpha, \beta]$ with $f^{(n+1)}$ being positive, such that f has at least two best L_p-approximations in the set $S_{n,k}$ of polynomial monosplines (Nürnberger and Braess 1982). The essential tool will again be a symmetry argument. Therefore, we assume that the underlying interval is $[\alpha, \beta] = [-2, +2]$.

As a first step, we construct a function $f \in S_{n,k+1}^+$ with at least two nearest points from $S_{n,k}$. We choose a function $s_0 \in S_{n,k-1}^+$ with $k-1$ distinct nodes in $(-1, +1)$ such that

$$s_0(-x) = (-1)^{n+1} s_0(x), \qquad 0 \leq x \leq 2.$$

Let $d = d(s_0|_{[-1,+1]}, S_{n,k-2})$ if $k \geq 2$ and $d = 1$ otherwise. Obviously, given $0 < \delta < 1/2$, the function

$$h(x) = \beta[(x - 2 + \delta)_+^n + (-1)^{n+1}(-x - 2 + \delta)_+^n] \tag{4.1}$$

can be normalized according to $\|h\|_p = d$ by choosing $\beta > 0$ appropriately. Note that $\tilde{f} = s_0 + h$ has the same symmetry behaviour as s_0.

We claim that each best L_p-approximation to $\tilde{f}$ from $S_{n,k}$ has exactly $k-1$ nodes in $(-1, +1)$ and exactly one node in $(-2, -1) \cup (1, 2)$, whenever δ is sufficiently small. To prove this, we consider

$$s_1(x) := s_0(x) + \beta(x - 2 + \delta)_+^n.$$

Then $s_1 \in S_{n,k}$ and

$$\|\tilde{f} - s_1\|_p^p = \int_{-2}^{-2+\delta} |\tilde{f} - s_1|^p \, dx = \int_{-2}^{0} |h|^p \, dx = \frac{1}{2} \|h\|_p^p.$$

Hence,

$$\|\tilde{f} - s_1\|_p = 2^{-1/p} d < d. \tag{4.2}$$

All norms on the finite dimensional space Π_n are equivalent. Hence, for some $c > 0$,

$$\|u\|_{\infty,[1,2]} \leq c\|u\|_{p,[1,3/2]} \text{ for } u \in \Pi_n.$$

From this and $\int_{2-\delta}^{2} |u|^p \, dx \leq \delta \|u\|_{\infty,[1,2]}^p$, we obtain

$$\|u\|_{p,[2-\delta,2]} \leq c \cdot \delta^{1/p} \|u\|_{p,[1,2-\delta]} \text{ for } u \in \Pi_n.$$

Assume that $s \in S_{n,k}$ has fewer than $k-1$ nodes in $(-1,+1)$. Then

$$\|\tilde{f} - s\|_p \geq \|s_0 - s\|_{p,[-1,+1]} \geq d_{L_p[-1,+1]}(s_0, S_{n,k-2}) = d > \|\tilde{f} - s_1\|_p.$$

Therefore, in this case s is not a nearest point in $S_{n,k}$.

Next assume that $s \in S_{n,k}$ has no node in $(1,2)$. Then the restriction of the function $u = s - s_0$ to $[1,2]$ is a polynomial. We consider two cases:

1. If $\|u\|_{p,[1,2-\delta]} \geq 3^{-1/p} d$, it follows that

$$\|\tilde{f} - s\|_{p,[1,2-\delta]} = \|u\|_{p,[1,2-\delta]} \geq 3^{-1/p} d.$$

2. If $\|u\|_{p,[1,2-\delta]} < 3^{-1/p} d$, then $\|u\|_{p,[2-\delta,2]} < c\delta^{1/p} 3^{-1/p} d$, and recalling that $\tilde{f} - s = h - u$, one has

$$\begin{aligned} \|\tilde{f} - s\|_{p,[0,2]} &\geq \|h\|_{p,[2-\delta,2]} - \|u\|_{p,[2-\delta,2]} \\ &\geq [2^{-1/p} - c(\delta/3)^{1/p}] d > 3^{-1/p} d, \end{aligned}$$

provided that δ is sufficiently small. In the same way, one obtains $\|\tilde{f} - s\|_{p,[-2,0]} \geq 3^{-1/p} d$ if s has no node in $(-2,-1)$. Therefore, for every spline function without nodes in $(-2,-1) \cup (1.2)$ it follows that $\|\tilde{f} - s\|_p \geq 2 \cdot 3^{-1/p} d > 2^{-1/p} d = \|\tilde{f} - s_1\|_p$.

Consequently, each best L_p-approximation s to $\tilde{f}$ from $S_{n,k}$ has exactly $k-1$ nodes in $(-1,+1)$ and exactly one node in $(-2,-1) \cup (1,2)$. Setting $\tilde{s}(x) = (-1)^{n+1} s(-x)$, we obtain another spline function $\tilde{s} \neq s$, which by the invariance principle is also a best approximation.

Finally, we note that $\tilde{g} := \tilde{f}^{(n)}$ is a piecewise constant nondecreasing function such that $\tilde{g}(-x) = -\tilde{g}(x)$. We may choose a C^∞-function g, $g'(x) > 0$, with the same symmetry as $\tilde{g}$ and with $\int_{-2}^{+2} |\tilde{g} - g| \, dx$ arbitrarily small. Using this, we get a C^∞-function f with $f^{(n)} = g$, $f(-x) = (-1)^{n+1} f(x)$ and $\|f - \tilde{f}\|_p$ arbitrarily small. Then the nodes of a best approximation to f are distributed as for $\tilde{f}$, and there are again at least two best approximations from $S_{n,k}$.

B. Duality

Let $K(t,x)$ be an extended totally positive kernel on $[t^-, t^+] \times [x^-, x^+]$. The domain for (interior) nodes is $\Delta_m[t^-, t^+]$, cf. (3.19). Given multiplicities $\mathbf{v} = (v_0, v_1, \ldots, v_m, v_{m+1})$ and nodes $\mathbf{t} \in \Delta_m[t^-, t^+]$, $t_0 = t^-$, $t_{m+1} = t^+$, the linear space $U_{\mathbf{t},\mathbf{v}}$ consists of the K-polynomials

$$\sum_{i=0}^{m+1}\sum_{j=0}^{\nu_i-1} a_{ij}K_j(t_i,.)$$

with $K_j = (\partial/\partial t)^j K$. Moreover, let $d\mu$ be a nonnegative measure. In this framework of extended total positivity, (generalized) monosplines are functions of the form $M = f - u$ where

$$f(x) = f_{\mathbf{t},\mathbf{v}}(x) = \int_{t^-}^{t^+} K(t,x)\sigma_{\mathbf{t},\mathbf{v}}(t)\,d\mu(t) \tag{4.3}$$

and $u \in U_{\mathbf{t},\mathbf{v}}$. The sign function $\sigma(t) = \sigma_{\mathbf{t},\mathbf{v}}(t)$, whose definition is found in (I.4.5), depends on the multiplicities of the nodes on the left hand side of t. We recall that the case where $\nu_1, \nu_2, \ldots, \nu_m$ are even (and hence $\sigma_\nu \geq 0$) corresponds to the classical monosplines.

It follows from Remark VII.1.11 that each monospline M has at most $N = \sum_i \nu_i$ zeros. If N is even and M has exactly $N/2$ double zeros, then M does not change its sign. More generally, arbitrary multiplicities may be prescribed in the form of a multiplicity vector $\boldsymbol{\omega} = (\omega_0, \omega_1, \ldots, \omega_n, \omega_{n+1})$, where ω_0 and ω_{n+1} refer to zeros at the end points. Corresponding subsets of K-polynomials $U_{\mathbf{t},\mathbf{v}}(f,\boldsymbol{\omega})$ are defined by (I.4.17). Naturally, the cases $\boldsymbol{\omega} = \mathbf{1} := (0,1,1,\ldots 1,0)$ and $\boldsymbol{\omega} = \mathbf{2} := 2\cdot\mathbf{1}$ are of particular interest.

The counterexamples in the preceding section with more than one solution refer to measures of which a large proportion is concentrated close to both endpoints of the interval $[t^-, t^+]$. Therefore uniqueness can only be expected if the measure is uniform or if a sufficiently large amount has its support in the middle of the interval. Indeed uniqueness has been proven by Dyn (1986) for logarithmically concave measures, i.e., if

$$d\mu(t) = w(t)\,dt \quad \text{with } w(t) > 0, \log w \text{ concave.} \tag{4.4}$$

In addition, assumptions on the kernel are necessary, otherwise one could transfer weights from the measure $d\mu$ to the kernel function because the change $K(t,x) \to \tilde{K}(t,x) := K(t,x)w(t)$ does not affect total positivity. It will not violate (4.4) to match the assumption that K is generated by a *Polya frequency function*, i.e.

$$K(t,x) = g(t-x) \tag{4.5}$$

where g is defined on an appropriate real domain.

4.1 Uniqueness Theorem. *Assume that $K(t,x)$ is extended totally positive on $[t^-,t^+] \times [x^-,x^+]$ and stems from a Polya frequency function. Let $m, n \geq 1$ and $\mathbf{v}$ and $\boldsymbol{\omega}$ be multiplicities such that*

$$\sum_{i=0}^{m+1} \nu_i = N = \sum_{l=0}^{n+1} \omega_l. \tag{4.6}$$

Furthermore, assume that μ and ρ are logarithmically concave measures on $[t^-, t^+]$ and $[x^-, x^+]$ respectively. Then among all monosplines of the form $f_{\mathbf{t},\mathbf{v}} - u$,

$u \in U_{\mathbf{t},\mathbf{v}}(f_{\mathbf{t},\mathbf{v}}, \boldsymbol{\omega})$ with $\mathbf{t} \in \Delta_m[t^-, t^+]$ there is a unique one with least $L_{1,\rho}$-norm. Moreover, the best one is of the form $f_{\mathbf{t},\mathbf{v}} - u$ with $u \in U_{\mathbf{t},\mathbf{v}-\mathbf{1}}$.

We will prove the uniqueness only for the Lebesgue measures $d\mu(t) = dt$ and $d\rho(x) = dx$ (see Braess and Dyn 1983). For the more general theorem, the reader is referred to Dyn (1986).

First we want to describe the basic idea of the proof and we demonstrate why the simplest case already leads to a framework with multiplicities as formulated in the theorem. Given f, let

$$f - \sum_{i=1}^{m} a_i K(t_i, .)$$

be a monospline with least $L_{1,\rho}$-norm. Computing derivatives with respect to the parameters a_i and t_i, we conclude from critical point theory that we have a best approximation in $U_{\mathbf{t},\mathbf{2}}$, i.e. where the nodes occur twice and the error curve has $n = N = 2m$ simple nodes. For this reason, we set $\mathbf{v} = \mathbf{2}$ and $\boldsymbol{\omega} = \mathbf{1}$.

The basic idea goes back to Jetter and Lange (1978). Let $u \in U_{\mathbf{t},\mathbf{1}}$ be a candidate for the spline part of a best monospline. The approximation will not deteriorate if u is replaced by its best approximation u_1 from the linear space $U_{\mathbf{t},\mathbf{2}}$. Then u_1 interpolates f at the canonical points $\mathbf{x} = \{x_l\}_{l=1}^{N}$ for the space $U_{\mathbf{t},\mathbf{2}}$, see Theorem I.4.5. The associated multiplicity vector is $\boldsymbol{\omega} = \mathbf{1}$. The optimality of u_1 implies that

$$\int_{x^-}^{x^+} \operatorname{sgn}(f - u_1) v \, d\rho(x) = 0 \quad \text{for } v \in U_{\mathbf{t},\mathbf{2}}. \tag{4.7}$$

After introducing the *Rodriguez function*

$$f^*(t) := \int_{x^-}^{x^+} K(t, x) \operatorname{sgn}(f - u_1) \, d\rho(x), \tag{4.8}$$

we rewrite (4.7) as

$$f^*(t_i) = f^{*\prime}(t_i) = 0, \qquad i = 1, 2, \ldots, n. \tag{4.9}$$

This means that the approximation problem from which we started is equivalent to a (nonlinear) interpolation problem. We are looking for $\mathbf{x}$ such that the perfect spline f^* with nodes $\{x_l\}$ solves (4.9). By Descartes' rule, f^* has no more than the specified $N = 2m$ zeros. Hence $f^*(t) \geq 0$. From this (cf. Remark VII.1.11), (4.7) and Fubini's theorem, we obtain in the spirit of (I.4.21)

$$\begin{aligned} \|f - u_1\|_{1,\rho} &= \int \operatorname{sgn}(f - u_1)(f - u_1) \, d\rho(x) \\ &= \int \operatorname{sgn}(f - u_1) f \, d\rho(x) \\ &= \int \operatorname{sgn}(f - u_1) K(t, x) \, d\mu(t) \, d\rho(x) \\ &= \int f^*(t) \, d\mu(t) = \|f^*\|_{1,\mu}. \end{aligned} \tag{4.10}$$

Next, we return from the consideration of f^* to a problem again involving f. Given f^* with nodes $\{x_i\}$, or for short, given $\mathbf{x} \in \Delta_N$, let

$$\tilde{u} = \sum_{i=1}^{m} \tilde{a}_i K(\tilde{t}_i, .)$$

be the spline function which interpolates f at $x_1, x_2, \ldots, x_N$. Then $f - \tilde{u}$ has no more sign changes than specified, and $\operatorname{sgn}(f - \tilde{u}) = \operatorname{sgn}(f - u_1)$. From (4.10) and Fubini's theorem, we obtain in the spirit of (I.4.22)

$$\begin{aligned} \|f - \tilde{u}\|_{1,\rho} &= \int \operatorname{sgn}(f - u_1)[f - \tilde{u}]\, d\rho \\ &= \int \operatorname{sgn}(f - u_1)\left[f - \sum_{i=1}^{m} \tilde{a}_i K(\tilde{t}_i, x)\right] d\rho(x) \qquad (4.11) \\ &= \|f^*\|_{1,\mu} - \sum_{i=1}^{m} \tilde{a}_i f^*(\tilde{t}_i). \end{aligned}$$

Since $f^*(t) \geq 0$, $\tilde{a}_i \geq 0$ and $\|f^*\|_{1,\mu} \leq \|f - u\|_{1,\rho}$, it follows that $\tilde{u}$ is at least as good an approximation as u. Moreover, if $\tilde{t}_i \neq t_i$ for some i, then $\tilde{u}$ is actually a better approximation. Therefore, each best approximation must be a fixed point of the mapping which sends u to $\tilde{u}$. It will be shown later that the corresponding mapping for the nodes is a contraction. Therefore, there is only one fixed point and a unique best approximation.

The improvement has been constructed in two steps. To make the duality more obvious, let us introduce the adjoint kernel

$$K^*(x, t) = K(t, x)$$

on $[x^-, x^+] \times [t^-, t^+]$. The linear spaces $U^*_{\mathbf{x},\boldsymbol{\omega}}$ of K^*-polynomials are defined analogously to $U_{\mathbf{t},\mathbf{v}}$. We recall that the first step consisted of an interpolation of f^*, while the second one was an interpolation of f. Likewise, in the first step canonical points for $U_{\mathbf{t},2}$ were required, while in the second step canonical points for $U^*_{\mathbf{x},1}$ were involved.

Finally, we note that $f - u$ is a classical monospline, while f^* in the adjoint problem is a perfect spline. It was the unified treatment of both cases which has led to the concept of generalized monosplines.

C. The Improvement Operator

The idea of successive improvement can be formulated for generalized monosplines in such a way that it is no longer necessary to distinguish between the given problem and the adjoint one.

Let $\mathbf{v} = (v_0, v_1, \ldots, v_{m+1})$ and $\boldsymbol{\omega} = (\omega_0, \omega_1, \ldots, \omega_{n+1})$ be multiplicities such that (4.6) holds. Furthermore, let K be an extended totally positive kernel on $[t^-, t^+] \times [x^-, x^+]$. Given $\mathbf{t} \in \Delta_m[t^-, t^+]$, the N-dimensional Haar subspace $U_{\mathbf{t},\mathbf{v}} \subset L_{1,\rho}$ is spanned by an ET-system. By Theorem I.4.1, there is a set $\mathbf{x} \in$

$\Delta_n[x^-, x^+]$ of canonical points with multiplicities $\omega_0, \omega_1, \ldots, \omega_{n+1}$ for $U_{\mathbf{t},\mathbf{v}}$. With this, a mapping

$$T = T(K, \mathbf{v}, \boldsymbol{\omega}, \rho): \Delta_m[t^-, t^+] \to \Delta_n[x^-, x^+] \tag{4.12}$$

is defined. The corresponding weight factors $\mathbf{a} = (a_{ij})$ will be denoted by $T_a(K, \mathbf{v}, \boldsymbol{\omega}, \rho)\mathbf{t}$.

The improvement nature of the mapping T will now be described.

4.2 Lemma. *Assume that K is ETP on $[t^-, t^+] \times [x^-, x^+]$. Let $\mathbf{v}$ and $\boldsymbol{\omega}$ satisfy* (4.6). *For $\mathbf{t} \in \Delta_n[t^-, t^+]$ let $\mathbf{x} = T(K, \mathbf{v}, \boldsymbol{\omega}, \rho)\mathbf{t}$ and*

$$f_{\mathbf{t},\mathbf{v}} = \int K(t, x)\sigma_{\mathbf{t},\mathbf{v}}(t)\, d\mu(t), \tag{4.13a}$$

$$f^*_{\mathbf{x},\boldsymbol{\omega}} = \int K^*(x, t)\sigma_{\mathbf{x},\boldsymbol{\omega}}(x)\, d\rho(x). \tag{4.13b}$$

Then

$$d_{L_{1,\rho}}(f_{\mathbf{t},\mathbf{v}}, U_{\mathbf{t},\mathbf{v}}(f_{\mathbf{t},\mathbf{v}}, \boldsymbol{\omega})) = \|f^*_{\mathbf{x},\boldsymbol{\omega}} - v^*\|_{1,\mu} \geq d_{L_{1,\mu}}(f^*_{\mathbf{x},\boldsymbol{\omega}}, U^*_{\mathbf{x},\boldsymbol{\omega}}(f^*_{\mathbf{x},\boldsymbol{\omega}}, \mathbf{v})), \tag{4.14}$$

where $v^* \in U^*_{\mathbf{x},\boldsymbol{\omega}-1}$ satisfies the interpolation conditions:

$$(f^*_{\mathbf{x},\boldsymbol{\omega}} - v^*)^{(j)}(t_i) = 0, j = 0, 1, \ldots, v_i, i = 0, 1, \ldots, n + 1. \tag{4.15}$$

Moreover, equality in (4.14) *holds only if* $\mathbf{t} = T(K^*, \boldsymbol{\omega}, \mathbf{v}, \mu)\mathbf{x}$.

Proof. Let $\hat{u}$ be the best $L_{1,\rho}$-approximation to $f_{\mathbf{t},\mathbf{v}}$ from $U_{\mathbf{t},\mathbf{v}}(f_{\mathbf{t},\mathbf{v}}, \boldsymbol{\omega})$. By Descartes' rule, $f_{\mathbf{t},\mathbf{v}}$ is in the convexity cone of $U_{\mathbf{t},\mathbf{v}}$ and $\hat{u}$ interpolates at the canonical points with multiplicities $\omega_0, \omega_1, \ldots, \omega_{n+1}$. Moreover, $f_{\mathbf{t},\mathbf{v}}$-$\hat{u}$ has no more zeros than specified. With the associated generalized Gaussian quadrature formula, we obtain

$$\begin{aligned}
\|f_{\mathbf{t},\mathbf{v}} - \hat{u}\|_{1,\rho} &= \int (f_{\mathbf{t},\mathbf{v}} - \hat{u})\sigma_{\mathbf{x},\boldsymbol{\omega}}\, d\rho(x) \\
&= \int f_{\mathbf{t},\mathbf{v}}\sigma_{\mathbf{x},\boldsymbol{\omega}}\, d\rho - \sum_{l=0}^{n+1} \sum_{j=0}^{\omega_l - 1} a_{lj}\hat{u}^{(j)}(x_l) \\
&= \int f_{\mathbf{t},\mathbf{v}}\sigma_{\mathbf{x},\boldsymbol{\omega}}\, d\rho - \sum_{l=0}^{n+1} \sum_{j=0}^{\omega_l - 1} a_{lj} f^{(j)}_{\mathbf{t},\mathbf{v}}(x_l).
\end{aligned} \tag{4.16}$$

Set $v^*(t) = \sum_{lj} a_{lj}(\partial/\partial x)^j K^*(x_l, t)$. We note that the integration rule

$$\int u\sigma_{\mathbf{x},\boldsymbol{\omega}}\, d\rho = \sum_{l=0}^{n+1} \sum_{j=0}^{\omega_l - 1} a_{lj} u^{(j)}(x_l) \text{ for } u \in U_{\mathbf{t},\mathbf{v}}$$

is equivalent to the interpolation condition (4.15) for the adjoint monospline.

In particular, $f^*_{\mathbf{x},\boldsymbol{\omega}} - v^*$ has no other zeros than those specified. Recalling the integral representation for $f_{\mathbf{t},\mathbf{v}}$ and applying Fubini's theorem, (4.16) becomes

$$\int \left\{ \int K(t,x)\sigma_{\mathbf{x},\boldsymbol{\omega}}\,d\rho - v^*(t) \right\} \sigma_{\mathbf{t},\mathbf{v}}\,d\mu = \int \{ f^*_{\mathbf{x},\boldsymbol{\omega}} - v^* \} \sigma_{\mathbf{t},\mathbf{v}}\,d\mu = \| f^*_{\mathbf{x},\boldsymbol{\omega}} - v^* \|_{1,\mu}.$$

Therefore we have $\|f_{\mathbf{t},\mathbf{v}} - \hat{u}\|_{1,\rho} = \|f^*_{\mathbf{x},\boldsymbol{\omega}} - v^*\|_{1,\mu}$. Finally by Theorem I.4.5, v^* is optimal in $U^*_{\mathbf{x},\boldsymbol{\omega}}(f^*_{\mathbf{x},\boldsymbol{\omega}}, \mathbf{v})$ if and only if it interpolates at the canonical points of $U_{\mathbf{x},\boldsymbol{\omega}}$, i.e. if and only if $\mathbf{t} = T(K^*, \boldsymbol{\omega}, \mathbf{v}, \mu)\mathbf{x}$. □

The contractive nature of the mapping T is established by the next lemma.

4.3 Lemma. *Let the conditions of the preceding lemma prevail. Then for* $\mathbf{x} = T(K, \mathbf{v}, \boldsymbol{\omega}, \rho)\mathbf{t}$,

$$\partial x_l/\partial t_i > 0 \quad \text{for } l = 1, 2, \ldots, n, i = 1, 2, \ldots, m. \tag{4.17}$$

If moreover $K(t,x) = g(t-x)$ *and* $d\rho(x) = dx$, *then*

$$\sum_{i=1}^{m} \frac{\partial x_l}{\partial t_i} < 1 \quad \text{for } l = 1, 2, \ldots, n. \tag{4.18}$$

The proof will be given in the next section.

Proof of Theorem 4.1. Given $t \in \Delta_m[t^-, t^+]$, let $\mathbf{x} = T(K, \mathbf{v}, \boldsymbol{\omega}, \rho)\mathbf{t}$ and $\boldsymbol{\tau} = T(K^*, \mathbf{v}, \boldsymbol{\omega}, \mu)\mathbf{x}$. By applying Lemma 4.2 twice, we get

$$d(f_{\mathbf{t},\mathbf{v}}, U_{\mathbf{t},\mathbf{v}}(f_{\mathbf{t},\mathbf{v}}, \boldsymbol{\omega})) \geq d(f_{\boldsymbol{\tau},\mathbf{v}}, U_{\boldsymbol{\tau},\mathbf{v}}(f_{\boldsymbol{\tau},\mathbf{v}}, \boldsymbol{\omega}))$$

with equality only if $\boldsymbol{\tau} = \mathbf{t}$. Therefore the set of nodes $\mathbf{t}$ of a generalized monospline with least $L_{1,\rho}$-norm is a fixed point of

$$T(K^*, \boldsymbol{\omega}, \mathbf{v}, \mu) \circ T(K, \mathbf{v}, \boldsymbol{\omega}, \rho).$$

By Lemma 4.3, the product map is a contractive mapping from Δ_m into itself. Hence there is at most one fixed point and at most one solution.

In settling the existence problem, we must again consider the optimization problem on the closure $\overline{\Delta_m}$. Assume that there is a solution in $\partial\Delta_m$. Let $\hat{\mathbf{t}}$ denote the set of distinct nodes and $\hat{\mathbf{v}}$ the corresponding multiplicities. Then the double step $T(K^*, \boldsymbol{\omega}, \mathbf{v}, \mu) \circ T(K, \hat{\mathbf{v}}, \boldsymbol{\omega}, \rho)\hat{t}$ would lead to a better approximation and a contradiction. Thus there is a unique solution with m distinct nodes in (t^-, t^+).

Finally, from the construction of v^* in the proof of Lemma 4.2, we know that the spline portion is obtained from the weights of a Gaussian quadrature formula and belongs to some space $U_{\mathbf{t},\mathbf{v}-\mathbf{1}}$. □

We note that Jetter and Lange (1978), Jetter (1978) and Strauß (1979) needed extra arguments for polynomial monosplines: The kernel $K(t,x) = (x-t)^n_+$ is generated by a Polya frequency function but it is only totally positive and not strictly totally positive. Therefore the improvement operators are only non-expansive in general, that is

$$\sum_l \frac{\partial x_i}{\partial t_l} \leq 1, \quad \text{for } i = 1, 2, \ldots, n.$$

Fortunately, in the neighborhood of any solution the kernel behaves like a strictly and even extended totally positive kernel, since for any monospline with a maximal number of nodes the interlacing conditions are satisfied.

D. Proof of Lemma 4.3

Let $\mathbf{x} = T(K, \mathbf{v}, \boldsymbol{\omega}, \rho)\mathbf{t}$ and $\mathbf{a} = T_a(K, \mathbf{v}, \boldsymbol{\omega}, \rho)\mathbf{t}$. Define the adjoint generalized monospline $M^* = M^*_{\mathbf{x},\boldsymbol{\omega}} = f^* - v^*$ with f^* and v^* as in Lemma 4.2.

$$M^*(t) = \int K^*(x,t)\sigma_{\mathbf{x},\boldsymbol{\omega}}(x)\,d\rho(x) - \sum_{l=0}^{n+1}\sum_{j=0}^{\omega_l-1}{}' a_{lj}K_j^*(x_l,t). \tag{4.19}$$

The summation symbol with the prime indicates that the terms for $j = \omega_l - 1$, $1 \le l \le n$ vanish. Since all zeros of M^* are specified by (4.15) and K^* is ETP, the sign of M^* is known from Descartes' rule of signs. The normalization of the sign in definition (I.4.5) implies that

$$\operatorname{sgn} M^*(t) = (-1)^{\nu_0+\omega_0}\sigma_{\mathbf{t},\mathbf{v}}(t). \tag{4.20}$$

The N equations (4.15) define $\mathbf{x} = (x_1, x_2, \ldots, x_n)$ and the coefficient vector $\mathbf{a}$ implicitly as a function of $\mathbf{t}$:

$$F_{rs}(\mathbf{x},\mathbf{a},\mathbf{t}) := M^{*(s)}(t_r) = 0,\ s = 0, 1, \ldots, \nu_r - 1,\ r = 0, 1, \ldots, m+1. \tag{4.21}$$

The Jacobian of (4.21) regarded as equation in $\mathbf{x}$ and $\mathbf{a}$ is nonsingular, since K is ETP and

$$\begin{aligned}\frac{\partial F_{rs}}{\partial a_{lj}} &= -\frac{\partial^s}{\partial t^s}K_j^*(x_l,t_r),\\ \frac{\partial F_{rs}}{\partial x_l} &= -\frac{\partial^s}{\partial t^s}\sum_{j=1}^{\omega_l-1} a_{l,j-1}K_j^*(x_l,t_r).\end{aligned} \tag{4.22}$$

Here we have adopted the abbreviation $a_{l,-1}$ of (I.4.14).

Now, fix $i \le m$ and consider the following nonsingular system of N equations for $\partial x_l/\partial t_i$, $i = 1, 2, \ldots, m$, and for $\partial a_{lj}/\partial t_i$:

$$\sum_{l=1}^{m}\frac{\partial F_{rs}}{\partial x_l}\frac{\partial x_l}{\partial t_i} + \sum_{l=0}^{m+1}\sum_{j=0}^{\omega_l-1}{}'\frac{\partial F_{rs}}{\partial a_{lj}}\frac{\partial a_{lj}}{\partial t_i} = -\frac{\partial F_{rs}}{\partial t_i}. \tag{4.23}$$

Using (4.22), we may rewrite the left-hand side of (4.23) in terms of the following K^*-polynomial:

$$\begin{aligned}v_i &:= \sum_{l=0}^{m+1}\sum_{j=0}^{\omega_l-1}{}'\frac{\partial a_{lj}}{\partial t_i}K_j^*(x_l,.) + \sum_{l=1}^{m}\frac{\partial x_l}{\partial t_i}\sum_{j=0}^{\omega_l-1} a_{l,j-1}K_j^*(x_l,.)\\ &=: \sum_{l=1}^{m} a_{l,\omega_l-2}\frac{\partial x_l}{\partial t_i}K^*_{\omega_l-1}(x_l,.) + \sum_{l=0}^{m+1}\sum_{j=0}^{\omega_l-1}{}'\beta_{lj}K_j^*(x_l,.).\end{aligned} \tag{4.24}$$

Note that $\partial F_{rs}/\partial t_i$ is easily obtained from (4.21) and the system (4.23) becomes

$$v_i^{(s)}(t_r) = \frac{\partial F_{rs}}{\partial t_i} = \begin{cases} M^{*(v_i)} & \text{for } s = v_i - 1 \text{ and } r = i, \\ 0 & \text{otherwise.} \end{cases} \tag{4.25}$$

Since $v_i \in U^*_{\mathbf{x},\boldsymbol{\omega}}$, it has no other zeros in $[t^-, t^+]$ than the $N-1$ ones specified. Hence, v_i has the same zeros with the same multiplicities as M^*, except at t_i where the multiplicity is one less. Moreover, we conclude from (4.25) that v_i and M^* have the same sign for $t > t_i$. This, together with (4.20) implies that $\operatorname{sgn} v_i(t) = \operatorname{sgn} M^*(t) = (-1)^{N-v_{n+1}-\omega_a}$ for $t_m < t < t^+$ and

$$\operatorname{sgn} v_i^{(v_{m+1})}(t^+) = (-1)^{N-\omega_0}. \tag{4.26}$$

By applying Descartes' rule VII.1.10 for K^*-polynomials, the signs of the leading terms in (4.24), i.e. the signs of the coefficients in the first sum, may be determined. Since v_i is an extended K^*-polynomial of order N with the maximal number $N-1$ of zeros, the signs of the vector of generalized signs alternate. Moreover, in view of (4.26) the last component is $(-1)^{N-\omega_0}$. By counting backwards and recalling (4.6), we get

$$\operatorname{sgn}\left[a_{l,\omega_l-2}\frac{\partial x_l}{\partial t_i}\right] = (-1)^{\omega_1+\omega_2+\cdots+\omega_l}, \qquad l = 1, 2, \ldots, \text{n}.$$

From Descartes' rule for generalized monosplines (or from Lemma I.4.4), we know that a_{l,ω_l-2} has the same sign. Hence (4.17) holds.

In order to verify the second claim in the lemma consider the auxiliary function

$$u := \frac{d}{dt}M^* - \sum_{i=1}^{n} v_i. \tag{4.27}$$

If $K(t,x) = g(t-x)$ and $d\rho(x) = dx$, then by putting $\psi(t) = \int^t g(\tau)\,d\tau$ we have $\int_{t_i}^{t_{i+1}} K^*(x,t)\,d\rho(x) = \psi(t_{i+1}-x) - \psi(t_i - x)$ and

$$\frac{d}{dt}K_j^*(x,t) = -\frac{d}{dx}K_j^*(x,t) = -K_{j+1}^*(x,t).$$

With this, the derivative of M^* can be calculated from (4.19):

$$\frac{d}{dt}M^* = \sum_{l=0}^{n+1} a_{l,-1}K^*(x_l, .) + \sum_{l=0}^{n+1}\sum_{j=1}^{\omega_l}{}' a_{l,j-1}K_j^*(x_l, .),$$

where $a_{l,-1}$ for $l = 1, 2, \ldots, n$, are as in (I.4.14) and

$$a_{0,-1} = +1,\; a_{n+1,-1} = -\sigma_{\mathbf{x},\boldsymbol{\omega}}(x^+) = (-1)^{N-\omega_0-\omega_{n+1}}. \tag{4.28}$$

Hence u is a K^*-polynomial with nodes $x_0, x_1, \ldots, x_n, x_{n+1}$ of multiplicities $\omega_0 + 1, \omega_1, \ldots, \omega_n, \omega_{n+1} + 1$:

$$\begin{aligned} u = \sum_{l=1}^{n} a_{l,\omega_l-2}\left[1 - \sum_{i=1}^{m} \frac{\partial x_l}{\partial t_i}\right] K^*_{\omega_l-1}(x_l, .) \\ + a_{0,\omega_0-1} K^*_{\omega_0}(x^-, .) + a_{n+1,\omega_{n+1}-1} K^*_{\omega_{n+1}}(x^+, .) \\ + \sum_{l=0}^{n+1} \sum_{j=0}^{\omega_l-1}{}' \gamma_{lj} K^*_l(x_l, .). \end{aligned} \tag{4.29}$$

Specifically, u is a K^*-polynomial of order $N + 2$. In view of (4.25) the function u has N zeros counting multiplicities. By Descartes' rule VII.1.10, there are at least N sign changes in the sequence of the generalized signs $\{s_1, s_2, \ldots, s_{N+2}\}$ of u. In order to determine the signs, we recall that we know the signs of the leading coefficients in (4.19). In particular

$$\begin{aligned} \operatorname{sgn} a_{0,\omega_0-1} &= +1 && \text{if } \omega_0 > 0, \\ \operatorname{sgn} a_{n+1,\omega_{n+1}-1} &= (-1)^{\omega_1+\omega_2+\cdots+\omega_{n+1}} && \text{if } \omega_{n+1} > 0, \end{aligned}$$

In view of (4.28) these relations are also true if $\omega_0 = 0$ or $\omega_{n+1} = 0$. Since x^- is a node of multiplicity $\omega_0 + 1$ for u, it follows that $s_{N+2} = (-1)^{N-\omega_0} = (-1)^{N-1} s_1$. Hence there are $N + 1$ (mod 2) sign changes in the sequence. Since the number of sign changes is at least N, it equals $N + 1$ and

$$s_r = (-1)^{\omega_0+1+r}, \qquad r = 1, 2, \ldots, N + 2.$$

For the signs of the leading terms in (4.29) we conclude that

$$\operatorname{sgn}\left\{a_{l,\omega_l-2}\left[1 - \sum_{i=1}^{m} \frac{\partial x_l}{\partial t_i}\right]\right\} = (-1)^{\omega_1+\omega_2+\cdots+\omega_l}, \qquad l = 1, 2, \ldots n.$$

Substituting the signs of a_{l,ω_l-2}, which we have already used in the first part of the proof, we get (4.18). □

§ 5. Monosplines of Least L_p-Norm

In 1979, Bojanov showed the uniqueness of the polynomial monospline $x^{n+1} - s_{n,k}(x)$ of least L_p-norm by using degree theory. Later, the uniqueness of monosplines for extended totally positive (ETP) kernels was investigated in a more general context by Braess and Dyn (1986). In both cases, uniqueness is established by showing that there is only one critical point. A different approach is due to Zhensykbaev (1979, 1981). He showed that there is a unique global best approximation among the critical points if the weight function belongs to a certain class of functions.

Here, the L_p-approximation is studied in the framework of monosplines for ETP kernels. Since we have already seen how to treat multiplicities in the last section, we will restrict ourselves to a simple case. In this section, all nodes are assumed to have the same multiplicity $m \geq 1$ and only (generalized) monosplines with simple zeros are considered.

Let $K(t,x)$ be extended totally positive (ETP) on $[t^-,t^+] \times [x^-,x^+]$ and let

$$f_{\mathbf{t}}(x) = \int K(t,x)\sigma_{\mathbf{t}}(t)\,d\mu(t), \tag{5.1}$$

where $\sigma_{\mathbf{t}}(\xi) := \sigma_{\mathbf{t},m\cdot\mathbf{1}}(\xi) = \operatorname{sgn} \Pi_{1\le i\le n}(t_i - \xi)^m$ and μ is a non-negative measure. We consider monosplines of the form

$$M = f_{\mathbf{t}} - u \text{ where } u \in U_{\mathbf{t},m-1}, \tag{5.2}$$

and

$$U_{\mathbf{t},m} := \left\{ u(x) = \sum_{i=1}^{n} \sum_{j=0}^{m-1} a_{ij} K_j(t_i,x) \colon a_{ij} \in \mathbb{R} \right\}.$$

The main idea is as follows. Let $\mathbf{t} \in \Delta_n[t^-,t^+]$. If M is a non-degenerate monospline (i.e., $a_{i,m-2} \neq 0$ for $i = 1,2,\ldots,n$), then the derivatives of M with respect to the parameters $\{a_{ij}\}$ and $\{t_i\}$ span the linear space $U_{\mathbf{t},m}$. From critical point theory we know that a monospline of least norm with respect to the free parameters $\{a_{ij}\}$ and $\{t_i\}$ is also of least norm among all functions of the form

$$f_{\mathbf{t}} - v,\ v \in U_{\mathbf{t},m},$$

with $\mathbf{t}$ being the set of nodes of the monospline under consideration.

Therefore, given $\mathbf{t}$, we may define $\mathbf{a} = \mathbf{a}(\mathbf{t})$ as the set of unique parameters $\{a_{ij}\}$ such that

$$\left\| f_{\mathbf{t}} - \sum_{i=1}^{n} \sum_{j=0}^{m-1} a_{ij}(\mathbf{t}) K_j(t_i,.) \right\| \le \| f_{\mathbf{t}} - v \| \text{ for all } v \in U_{\mathbf{t},m}.$$

In particular, we are interested in the parameters corresponding to the augmented terms and define $b_i(\mathbf{t}) := a_{i,m-1}(\mathbf{t})$, for $1 \le i \le n$. The equivalent problem to the uniqueness of the least-norm monospline is the uniqueness of the solution of the system

$$\Phi(\mathbf{t}) := \mathbf{b}(\mathbf{t}) = \mathbf{0}. \tag{5.3}$$

The components of a solution $\mathbf{t}$ of (5.3) will be denoted as *critical nodes*, since a monospline with critical nodes $\mathbf{t}$ and coefficients $a_{ij}(\mathbf{t})$ is a critical point.

We shall apply degree theory to the nonlinear equation (5.3), and show that the degree of Φ with respect to $\Phi(\mathbf{t}) = \mathbf{0}$ is independent of the measure μ in (5.1) and equals $\Pi_{i=1}^{n}(-1)^{1+im}$ for all ETP kernels. We note that this result applies not only to L_p-norms but to a wider class of norms (cf. Braess and Dyn 1986). The same is true for interpolation problems, and thus the fundamental theorem of algebra for generalized monosplines can also be proved in this framework.

A. The Rodriguez Function for the L_p-Norms

Let $1 < p < \infty$, $d\rho(x)$ be a positive measure, and let $\|f\| := \|f\|_{p,\rho} = (\int |f|^p\,d\rho(x))^{1/p}$. Given a set $\mathbf{t}$ of n nodes, let $v_{\mathbf{t}}$ be the best L_p-approximation to $f_{\mathbf{t}}$ from $U_{\mathbf{t},m}$. The remainder $R := R_{\mathbf{t}} = f_{\mathbf{t}} - v_{\mathbf{t}}$ is a monospline at a critical point $\mathbf{t}$.

The solution of the linear L_p-approximation problem is characterized by the orthogonality relations

$$\int |R|^{p-2} R v \, d\rho(x) = 0 \quad \text{for each } v \in U_{\mathbf{t},m}. \tag{5.4}$$

Since $U_{\mathbf{t},m}$ is a Haar space, the function R has $N = n \cdot m$ simple zeros in $[x^-, x^+]$. Following Schoenberg, the auxiliary function

$$F(y) := \int |R|^{p-2} R K(y, .) \, d\rho(x) \quad \text{for } y \in [t^-, t^+]$$

may be called the *Rodriguez function* for the L_p-problem (Jetter 1978). Obviously, (5.4) can be rewritten in terms of F as

$$F^{(j)}(t_i) = 0 \quad \text{for } i = 1, 2, \ldots, n, j = 0, 1, \ldots, m-1. \tag{5.5}$$

For the study of the Rodriguez function a composite kernel is defined on $[t^-, t^+]^2$:

$$G(t, y) = G(y, t) := \int |R|^{p-2} K(t, .) K(y, .) \, d\rho. \tag{5.6}$$

Here $|R|^{p-2}$ may be considered as a weight function. The following Lemma asserts that G is *ETP* whenever K is.

5.1 Lemma. *Let the kernels K and L be TP(ETP) on $[t^-, t^+] \times [x^-, x^+]$ and $[z^-, z^+] \times [x^-, x^+]$ respectively, and let $d\rho(x)$ be a nonnegative measure which is not concentrated on a finite set. Then the composite kernel*

$$G(t, y) = \int_{x^-}^{x^+} K(t, x) L(z, x) \, d\rho(x)$$

is TP(ETP) on $[t^-, t^+] \times [z^-, z^+]$.

Proof. The essential step is the proof of the following composition rule which is given in the notation of (VII.1.5) and may be viewed as a continuous version of the Cauchy-Binet formula (cf. Karlin 1968), p. 17):

$$G\begin{pmatrix} t_1, t_2, \ldots, t_n \\ z_1, z_2, \ldots, z_n \end{pmatrix} = \int_{x^- \le x_1 \le \cdots \le x_n \le x^+} \cdots \int K\begin{pmatrix} t_1, \ldots, t_n \\ x_1, \ldots, x_n \end{pmatrix} L\begin{pmatrix} z_1, \ldots, z_n \\ x_1, \ldots, x_n \end{pmatrix} d\rho(x_1) \ldots d\rho(x_n). \tag{5.7}$$

To prove this formula, let π denote a permutation of $\{1, 2, \ldots, n\}$. By definition,

$$G(t_1, z_{\pi(1)}) \cdots G(t_n, z_{\pi(n)}) = \int_{x^- \le x_i \le x^+} \cdots \int K(t_1, x_1) \ldots K(t_n, x_n) L(z_{\pi(n)}, x_1) \ldots \times L(z_{\pi(n)}, x_n) \, d\rho(x_1) \ldots d\rho(x_n).$$

By summing with respect to all permutations π according to the definition of determinants, we obtain

$$G\begin{pmatrix} t_1, \dots, t_n \\ z_1, \dots, z_n \end{pmatrix}$$

$$= \int_{x^- \le x_i \le x^+} \cdots \int K(t_1, x_1) \dots K(t_n, x_n) L\begin{pmatrix} z_1, \dots, z_n \\ x_1, \dots, x_n \end{pmatrix} d\rho(x_1) \dots d\rho(x_n).$$

The region of integration can be decomposed into non-overlapping simplices Δ_v characterized by $x_{v^{-1}(1)} < x_{v^{-1}(2)} < \dots < x_{v^{-1}(n)}$, where v is again a permutation of $\{1, 2, \dots, n\}$. Transforming Δ_v by means of the change of variable $x_{v(i)} \mapsto x_i$ for $i = 1, 2, \dots, n$, we obtain for the last integral

$$\int_{x_1 < \dots < x_n} \cdots \int \sum_v (-1)^{|v|} K(t_1, x_{v(1)}) \dots K(t_n, x_{v(n)}) \cdot$$

$$\times L\begin{pmatrix} z_1, \dots, z_n \\ x_1, \dots, x_n \end{pmatrix} d\rho(x_1) \dots d\rho(x_n).$$

The sign factor $(-1)^{|v|}$ results from the permutation of the arguments of $L(:::)$. With this, the formula (5.7) has been proved.

From (5.7), the statement of the lemma for *TP* kernels is obvious. The extensions to the modified determinants (VII.1.8) with coalescing arguments are straightforward. □

Recalling (5.1), (5.6), and $R = f_{\mathbf{t}} - \sum a_{ij} K_j(t_i, .)$, we rewrite the Rodriguez function in terms of the kernel G:

$$F(y) = \int G(t, y) \sigma_{\mathbf{t}}(t) \, d\mu(t) - \sum_{i=1}^{n} \sum_{j=0}^{m-1} a_{ij} G_{j0}(t_i, y), \tag{5.8}$$

with $G_{jl} := (\partial/\partial t)^j (\partial/\partial y)^l G(t, y)$. In particular, if $\mathbf{t}$ is critical, then F is a generalized monospline with kernel G which by (5.7) has $N = nm$ simple zeros $\{x_i\}_{i=1}^N$. From Descartes' rule of signs (Remark VII.1.11) we know that

$$\sigma_i a_{i,m-2} > 0 \quad \text{for } i = 1, 2, \dots, n. \tag{5.9}$$

Here we have adopted the conventions of Lemma I.4.4:

$$\sigma_i := \sigma_{\mathbf{t}}(t_i + 0),$$

$$a_{i,-1} = [\sigma_{\mathbf{t}}(t_i + 0) - \sigma_{\mathbf{t}}(t_i - 0)] w(t_i), \qquad w(t) \, dt = d\mu(t). \tag{5.10}$$

Furthermore, F cannot have more zeros than specified by (5.5). This and the normalization of $\sigma_{\mathbf{t}}$ imply that $F(t) > 0$ for $t < t_1$. Hence,

$$(-1)^{mk} F^{(m)}(t_k) > 0 \quad \text{for } k = 1, 2, \dots, n. \tag{5.11}$$

B. The Degree of the Mapping Φ

For the sake of completeness, we recall some facts from degree theory (see e.g. Amann 1974 or Schwartz 1969).

Let D be an open bounded set in n-space, and let $\Phi\colon \bar{D} \to \mathbb{R}^n$ be continuous. Then, given $\mathbf{c} \in \mathbb{R}^n$ with $\mathbf{c} \notin \Phi(\partial D)$, the *degree* of Φ with respect to D and $\mathbf{c}$ is defined, has an integer value and is denoted by $\deg(\Phi, D, \mathbf{c})$.

Properties of the Degree of a Map.

(i) (*Decomposition in the non-degenerate case*). Suppose that Φ is differentiable at $\mathbf{x}$ and that $\det(\Phi'(\mathbf{x})) \neq 0$ whenever $\mathbf{x} \in D$ and $\Phi(\mathbf{x}) = \mathbf{c}$. Then there exists a finite number of points, say $\mathbf{x}^i \in D$, $i \in I$, where $\Phi(\mathbf{x}^i) = \mathbf{c}$ and

$$\deg(\Phi, D, \mathbf{c}) = \sum_{i \in I} \operatorname{sgn} \det(\Phi'(x^i)).$$

(ii) If $\deg(\Phi, D, \mathbf{c}) \neq 0$, then $\Phi(\mathbf{x}) = \mathbf{c}$ holds for at least one point $\mathbf{x} \in D$.

(iii) (*Invariance under homotopy*). Let $\Phi(\lambda, \mathbf{x})$ be continuous on $[0,1] \times D$. Moreover, suppose that $\Phi(\lambda, \mathbf{x}) \neq \mathbf{c}$ for all $x \in \partial D$, $0 \leq \lambda \leq 1$. Then $\deg(\Phi(\lambda, .), D, \mathbf{c})$ is an integer which is independent of λ.

(iv) (*Multiplication property*). Let $\Phi\colon \bar{D} \to \mathbb{R}^n$, $\Psi\colon \mathbb{R}^n \to \mathbb{R}^n$ be two continuous mappings and let Δ_i be the bounded components of $\mathbb{R}^n \setminus \Phi(\partial D)$. Suppose that $\mathbf{c} \notin \Psi \circ \Phi(\partial D)$. Then

$$\deg(\Psi \circ \Phi, D, \mathbf{c}) = \sum_{\Delta_i} \deg(\Psi, \Delta_i, \mathbf{c}) \deg(\Phi, D, \Delta_i \cap \Psi^{-1}(\mathbf{c})).$$

In order to apply degree theory to Equ. (5.3), we have to choose a set D with compact closure and we must make sure that no solution of $\Phi(\mathbf{t}) = \mathbf{0}$ lies on its boundary. To this end, we establish a compactness property of generalized monosplines for *ETP* kernels which is analogous to Lemma 3.9.

For technical reasons, we will allow the measure μ in (5.1) to depend on $\mathbf{t}$. No complications will arise from this generalization, since on the other hand we assume that the mapping $\Delta_n[t^-, t^+] \to C^N[x^-, x^+]$ defined by

$$\mathbf{t} \mapsto f_{\mathbf{t}} = \int K(t, .)\sigma_{\mathbf{t}}(t)\, d\mu_{\mathbf{t}}(t) \tag{5.12}$$

is continuous. This includes the cases

$$f_{\mathbf{t}}(x) = \int K(t, x)\sigma_{\mathbf{t}}(t) w(t)\, dt, \quad w \in C[t^-, t^+],\ w \geq 0,$$

and

$$f_{\mathbf{t}}(x) = \sum_{i=1}^{n} \left[\prod_{j=1}^{n} (t_j - \tau_i)^m \right] K(t_i, x), \qquad \tau \in \Delta_n[t^-, t^+]. \tag{5.13}$$

Moreover, when applying the invariance of the degree under homotopy, μ will be assumed to be a convex combination of two given measures μ_0 and μ_1 with the above properties.

5.2 Lemma. *Let $K(t,x)$ be ETP on $[t^-, t^+] \times [x^-, x^+]$, let $m, n \geq 1$, $N = mn$ and let μ_0, μ_1 be nonnegative measures such that the map (5.12) is continuous. Then* *the set of generalized monosplines with* $\mu = \lambda\mu_1 + (1-\lambda)\mu_0$, $\lambda \in [0, 1]$ *which have* N *zeros in* $[x^-, x^+]$, *counting multiplicities, is compact.*

Remark. Actually we will prove compactness for the topology induced by the parameters a_{ij}, t_i and λ. This implies compactness in L_p for $1 \le p \le \infty$.

Proof of Lemma 5.2. Assume that $M = f_{\mathbf{t}} - u$ with $u \in U_{\mathbf{t},m-1}$ has N zeros $x_1 \le x_2 \le \cdots \le x_N$. In order to have a suitable auxiliary linear interpolation problem, we also consider u as an element of the N-dimensional space $U_{\mathbf{t},m}$. We choose a basis of $U_{\mathbf{t},m}$:

$$v_r(x) = K(\underbrace{t_1, \ldots, t_n, t_1, \ldots, t_n, \ldots, t_1, \ldots, t_n}_{j-1 \text{ complete groups}}, t_1, \ldots, t_i; x), \tag{5.14}$$

$r = (j-1)n + i$, for $1 \le i \le n$, $1 \le j \le m$. Here, and in (5.15), repeated arguments refer to divided differences. Consider the interpolation problem with N Hermite data. Given $\{x_l\}_{l=1}^N$ and $\mathbf{t}$, determine $u = \sum_{r=1}^N a_r v_r \in U_{\mathbf{t},m}$ such that

$$u(x_1, x_2, \ldots, x_l) = f_{\mathbf{t}}(x_1, x_2, \ldots, x_l) \quad \text{for } l = 1, 2, \ldots, N. \tag{5.15}$$

The matrix of the corresponding linear system for the unknowns $\{a_r\}_{r=1}^N$ is nonsingular, since K is *ETP*. Moreover, the matrix and the right hand side are continuous in $\mathbf{x}$, $\mathbf{t}$ and λ.

Let $(M_k)_{k\in\mathbb{N}}$ be a sequence of monosplines with nodes $\mathbf{t}^{(k)}$, $N = nm$ zeros $\mathbf{x}^{(k)}$ and measure parameters $\lambda^{(k)}$. After passing to a subsequence if necessary, we may assume that $\lim_{k\to\infty} t^{(k)} = t^*$, $\lim_{k\to\infty} \mathbf{x}^{(k)} = x^*$, and $\lim_{k\to\infty} \lambda^{(k)} = \lambda^*$. We know from the preceding discussion that the interpolation problems (5.15) with $\mathbf{t} = \mathbf{t}^*$, $\mathbf{x} = \mathbf{x}^*$, $\lambda = \lambda^*$, or with $\mathbf{t} = \mathbf{t}^{(k)}$, $\mathbf{x} = x^{(k)}$, $\lambda = \lambda^{(k)}$, respectively, are solvable.

Since (M_k) is a sequence of monosplines, we have $a_r^{(k)} = 0$ for $r > N - m$. From the continuity arguments above, it follows that the same is true for the limit case and that the limit function u^* interpolates $f_{\mathbf{t}}^*$. In particular, $f_{\mathbf{t}^*} - u^*$ has $N = nm$ zeros counting multiplicities. By Descartes' rule this is only possible if $t^- < t_1^* < t_2^* < \cdots < t_n^* < t^+$. Therefore, $f_{\mathbf{t}^*} - u^*$ is a monospline. Moreover, we may abandon the basis (5.14) with divided differences and return to the natural basis. □

Since the compactness result refers to the topology given by the parameters, we have

5.3 Corollary. *Let the conditions of Lemma 5.2 prevail. Then there is an* $\varepsilon > 0$ *which depends only on* K, n, m, μ_0 *and* μ_1 *such that the nodes* $t_1 < t_2 < \cdots < t_n$ *of each monospline of the form* (5.2) *with* nm *zeros satisfy*

$$t_{i+1} - t_i \ge 2\varepsilon \quad \textit{for } i = 0, 1, \ldots, n.$$

By the corollary, all critical nodes are contained in some

$$\Delta_{n,\varepsilon} := \{\mathbf{t} \in \Delta_n[t^-, t^+]: t_{i+1} - t_i > \varepsilon \text{ for } i = 0, 1, \ldots, n\}$$

with $t_0 = t^-$, $t_{n+1} = t^+$. Therefore, we set

$$\deg(\Phi, \Delta_n, \mathbf{0}) := \lim_{\varepsilon \to 0} \deg(\Phi, \Delta_{n,\varepsilon}, \mathbf{0}).$$

At this moment we should note that the mapping Φ is continuous. In fact, since $\mathbf{a}(\mathbf{t})$ is the parameter vector of a strict best approximation in $U_{\mathbf{t},m}$, the continuity follows by the same arguments as in the proofs of the continuity lemma III.1.13 and the perturbation lemma IV.2.8. [Further, we recall (5.10).]

5.4 Proposition. *Let $K(t, x)$ be an ETP kernel and let $n, m \geq 1$. Assume that the mapping* (5.12) $\mathbf{t} \to f_{\mathbf{t}}$ *is continuous. Then for any L_p-approximation problem* $(1 \leq p \leq \infty)$ *the degree of the mapping* (5.3) *is given by*

$$\deg(\Phi, \Delta_n, \mathbf{0}) = \prod_{i=1}^{n} (-\sigma_i). \tag{5.16}$$

Remark. The formula holds for a larger class of norms. Actually, we only make use of the uniqueness of the best approximations in the linear Haar spaces $U_{\mathbf{t},m}$ and of the $N = nm$ zeros of the corresponding error functions.

Proof of the Proposition. Given two measures μ_0 and μ_1 which satisfy the assumptions of the proposition, let $\Phi(\lambda, \mathbf{t})$ be the mapping associated to the measure $\lambda\mu_1 + (1 - \lambda)\mu_0$. By Corollary 5.3, we have $\Phi(\lambda, \mathbf{t}) \neq 0$ for $0 \leq \lambda \leq 1$ and $\mathbf{t} \in \partial\Delta_{n,\varepsilon}$ for ε sufficiently small. The invariance of the degree under homotopy yields $\deg(\Phi(0, .), \Delta_n, \mathbf{0}) = \deg(\Phi(1, .), \Delta_n, \mathbf{0})$. Therefore, the degree is the same for all measures under consideration.

The degree is now determined from a suitable example. Following an idea of Barrow (1978), we choose an atomic measure. Specifically, we fix $\boldsymbol{\tau} \in \Delta_n$ and consider $f_{\mathbf{t}}$ defined by (5.13).

To see that $\mathbf{t} = \boldsymbol{\tau}$ is the only solution of $\Phi(\mathbf{t}) = \mathbf{0}$, observe that the monospline $f_{\mathbf{t}} - u$ with $u \in U_{\mathbf{t},m-1}$ degenerates to a non-zero K-polynomial of order $N = nm$ whenever $\mathbf{t} \neq \boldsymbol{\tau}$. Thus $f_{\mathbf{t}} - u$ has at most $N - 1$ zeros and cannot be a critical point.

Next, we expand $K(\tau_i, .)$ at $\tau_i + h_i$. Using the abbreviation $Q_{\mathbf{t}}(\xi) := \prod_{j=1}^{n} (t_j - \xi)^m$, we obtain for $f_{\boldsymbol{\tau}+\mathbf{h}} = \sum_{i=1}^{n} Q_{\boldsymbol{\tau}+\mathbf{h}}(\tau_i) K(\tau_i, .)$ the expansion

$$\begin{aligned} f_{\boldsymbol{\tau}+\mathbf{h}} &= \sum_{i=1}^{m} \Big\{ [Q_{\boldsymbol{\tau}}^{(m)}(\tau_i) + o(1)](-h_i)^m \frac{1}{m!} \\ &\qquad \times \Big[\sum_{j=0}^{m-1} \frac{(-h_i)^j}{j!} K_j(\tau_i + h_i, .) + o(h_i^{m-1}) \Big] \Big\} \\ &= v_1 - \sum_{i=1}^{n} Q_{\boldsymbol{\tau}}^{(m)}(\tau_i) \frac{h_i^{2m-1}}{m!(m-1)!} K_{m-1}(\tau_i + h_i, .) + g, \end{aligned} \tag{5.17}$$

where $v_1 \in U_{\mathbf{t},m-1}$, $\mathbf{t} = \boldsymbol{\tau} + \mathbf{h}$ and $\|g\| = o(\sum_{i=1}^{n} |h_i|^{2m-1})$. Note that the best approximation Pg to g in $U_{\mathbf{t},m}$ has norm $\|Pg\| \leq 2\|g\|$. Moreover, the best approximation to $f_{\boldsymbol{\tau}+\mathbf{h}}$ is Pg + the K-polynomial—part of (5.17). Hence, we obtain for the leading coefficient of the best approximation to $f_{\boldsymbol{\tau}+\mathbf{h}}$ the formula

$$b_i(\boldsymbol{\tau} + \mathbf{h}) = -Q_{\boldsymbol{\tau}}^{(m)}(\tau_i) h_i^{2m-1}/m!(m-1)! + o(\textstyle\sum |h_i|^{2m-1}).$$

Next, the order preserving map $\Psi: \mathbf{t} \to \tilde{\mathbf{h}}$ is applied, where

$$t_i = \tau_i + h_i \text{ and } \tilde{h}_i = h_i^{2m-1} \quad \text{for } i = 1, 2, \ldots, n.$$

Hence, (5.17) yields the derivatives

$$\frac{\partial b_i}{\partial \tilde{h}_k} = -\delta_{ik} Q_\tau^{(m)}(\tau_i)/m!(m-1)! \quad \text{for } i, k = 1, 2, \ldots, n,$$

and

$$\operatorname{sgn} \det((\Phi \circ \Psi^{-1})') = \operatorname{sgn} \prod_{i=1}^{n} (-Q_\tau^{(m)}(\tau_i)) = \prod_{i=1}^{n} (-\sigma_i).$$

Since Ψ is an order preserving map, we have $\deg(\Psi^{-1}, \Psi(\Delta_n), \mathbf{t}) = +1$ for $\mathbf{t} \in \Delta_n$. The multiplication rule of the degree is simple here, since Ψ is one-one. It yields $\deg(\Phi, \Delta_n, \mathbf{0}) = \deg(\Phi \circ \Psi^{-1}, \Psi(\Delta_n), \mathbf{0}) = \operatorname{sgn} \det((\Phi \circ \Psi^{-1})')$, which completes the proof of (5.16). □

C. The Uniqueness Theorem

From Proposition 5.4 we know that $|\deg(\Phi, \Delta_n, \mathbf{0})| = 1$ holds for the mapping defined in (5.3). If we show that

$$\det(\Phi') = \det\left(\frac{\partial b_i}{\partial t_k}\right)_{i,k=1}^{n}$$

does not vanish and has the same sign at all critical points, then Property (i) of the degree implies uniqueness.

Now we return to L_p-approximation ($1 < p < \infty$). In order to determine the partial derivatives $\partial b_i/\partial t_k$, we rewrite (5.5) as

$$\int |R|^{p-1} \operatorname{sgn} R K_l(t_\nu, .)\, d\rho(x) = 0, \; l = 0, 1, \ldots, m-1, \; \nu = 1, 2, \ldots, n. \tag{5.18}$$

Differentiating with respect to t_k, we obtain

$$\begin{aligned} -(p-1) \int |R|^{p-2} \left\{ \sum_{i=1}^{n} \sum_{j=0}^{m-1} \frac{\partial a_{ij}}{\partial t_k} K_j(t_i, .) + \sum_{j=0}^{m} a_{k,j-1} K_j(t_k, .) \right\} K_l(t_\nu, .)\, d\rho \\ + \delta_{k\nu} \int |R|^{p-2} R K_{l+1}(t_\nu, .)\, d\rho = 0, \end{aligned} \tag{5.19}$$

with $a_{k,-1}$ defined as in (5.10). The second integral equals $F^{(l+1)}(t_\nu)$ and is nonzero only if $l = m-1$. Now we introduce the G-polynomials:

$$\begin{aligned} v_k(y) := \sum_{i=1}^{n} \Bigg\{ \sum_{j=0}^{m-2} \left[\frac{\partial a_{ij}}{\partial t_k} + \delta_{ik} a_{k,j-1} \right] G_{j0}(t_i, y) \\ + \left[\frac{\partial b_i}{\partial t_k} + \delta_{ik} a_{k,m-2} \right] G_{m-1,0}(t_i, y) \Bigg\}. \end{aligned}$$

The first integral in (5.19) is easily expressed in terms of v_k or its derivatives (if $l > 0$), and equs. (5.19) become interpolation conditions:

$$v_k^{(j)}(t_\nu) = \begin{cases} 0 & j = 1, 2, \ldots, m-2, \\ \dfrac{\delta_{k\nu}}{p-1} F^{(m)}(t_k), & j = m-1, \end{cases} \qquad \nu = 1, 2, \ldots, n. \tag{5.20}$$

Recalling (5.10) for the signs σ_i, we have

5.5. Lemma. *Let* $\mathbf{t}^*$ *be critical nodes. Then*

$$\frac{\partial b_i}{\partial t_k}(\mathbf{t}^*) > 0 \quad \textit{for } i \neq k,\ 1 \le i, k \le n.$$

Proof. The function v_k is a G-polynomial of degree nm. Since G is ETP, v_k has no other than the $nm - 1$ zeros given by (5.20). Here as usually zeros are counted with their multiplicities. Also by (5.20) and (5.12),

$$\operatorname{sgn} v_k(t_k + 0) = \operatorname{sgn} F^{(m)}(t_k + 0) = \sigma_k.$$

Therefore, by Descartes' rule of signs the leading coefficients of v_k are known:

$$\operatorname{sgn}\left[\frac{\partial b_i}{\partial t_k} + \delta_{ik} a_{k,m-2}\right] = \sigma_i.$$

In particular, for $i \neq k$ the statement of the lemma is obtained. □

The next step is to determine the sign of $\sum_{k=1}^n \partial b_i/\partial t_k$. To this end, we introduce the auxiliary function

$$u := \frac{1}{p-1} F' - \sum_{k=1}^{n} v_k.$$

By (5.5) and (5.20),

$$u^{(j)}(t_i) = 0, \qquad j = 0, 1, \ldots, m-1, \qquad i = 1, 2, \ldots, n. \tag{5.21}$$

In the following, the method of proof applies only to kernels of the form $K(t,x) = g(t-x)$ and to the Lebesgue measures $d\rho = dx$, $d\mu = dt$. Integration by parts of F' (at a critical node $\mathbf{t}$) yields

$$\begin{aligned} F'(y) &= \int |M|^{p-2} M \frac{\partial}{\partial y} K(y, .)\, dx \\ &= -\int |M|^{p-1} \operatorname{sgn} M \frac{\partial}{\partial x} K(y, .)\, dx \\ &= (p-1) \int |M|^{p-2} M' K(y, .)\, dx \\ &\quad + |M(x^-)|^{p-2} M(x^-) K(y, x^-) - |M(x^+)|^{p-2} M(x^+) K(y, x^+). \end{aligned}$$

The boundary terms in the last expression will be abbreviated by BT. We proceed by writing M' explicitly and again using $(\partial/\partial t)\, K = -(\partial/\partial x)K$:

$$F'(y) = (p-1)\int |M|^{p-2}\left\{\int_{t^-}^{t^+} \sigma_t(t)\frac{\partial}{\partial t}K(t,x)\,dt\right.$$

$$\left. - \sum_{i=1}^{n}\sum_{j=0}^{m-2} a_{ij}K_{j+1}(t_i,x)\right\} K(y,x)\,dx + BT$$

$$= (p-1)\left\{(-1)^{nm}G(t^+,y) - G(t^-,y) - \sum_{i=1}^{n}\sum_{j=0}^{m-1} a_{i,j-1}G_{j,0}(t_i,y)\right\} + BT$$

with $a_{i,-1}$ as defined in (5.10) and $w(x) \equiv 1$. Combining the expressions for F' and v_k, we obtain

$$\begin{aligned} u(y) = \beta^- K(y,x^-) + G(t^-,y) - \sum_{i=1}^{n}\left\{\sum_{j=0}^{m-2}\left[\sum_{k=1}^{n}\frac{\partial a_{ij}}{\partial t_k} + a_{i,j-1}\right]G_{j0}(t_i,y)\right. \\ \left. + \left[\sum_{k=1}^{n}\frac{\partial b_i}{\partial t_k}\right]G_{m-1,0}(t_i,y) - \sigma_n G(t^+,y) - \beta^+ K(y,x^-)\right. \end{aligned} \tag{5.22}$$

with

$$\beta^{\pm} = \frac{1}{p-1}|M(x^{\pm})|^{p-2}M(x^{\pm}).$$

We intend to derive from u and its zeros (5.21) the signs of the leading coefficients. Note that u is not a G-polynomial, since (5.21) also contains terms in $K(t^-,.)$ and $K(t^+,.)$. Fortunately, one may define a modified kernel $\tilde{G}$, which is ETP, such that u is a $\tilde{G}$-polynomial. Specifically, we add formal points $t^{(l)}$ (resp. $t^{(r)}$), which are taken to lie on the left (resp. right) hand side of $[t^-,t^+]$, and define $\tilde{G}$ on the set $(\{t^{(l)}\}\cup[t^-,t^+]\cup\{t^{(r)}\}) \times [t^-,t^+]$ by

$$\tilde{G}(t,y) = \begin{cases} G(t,y) & \text{if } t\in[t^-,t^+] \\ K(y,x^-) & \text{if } t = t^{(l)}, \\ K(y,x^+) & \text{if } t = t^{(r)}. \end{cases}$$

The kernel $\tilde{G}$ is ETP. Indeed, the composition formula

$$\tilde{G}\begin{pmatrix} t^{(l)}, t_1, \ldots, t_n, t^{(r)} \\ y_0, y_1, \ldots, y_n, y_{n+1} \end{pmatrix} = \int_{x^-\le x_1\le\cdots\le x_n\le x^+} \cdots \int K\begin{pmatrix} y_0, \ldots, y_{n+1} \\ x^-, x_1, \ldots, x_n, x^+ \end{pmatrix}$$

$$\times K\begin{pmatrix} t_1, \ldots, t_n \\ x_1, \ldots, x_n \end{pmatrix} d\tilde{\rho}(x_1)\ldots d\tilde{\rho}(x_n)$$

where $d\tilde{\rho} = |M|^{p-2}\,d\rho$, is easily derived in the same way as formula (5.7). Moreover, determinants with one argument $t^{(l)}$ or $t^{(r)}$ are similarly decomposed, and $\tilde{G}$ is ETP.

Now we are in a position to establish.

5.6 Lemma. *Assume that* $K(t,x) = g(t-x)$, $d\mu(t) = dt$ *and* $d\rho(x) = dx$. *Let* $\mathbf{t}^*$ *be critical nodes. Then*

$$\sigma_i \sum_{k=1}^{n} \frac{\partial b_i}{\partial t_k}(\mathbf{t}^*) < 0 \quad (i = 1, 2, \ldots, n). \tag{5.23}$$

Proof. The auxiliary function u is a $\tilde{G}$-polynomial of the form

$$u(y) = \beta^- \tilde{G}(t^{(l)}, y) + \tilde{G}(t^-, y) + \sum_{i=1}^{n} \sum_{j=0}^{m-1} c_{ij} \tilde{G}_{jo}(t_i, y)$$
$$- \sigma_n \tilde{G}(t^+, y) - \beta^+ \tilde{G}(t^{(r)}, y).$$

Now the sign pattern of M with $N = nm$ zeros and extended total positivity imply that $M(t^-) > 0$ and $(-1)^N M(t^+) > 0$. Therefore, $\beta^- > 0$ and $\sigma_n \beta^+ > 0$.

If we introduce the notation

$$\eta_{im} = \operatorname{sgn} c_{i,m-1}, \eta_{ij} = (-1)^{m-j} \eta_{im} (j = 1, 2, \ldots, m-1, i = 1, 2, \ldots, n),$$

the generalized signs VII.1.7 of the $N + 4$ coefficients of the $\tilde{G}$-polynomial u are given by the sequence

$$1, 1, \eta_{11}, \ldots, \eta_{1m}, \eta_{21}, \ldots, \eta_{nm}, -\sigma_n, -\sigma_n. \tag{5.24}$$

From Descartes' rule of signs and the nm zeros of u we conclude that there are at least N sign changes in the sequence (5.24). Observing that the number of sign changes equals $N + 1$ modulo 2, we conclude that this number equals $N + 1$. Therefore, all signs in the sequence are known. In particular, recalling (5.22) we obtain

$$\operatorname{sgn}\left[\sum_{k=1}^{n} \frac{\partial b_i}{\partial t_k}\right] = \operatorname{sgn}(-c_{i,m-1}) = -\eta_{im} = -\sigma_i \quad (i = 1, 2, \ldots, n)$$

which completes the proof. □

A consequence of this equation and Lemma 5.5 is

$$\sigma_i \frac{\partial b_i}{\partial t_i} < 0 \quad \text{for } i = 1, 2, \ldots, n. \tag{5.25}$$

5.7 Uniqueness Theorem (Braess and Dyn 1986). *Let* $K(t,x) = g(t-x)$ *be extended totally positive, and let* $d\mu = dt$ *and* $d\rho = dx$. *Then there is a unique generalized monospline* (5.1) *of least* L_p*-norm whenever* $1 < p < \infty$.

Proof. By Lemma 5.4 we have $\deg(\Phi, \Delta_n, \mathbf{0}) = \prod_{i=1}^{n}(-\sigma_i)$. In view of Property (i) of the degree, it is sufficient to show that at each $\mathbf{t}^* \in \Delta_n$ with $\Phi(\mathbf{t}^*) = \mathbf{0}$:

$$\prod_{i=1}^{n} (-\sigma_i) \det \Phi'(t^*) > 0 \tag{5.26}$$

in order to verify uniqueness.

Now, from the above lemmas and (5.25) we conclude that at each critical point t^*

$$\left|\frac{\partial b_i}{\partial t_i}\right| = -\sigma_i \frac{\partial b_i}{\partial t_i} > \sum_{k \neq i} \sigma_i \frac{\partial b_i}{\partial t_k} = \sum_{k \neq i} \left|\frac{\partial b_i}{\partial t_k}\right| \quad \text{for } i = 1, 2, \ldots, n.$$

Hence $\left(\frac{\partial b_i}{\partial t_k}(\mathbf{t}^*)\right)_{i,k=1}^n$ is a diagonally dominant matrix. By Gerschgorin's theorem (Gerschgorin 1931), the determinant has the same sign as the product of its diagonal elements. Hence, the claim (5.26) follows from (5.25), and the proof is complete. □

The uniqueness theorem can be extended to $d\mu = w_1(t)\,dt$, $d\rho = w_2(x)\,dx$ with w_1, w_2 logarithmically concave. For this case (5.23) can be proved by the method developed by Dyn (1986) for the L_1-problem. A special case with logarithmically concave weight functions was also considered by Barrow, Chui, Smith and Ward (1978).

Finally we note that $|\deg(\Phi, \Delta_n, \mathbf{0})| = 1$ also holds in the case where more than one solution exists. But then sgn det $\Phi'(\mathbf{t})$ is not the same at all critical points.

Appendix. The Conjectures of Bernstein and Erdös

Linear approximation problems often give rise to nonlinear problems, for example the uniqueness of the canonical points for linear L_1-approximation has been treated by methods from nonlinear analysis. A famous problem of the same kind is the uniqueness and the characterization of best points for Lagrangian interpolation. It took almost 50 years until the problem raised by Bernstein could be proved.

Let $n \geq 2$. To each point $\mathbf{t}$ in $\Delta_{n-1} := \Delta_{n-1}(\alpha, \beta) = \{\mathbf{t} \in \mathbb{R}^{n-1}: \alpha < t_1 < t_2 < \cdots < t_{n-1} < \beta\}$, there is a linear map $P := P_\mathbf{t}: C[\alpha, \beta] \to \Pi_n$ of polynomial interpolation at the $n + 1$ knots $\alpha = t_0, t_1, \ldots, t_n = \beta$. Specifically, in its Lagrange form,

$$P_\mathbf{t} f = \sum_{i=0}^{n} f(t_i) l_i$$

with the polynomials

$$l_i(x) = \prod_{j \neq i} \frac{x - t_j}{t_i - t_j} \quad \text{for } 0 \leq i \leq n. \tag{1}$$

One is interested in optimal knots, i.e. in a $\mathbf{t} \in \Delta_{n-1}$ for which $\|P_\mathbf{t}\|$ is minimal. Here, $C[\alpha, \beta]$ is endowed with the uniform norm and as usual $\|P_\mathbf{t}\| = \sup\{\|P_\mathbf{t} f\|: f \in C[\alpha, \beta], \|f\| = 1\}$. The problem is motivated by the fact that $P_\mathbf{t}$ is a projection onto Π_n. Therefore by a well known lemma of Lebesgue, one has $\|f - P_\mathbf{t} f\| \leq (1 + \|P_\mathbf{t}\|) d(f, \Pi_n)$.

Obviously, for any $\|f\| \leq 1$, one has

$$|(P_\mathbf{t} f)(x)| \leq \Lambda_\mathbf{t}(x) := \sum_{i=0}^{n} |l_i(x)|,$$

where Λ_t is called the Lebesgue function of the interpolation process. Consequently,

$$\|P_\mathbf{t}\| = \|\Lambda_\mathbf{t}\|.$$

Let $1 \leq i \leq n$. On the subinterval $[t_{i-1}, t_i]$, the Lebesgue function coincides with the polynomials $p_i = p_i(., \mathbf{t})$ given by

$$p_i(t_k) = \begin{cases} (-1)^{i-k-1} & \text{for } 0 \leq k \leq i-1, \\ (-1)^{i-k} & \text{for } i \leq k \leq n. \end{cases} \tag{2}$$

The polynomial p_i has a unique maximum in (t_{i-1}, t_i), say at z_i. From this, it is natural to consider the numbers

$$\lambda_i(\mathbf{t}) := \max_{t_{i-1} \le x \le t_i} p_i(x) \quad \text{for } 1 \le i \le n.$$

In 1931, Bernstein conjectured that $\|P_t\|$ is minimal if $\Lambda_\mathbf{t}$ equioscillates, i.e. if

$$\lambda_1(\mathbf{t}) = \lambda_2(\mathbf{t}) = \cdots = \lambda_n(\mathbf{t}). \tag{3}$$

Later, Erdös (1957) conjectured that even a stronger result is true.

Conjectures of Bernstein and Erdös. A polynomial projection $P_\mathbf{t}$ is the unique projection of minimal norm $\|P_\mathbf{t}\|$ with $\mathbf{t} \in \Delta_{n-1}(\alpha, \beta)$, if and only if $\Lambda_\mathbf{t}$ equioscillates. Moreover, for any $\mathbf{t} \in \Delta_{n-1}(\alpha, \beta)$ one has

$$\min_i \lambda_i(\mathbf{t}) \le \lambda^* := \inf\{\|P_\mathbf{s}\|, \mathbf{s} \in \Delta_{n-1}\}. \tag{4}$$

The conjecture has been an open problem for many years and many numerical computations like those in Ehlich and Zeller (1966) have substantiated the conjecture. In 1976, Kilgore and Cheney found that, for each $n > 1$ there is a set of knots with an equioscillating Lebesgue function. Finally, in 1978 Kilgore showed that the equioscillation property is a necessary condition. On the basis of this local result and by applying topological arguments, de Boor and Pinkus (1978) completed the proof of the conjectures. Notes on the interesting history of the conjectures, their proof, and related results are found in the cited literature.

To prove the conjecture, we start with some simple properties of the polynomials p_i. From (2), it follows that p_i has $n-1$ zeros in (α, β) and that in addition $(-1)^n p_i(\alpha) \cdot p_i(\beta) < 0$. Therefore p_i has one more zero in $(-\infty, \alpha) \cup (\beta, \infty)$. All zeros of p_i are real and have multiplicity 1. By Rolle's theorem, the same holds for p_i'. There is a unique z_i in (t_{i-1}, t_i) with

$$p_i'(z_i) = 0, \qquad p_i''(z_i) \neq 0.$$

By applying the implicit function theorem to $p_i'(z_i, \mathbf{t}) = 0$, we see that λ_i is a differentiable function of $\mathbf{t}$. Moreover, $p_i'(z_i) = 0$ implies that

$$\frac{\partial \lambda_i}{\partial t_k} = \frac{\partial p_i(z_i, \mathbf{t})}{\partial t_k}.$$

To evaluate the derivative, observe that $p_i(.,\tilde{\mathbf{t}}) - p_i(.,\mathbf{t})$ has the n zeros $\{t_j\}_{0 \le j \le n, j \ne k}$ whenever $\tilde{t}_j = t_j$ for $j \neq k$. Hence, the polynomial is a multiple of l_k. The factor can be determined from the value of the difference at $\tilde{t}_k$, which equals $-p_i'(t_k)(\tilde{t}_k - t_k) + o(|\tilde{t}_k - t_k|)$. This yields

$$\frac{\partial \lambda_i}{\partial t_k} = -p_i'(t_k) l_k(z_i). \tag{5}$$

The crucial point in Kilgore's proof will be the verification that the determinants of the Jacobians

$$J_r = \det\left(\frac{\partial \lambda_i}{\partial t_k}\right)_{\substack{1 \le i \le n, i \ne r, \\ 1 \le k \le n-1}} \quad \text{for } r = 1, 2, \ldots, n, \tag{6}$$

do not vanish. The first step in this direction is the observation (Braess 1973d), that (5) may be rewritten as

$$\frac{\partial \lambda_i}{\partial t_k} = \frac{p_i'(t_k)}{t_k - z_i} \prod_{j=0}^{n} (z_i - t_j) \Big/ \prod_{j \ne k} (t_k - t_j).$$

Since the multiplication of a row or a column of a matrix by a nonzero factor has a well known effect on the determinant, it suffices to prove that

$$\det(q_i(t_k))_{\substack{1 \le i \le n, i \ne r \\ 1 \le k \le n-1}} \ne 0 \quad \text{for } 1 \le r \le n \text{ and } \mathbf{t} \in \Delta_{n-1} \tag{7}$$

with

$$q_i(x) := \frac{p_i'(x)}{x - z_i}.$$

Since $q_i \in \Pi_{n-2}$, the inequalities (7) are in turn equivalent to the linear independence of any $n - 1$ of the n polynomials $q_1, q_2, \ldots, q_n$.

In this context, locations of the zeros of p_i are of interest. Let $y_1^{(i)}, y_2^{(i)}, \ldots, y_{n-1}^{(i)}$ denote those of p_i in $[\alpha, \beta]$

$$y_k^{(i)} \in \begin{cases} (t_{k-1}, t_k) & \text{for } k < i, \\ (t_k, t_{k+1}) & \text{for } k > i. \end{cases}$$

The additional zero in $(-\infty, \alpha)$ or (β, ∞) will be denoted by $y_0^{(i)}$ or $y_n^{(i)}$, respectively.

Lemma 1. *For $i < j$ the zeros of p_i and p_j strictly interlace and $y_k^{(j)} < y_k^{(i)}$ holds for all applicable k in $[0, n]$. Moreover, if p_i has a zero $y_n^{(i)} > \beta$, then p_j has a zero $y_n^{(j)} \in (\beta, y_n^{(i)})$. If p_j has a zero $y_0^{(j)} < \alpha$, then p_i has a zero $y_0^{(i)} \in (y_0^{(j)}, \alpha)$.*

Proof. The function $h := p_i - (-1)^{j-i} p_j$ satisfies

$$h(t_k) = \begin{cases} 0 & \text{for } 0 \le k \le i - 1 \text{ or } j \le k \le n, \\ 2(-1)^{k-i} & \text{for } i \le k \le j - 1. \end{cases}$$

Thus h has at least $j - i - 1$ zeros in $[t_i, t_{j-1}]$ and $n + 1 + i - j$ zeros outside that interval. Since $h \in \Pi_n$, it cannot have any additional zeros and all of them must be simple. From this and the fact that $h(t_i) = 2 > 0$, it follows that $(-1)^{i-k} h > 0$ on (t_{k-1}, t_k) for $k \le i$. (Here, $t_{-1} = -\infty$).

The signs of p_j and $(-1)^{j-i} p_i$ at t_{k-1} are known, as well as the difference in (t_{k-1}, t_k). Referring to Fig. 38, we conclude that

$$t_{k-1} < y_k^{(j)} < y_k^{(i)} < t_k \quad \text{for } k = 1, 2, \ldots, i - 1,$$

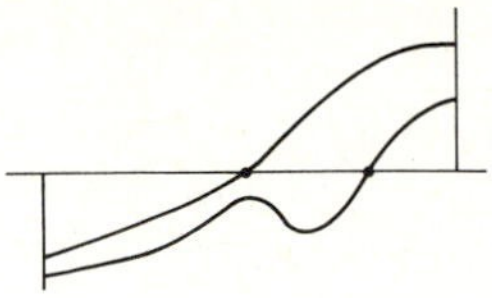

Fig. 38. Location of the zeros of two functions with a difference of constant sign

and

$$y_0^{(j)} < y_0^{(i)} < t_0 \quad \text{if } y_0^{(j)} \text{ exists.}$$

By a symmetry argument, we conclude that

$$t_k < y_k^{(j)} < y_k^{(i)} < t_{k+1} \quad \text{for } k = j, j+1, \ldots, n-1,$$

and also that

$$t_n < y_n^{(j)} < y_n^{(i)} \quad \text{if } y_n^{(i)} \text{ exists.}$$

Finally, the function $g := p_i + (-1)^{j-i} p_j$ satisfies

$$g(t_k) = \begin{cases} 2(-1)^{k-i-1} & \text{for } 0 \le k \le i-1, \\ 0 & \text{for } i \le k \le j-1, \\ 2(-1)^{k-i} & \text{for } j \le k \le n. \end{cases}$$

Thus g has at least $j - i$ zeros in $[t_{i-1}, t_j]$ and $(i-1) + (n-j)$ zeros outside that interval, giving a total of at least $n - 1$ zeros. Since $g(t_{i-1})g(t_j) = 4(-1)^{j-i}$, the number of zeros of g in $[t_{i-1}, t_j]$ must be of parity $j - i$. Hence $g \in \Pi_n$ implies that g has no other zeros in $[t_{i-1}, t_j]$. Thus $(-1)^{k-i} g > 0$ on (t_{k-1}, t_k) for $i \le k \le j$. With the arguments as above, we conclude that

$$t_{k-1} < y_{k-1}^{(i)} < y_k^{(j)} < t_k \quad \text{for } i < k < j.$$

Moreover $y_k^{(j)} < t_k < y_k^{(i)}$ for $i \le k \le j$, and the proof of the lemma is complete. □

Next, we note that p_1 has a zero in $(-\infty, \alpha)$, since p_1' does not change its sign for $x < z_1$. Similarly, p_n has a zero in (β, ∞). Summing up, the zeros of $p_1, p_2, \ldots, p_n$ on $(-\infty, +\infty)$ lie in the pattern

$$y_0^I, \ldots, y_0^{(1)}, y_1^{(n)}, \ldots, y_1^{(1)}, y_2^{(n)}, \ldots, y_{n-1}^{(1)}, y_n^{(n)}, \ldots, y_n^{(I+1)}, \tag{8}$$

where I is an integer with $1 \le I < n$.

The properties are inherited by the derivatives. Let $z_k^{(i)}$ denote the zero of p_i' which lies in $(y_{k-1}^{(i)}, y_k^{(i)})$. In particular, we have $z_i^{(i)} = z_i$. Due to results on polynomials of V.A. Markov (1916), the zeros of p_i' and p_j' interlace in the same manner. The elementary proof, which is based on the same ideas as the proof of Lemma 1, is omitted. The zeros of $p_1', p_2', \ldots, p_n'$ lie in the pattern

$$z_1^{(I)}, \ldots, z_1^{(1)}, z_2^{(n)}, \ldots, z_{n-1}^{(1)}, z_n^{(n)}, \ldots, z_n^{(I+1)},$$

where I is the integer defined with (8). To make the pattern more transparent, we restrict it to the interval $[z_i, z_n]$ and drop the lower indices. Writing i for a zero of p_i' and $\hat{i}$ for the point z_i, one gets

$$\hat{1}, n, n-1, \ldots, 3, \hat{2}, n, n-1, \ldots, \hat{3}, 3, 1, n, \ldots, 3, 2, 1, \hat{n}.$$

From this pattern, we see that $q_i = p_i'/(x - z_i)$ has no zero in $[z_{i-1}, z_{i+1}]$ but that it changes its sign in (z_k, z_{k+1}) whenever $k < i - 1$ or $k > i$. Therefore, after replacing q_i by $(-1)^i q_i$, one has

$$\operatorname{sgn} q_i(z_k) = \begin{cases} (-1)^k & \text{if } i \neq k, \\ (-1)^{k+1} & \text{if } i = k. \end{cases} \tag{9}$$

Now suppose that $\sum_{i=1}^n a_i q_i = 0$ for some $a \neq 0$. Set $P := \{i \in [1, n]: a_i \geq 0\}$ and $N := [1, n] \backslash P$. Consider the function

$$h := \sum_{i \in P} a_i q_i = -\sum_{i \in N} a_i q_i. \tag{10}$$

If $k \notin P$, then from (9) and the first sum in (10), we conclude that $(-1)^k h(z_k) \geq 0$. The same relation is obtained from the second sum in (10), if $k \notin N$. Thus h has $n - 1$ weak sign changes and $h \in \Pi_{n-2}$ implies that $h = 0$. Without loss of generality, we may assume that $N = \varnothing$. Note that

$$a_i q_i(z_i) = -\sum_{j \neq i} a_j q_j(z_i).$$

All the terms in the sum have the same sign, and from $a \neq 0$ we conclude that $a_i \neq 0$. Since this is true for each $i \in [1, n]$, no $n - 1$ of the polynomials $q_1, q_2, \ldots, q_n$ are linearly dependent. This proves (7) and

$$J_r \neq 0 \quad \text{for } 1, 2, \ldots, n. \tag{11}$$

Now we can show by standard arguments that the Lebesgue function for each optimal projection must have the equioscillation property. Assume that $\hat{\mathbf{t}} \in \Delta_{n-1}(\alpha, \beta)$ and that $\Lambda_{\hat{\mathbf{t}}}$ does not equioscillate. Then $\lambda_r(\hat{\mathbf{t}}) < \max_{1 \leq i \leq n} \lambda_i(\hat{\mathbf{t}})$ for some $r \in [1, n]$. Since $J_r(\hat{\mathbf{t}}) \neq 0$, by the inverse function theorem, some open neighborhood U of $\hat{t}$ in Δ_{n-1} is mapped homeomorphically onto an open set in $\mathbb{R}^{n-1}$ by $\mathbf{t} \mapsto (\lambda_i(t))_{1 \leq i \leq n, i \neq r}$. Consequently, $\max \lambda_i(\mathbf{t}) < \max \lambda_i(\hat{\mathbf{t}})$ holds for some $\mathbf{t} \in U \subset \Delta_{n-1}$ and $\|P_{\mathbf{t}}\|$ is not minimal. This proves the equioscillaltion property for minimal projections. □

For the treatment of the other parts of the conjectures, the map

$$\Gamma: \Delta_{n-1}(\alpha, \beta) \to \mathbb{R}^{n-1}, \qquad \mathbf{t} \mapsto (\lambda_{i+1}(\mathbf{t}) - \lambda_i(\mathbf{t}))_{i=1}^{n-1}$$

is considered.

Lemma 2. *If* $\mathbf{t} \to \partial \Delta_{n-1}$, *then* $\|\Gamma(\mathbf{t})\| \to \infty$.

Proof. Given $\mathbf{t} \in \Delta_{n-1}$ let $\delta = \min_j (t_j - t_{j-1}) = t_r - t_{r-1}$. From (1) we conclude that $|l_j(x)| \leq 1$ holds in $[t_{r-1}, t_r]$ for $j = 0, 1, \ldots, n$. Hence, $\lambda_r(\mathbf{t}) \leq n + 1$. An easy calculation yields $|l_i(x)| \geq 2^{-n}n^{-1}(\beta - \alpha)/\delta$ for some x in the largest subinterval (t_{i-1}, t_i). It follows that $\lambda_i(\mathbf{t}) \geq 2^{-n}n^{-1}(\beta - \alpha)/\delta$ for some i. Therefore $\|\Gamma(\mathbf{t})\| \to \infty$ if $\delta \to 0$, i.e. if $\mathbf{t} \to \partial \Delta_{n-1}(\alpha, \beta)$. □

The central part in verifying the conjectures will be the establishing of the properties of Γ, cf. Exercise I.3.20.

Theorem 3. *The map Γ is a homeomorphism of $\Delta_{n-1}(\alpha, \beta)$ onto $\mathbb{R}^{n-1}$. Moreover, Γ and Γ^{-1} are C^1-mappings.*

Proof. In view of Theorem IV.1.8 and the preceding lemma, for the first statement it suffices to prove that Γ is a local homeomorphism. Thus it suffices to show that

$$\det(\partial(\lambda_{i+1} - \lambda_i)/\partial t_k)_{i,k=1}^{n-1} \neq 0 \quad \text{for all } \mathbf{t} \in \Delta_{n-1}. \tag{12}$$

The expansion of the determinant by rows yields

$$\det(\partial(\lambda_{i+1} - \lambda_i)/\partial t_k)_{i,k=1}^{n-1} = \sum_{r=1}^{n} (-1)^r J_r(\mathbf{t}).$$

From (11), it follows that the signs of the determinants $J_r(\mathbf{t})$ are constant in Δ_{n-1}, and it suffices to verify that

$$(-1)^k J_k(\mathbf{t})/J_1(\mathbf{t}) < 0 \quad \text{for } 2 \leq k \leq n \tag{13}$$

holds for some point $\mathbf{t} \in \Delta_{n-1}$.

Let $\mathbf{s} \in \Delta_{n-1}$. Since $J_1(\mathbf{s}) \neq 0$, the vector $\mathbf{t}$ is a unique C^1-function of $(\lambda_i(\mathbf{t}))_{i=2}^n$ in some open neighborhood U of $(\lambda_i(\mathbf{s}))_{i=2}^n$. There is a C^1-function $G: U \to \mathbb{R}$ with

$$\lambda_1(\mathbf{t}) = G(\lambda_2(\mathbf{t}), \lambda_3(\mathbf{t}), \ldots, \lambda_{n-1}(\mathbf{t})).$$

By Cramer's rule (or Exercise 4), $d\lambda_1 = \sum_{k=2}^n (-1)^k (J_k/J_1)\, d\lambda_k$ and

$$\partial\lambda_1/\partial\lambda_k = (-1)^k J_k/J_1 \quad \text{for } k = 2, 3, \ldots, n. \tag{14}$$

Now consider (14) at the point $\mathbf{t}$ at which $\|P_{\mathbf{t}}\|$ is minimal. By Lemma 2, such an optimal $\mathbf{t}$ exists in Δ_{n-1} and from the discussion above it is known that the corresponding Lebesgue function equioscillates. It follows that

$$\partial\lambda_1/\partial\lambda_k < 0 \quad \text{for } k = 2, 3, \ldots, n. \tag{15}$$

Otherwise, the construction of an equally good projection without the oscillation property would be possible. From (14) and (15), one has (13) and Γ is a local and thus a global homeomorphism.

By the inverse function theorem Γ and Γ^{-1} are C^1-mappings. □

From Theorem 3, it follows that there is exactly one $\mathbf{t} \in \Delta_{n-1}$ with $\Gamma(\mathbf{t}) = 0$, i.e. exactly one set $\mathbf{t}$ of for which $\Lambda_{\mathbf{t}}$ equioscillates. Hence, the optimal projection is unique and the equioscillation property is also sufficient for optimality.

To conclude the proof of the conjectures, the inequality (4) of de la Vallée-Poisson type has to be verified. Given $\mathbf{s} \in \Delta_{n-1}$, consider the curve $(\mathbf{t}_\alpha)_{0 \leq \alpha \leq 1}$

such that $\Gamma(\mathbf{t}_\alpha) = (1 - \alpha)\Gamma(\mathbf{s})$, which by Theorem 3 exists and is of class C^1. Fix r by $\lambda_r(\mathbf{s}) = \min_j \lambda_j(\mathbf{s})$. Then for each $j \in [1, n]$, we have $\partial(\lambda_j - \lambda_r)(t_\alpha)/\partial\alpha = (\lambda_j - \lambda_r)(s) \leq 0$. Suppose that $\partial\lambda_r/\partial\alpha < 0$ for some $\alpha \in (0, 1)$. Then $\partial\lambda_j/\partial\alpha < 0$ for all $j \in [1, n]$. From $\sum_i (-1)^i J_i(\mathbf{t}_\alpha)\partial\lambda_i/\partial\alpha = 0$ and (13), we get a contradiction. Hence $\partial\lambda_r/\partial\alpha \geq 0$ holds for all $\alpha \in [0, 1]$ and $\lambda^* = \lambda_r(\mathbf{t}_1) \geq \lambda_r(\mathbf{t}_0) = \min_j \lambda_j(\mathbf{s})$. Consequently, (4) also holds. □

Exercise 4. Assume that

$$x = At, \qquad x \in \mathbb{R}^n, \qquad t \in \mathbb{R}^{n-1},$$

with A being an $n \times (n - 1)$ matrix. Show that

$$\sum_{r=1}^{n} (-1)^r J_r x_r = 0,$$

where J_r is the determinant of the matrix A with the r-th row deleted.
Hint: Apply Laplace's theorem to the $n \times n$ matrix which consists of A and the vector x.

Bibliography*

Achieser, N.I. (1930): On extremal properties of certain rational functions [Russian]. DAN *18*, 495–499

Achieser, N.I. (1947): Lectures on the Theory of Approximation. [Russian]. Gostekhizdat, Moscow, 1947 = Vorlesungen über Approximationstheorie. Akademie-Verlag, Berlin, 1953 = Theory of Approximation. Frederick Ungar, New York, 1956

Akhieser, N.I. [Achieser, N.I.] (1965): The Classical Moment Problem and some Related Questions in Analysis. Oliver & Boyd, Edinburgh-London

Akhlaghi, M., Wolfe, J.M. (1981): Functions with many best L_2-approximations. JAT *33*, 111–118

Amann, H. (1974): Lectures on some Fixed Point Theorems. IMPA, Rio de Janeiro

Ambrosetti, A., Rabinowitz, P.H. (1973): Dual variational methods in critical point theory and applications. J. Functional Anal. *14*, 349–381

Amir, D., Deutsch, F. (1972): Suns, moons and quasi-polyhedra. JAT *6*, 176–201

Anderson, J.E., Bojanov, B.D. (1984): A note on optimal quadrature in H^p. Numer. Math. *44*, 301–308

Arndt, H. (1974): On uniqueness of best spline approximations. JAT *11*, 118–125

Baker, Jr., G.A. (1975): Essentials of Padé Approximants. Academic Press, New York-San Francisco-London

Baker, Jr. G.A., Graves-Morris, P. (1981): Padé Approximants. I. Basic Theory. II. Extensions and Applications. Encyclopedia of Mathematics and its Applications. Addison-Wesley, London-Amsterdam

Barrar, R.B., Loeb, H.L. (1968): On N-parameter and varisolvent families. JAT *1*, 180–181

Barrar, R.B., Loeb, H.L. (1970): On the continuity of the nonlinear Tschebyscheff operator. Pacific J. Math. *32*, 593–601

Barrar, R.B., Loeb, H.L. (1978): On monosplines with odd multiplicity of least norm. J. d'Anal. Math. *33*, 12–38

Barrar, R.B., Loeb, H.L. (1980): Fundamental theorem of algebra for monosplines and related results. SIAM J. Numer. Anal. *17*, 874–882

Barrar, R.B., Loeb, H.L. (1981): Oscillating Tchebycheff systems. JAT *31*, 188–197

Barrar, R.B., Loeb, H.L., Werner, H. (1980): On the uniqueness of the best uniform extended totally positive monospline. JAT *28*, 20–29

Barrett, W. (1971): On the convergence of sequences of rational approximations to analytic functions of a certain class. J. Inst. Maths. Applics. 7, 308–323

Barrodale, I., Powell, M.J.D., Roberts, F.D.K. (1972): The differential correction algorithm for rational l_∞-approximation. SIAM J. Numer. Anal. *9*, 493–504

Barrow, D.L. (1976): On multiple node Gaussian quadrature formulae. Math. Comp. *32*, 431–439

Barrow, D.L., Chui, C.K., Smith, P.W., Ward, J.D. (1978): Unicity of best mean approximation by second order splines with variable knots. Math. Comp. *32*, 1131–1143

---

* Abbreviations

DAN SSSR = Dokl. Akad. Nauk SSSR

JAT = J. Approximation Theory

Bartke, K. (1978): Eine varisolvente Familie, welche das Phänomen der konstanten Fehlerkurve zuläßt. JAT *24*, 324–329

Bartke, K. (1984): Die Struktur der normalen rationalen Funktionen. Numer. Math. *43*, 379–388

Berens, H., Hetzelt, L. (1984): Die metrische Struktur der Sonnen in $l_\infty(n)$. aequationes math. *27*, 274–287

Bernstein, S.N. (1926): Leçons sur les propriétés extrémales et la meilleure approximation des fonctions analytiques d'une variable réelle. Gauthier-Villars, Paris

Bernstein, S.N. (1931): Sur la limitation des valeurs d'une polynome $P_n(x)$ de degré n sur tout un segment par ses valeurs en $(n + 1)$ points du segment. Izv. Akad. Nauk SSSR 7, 1025–1050

Billingsley, P. (1968): Convergence of Probability Measures. John Wiley & Sons, New York-London-Sydney-Toronto

Björck, Å. (1981): Least Squares Methods in Physics and Engineering. Lecture Notes. CERN-Report 81–16, Geneva

Blatt, H.P., Braess, D. (1980): Zur rationalen Approximation von e^{-x} auf $[0, \infty)$. JAT *30*, 169–172

Blatter, J. (1968): Approximative Kompaktheit verallgemeinerter rationaler Funktionen. JAT *1*, 85–93

Bojanov, B.D. (1979): Uniqueness of the monosplines of least deviation. In "Numerische Integration" (G. Hämmerlin, ed.), ISNM 45, pp. 67–97. Birkhäuser, Basel

Bojanov, B.D., Braess, D., Dyn, N. (1986): Generalized Gaussian quadrature formulas. JAT (to appear)

de Boor, C. (1969): On the approximation by γ-polynomials. In "Approximation with Special Emphasis on Spline Functions" (I.J. Schoenberg, ed.), pp. 157–183. Academic Press, New York-London

de Boor, C. (1978): A Practical Guide to Splines. Springer, New York-Heidelberg-Berlin

de Boor, C., Pinkus A. (1978): Proof of the conjectures of Bernstein and Erdös concerning the optimal nodes for polynomial interpolation. JAT *24*, 289–303

Borel, E. (1905): Leçons sur les Fonctions de Variables Réelles. Gauthier-Villars, Paris

Borosh, I., Chui, C.K., Smith, P.W. (1977): Best uniform approximation from a collection of subspaces. Math. Z. *156*, 13–18

Borsuk, K. (1933): Drei Sätze über die n-dimensionale euklidische Sphäre. Fund. Math. *20*, 177–191

Borwein, P. (1982): On a method of Newman and a theorem of Bernstein. JAT *34*, 37–41

Borwein, P. (1983): Uniform approximation by polynomials with variable exponents. Can. J. Math. *35*, 547–557

Braess, D. (1967): Approximation mit Exponentialsummen. Computing *2*, 309–321

Braess, D. (1967b): private communication, cf. G. Lamprecht (1967)

Braess, D. (1970): Die Konstruktion der Tschebyscheff-Approximierenden bei der Anpassung mit Exponentialsummen. JAT *3*, 261–273

Braess, D. (1971): Chebyshev approximation by spline functions with free knots. Numer. Math. *17*, 357–366

Braess, D. (1973a): Kritische Punkte bei der nichtlinearen Tschebyscheff-Approximation. Math. Z. *132*, 327–341

Braess, D. (1973b): Chebyshev approximation by γ-polynomials. JAT *9*, 20–43

Braess, D. (1973c): On a paper of C.B. Dunham concerning degeneracy in mean non-linear approximation. JAT *9*, 313–315

Braess, D. (1973d): private communication, cf. T.A. Kilgore (1978)

Braess, D. (1974a): Chebyshev approximation by γ-polynomials II. JAT *11*, 16–37

Braess, D. (1974b): Geometrical characterizations for nonlinear uniform approximation. JAT *11*, 260–274

Braess, D. (1974c): Rationale Interpolation, Normalität und Monosplines. Numer. Math. *22*, 219–232

Braess, D. (1974d): On varisolvency and alternation. JAT *12*, 230–233.

Braess, D. (1975a): On the number of best approximations in certain non-linear families of functions. aequationes math. *12*, 184–199

Braess, D. (1975b): An n-parameter Chebyshev set which is not a sun. Canad. Math. Bull. *18*, 489–492

Braess, D. (1975c): On the degree of approximation by spline functions with free knots. aequations math. 12, 80–81

Braess, D. (1976): On rational L_2-approximation. JAT *18*, 136–151

Braess, D. (1978): Chebyshev approximation by γ-polynomials III. On the number of solutions. JAT *24*, 119–145

Braess, D. (1984a): On the conjecture of Meinardus on rational approximation of e^x. II. JAT *40*, 375–379

Braess, D. (1984b): On rational approximation of the exponential and the square root function. In "Rational Approximation and Interpolation", (P.R. Graves-Morris, E.B. Saff, and R.S. Varga, eds.), pp. 89–99. Springer, Berlin-Heidelberg-New York-Tokyo

Braess, D. (1986): On nonuniqueness in rational L_p-approximation. JAT (to appear)

Braess, D., Dyn, N. (1982): On the uniqueness of monosplines and perfect splines of least L_1- and L_2-norm. J. d'Anal. Math. *41*, 217–233

Braess, D., Dyn, N. (1986): On the uniqueness of monosplines with least L_p-norm. Constr. Approx. 2, 79–99

Braess, D., Saff, E.B. (in prep.): On the degree of approximation of completely monotone functions by exponential sums.

Breckner, W.W. (1968): Bemerkungen über die Existenz von Minimal lösungen in normierten linearen Räumen. Mathematica(Cluj) *10*(33), 223–228

Breckner, W.W. (1970): Zur Charakterisierung von Minimallösungen. Mathematica(Cluj) *12*(35), 25–38

Breckner, W., Brosowski, B. (1971): Ein Kriterium zur Charakterisierung von Sonnen. Mathematica (Cluj) *13*(36), 181–188

Brezinski, C. (1980): Padé-Type Approximation and General Orthogonal Polynomials. Birkhäuser, Basel-Boston-Stuttgart.

Brezinski, C. (1983): Outline of Padé approximation. In "Computational Aspects of Complex Analysis" (H. Werner, L. Wuytack, E. Ng, and H.J. Bünger, eds.), pp. 1–50. D. Reidel, Dordrecht-Boston-Lancaster

Brink-Spalink, J. (1975): Tschebyscheff-Approximation durch γ-Polynome mit teilweise fixierten Frequenzen. JAT *15*, 60–77

Brink-Spalink, J. (1977): Eine Regularisierung der Splines mit freien Knoten. Report 23, Rechenzentrum Universität Münster

Brosowski, B. (1968): Nicht-lineare Tschebyscheff-Approximationen. Bibliographisches Institut, Mannheim

Brosowski, B. (1969): Einige Bemerkungen zum verallgemeinerten Kolmogoroffschen Kriterium. In "Funktionalanalytische Methoden der numerischen Mathematik" (L. Collatz and H. Unger, eds.), ISNM 12, 25 34. Birkhäuser, Basel-Stuttgart

Brosowski, B., Deutsch, F. (1974a): On some geometric properties of suns. JAT 10, 245–267

Brosowski, B., Deutsch, F. (1974b): Radial continuity of set-valued metric projections. JAT *11*, 236–253

Brosowski, B., Wegmann, R. (1970): Charakterisierung bester Approximationen in normierten Vektorräumen. JAT *3*, 369–397

Browder, F.E. (1983): Fixed point theory and nonlinear problems. Bull. Amer. Math. Soc. *9*, 1–39

Bulanov, A.P. (1968): Asymptotics for least deviation of $|x|$ from rational functions. Mat. Sbornik *76*(118), 288 – 303 = Math. USSR Sb. *5*, 275–290

Bunt, L.N. (1934): Bijdrage tot de theorie der convexe puntverzamelingen. Thesis, Univ. Groningen. publ. Amsterdam

Carpenter, A.J., Ruttan, A., Varga, R.S. (1984): Extended numerical computations on the "1/9" conjecture in rational approximation theory. In "Rational Approximation and Interpolation" (P.R. Graves-Morris, E.B. Saff, and R.S. Varga, eds.), pp. 383–411. Springer, Berlin-Heidelberg-New York-Tokyo

Cauchy, M. (1821): Cours d'Analyse de l'Ecole Royale Polytechnique; 1.re Partie. Analyse Algébraique. L'imprimerie Royal, Paris

Cavaretta, A.S. (1975): Oscillatory and zero properties for perfect splines and monosplines. J. d'Anal. Math. *28*, 41–59

Chebyshev, P.L. (1859): Sur les questions de minima qui se rattachent à la représentation approximative des fonctions. Oevres I, 273–378

Cheney, E.W. (1966): Introduction to Approximation Theory. McGraw-Hill, New York-London

Cheney, E.W., Goldstein, A.A. (1967): Mean-square approximation by generalized rational functions. Math. Z. *95*, 232–241

Chemey, E.W., Loeb, H.L. (1961): Two new algorithms for rational approximation. Numer. Math. *3*, 72–75

Cheney, E.W., Loeb, H.L. (1962): On rational Chebyshev approximation. Numer. Math. *4*, 124–127

Cheney, E.W., Loeb, H.L. (1964): Generalized rational approximation. SIAM J. (B), Numer. Anal. *1*, 11–25

Clarkson, J.A. (1936): Uniformly convex spaces. Trans. Amer. Math. Soc. *40*, 396–414

Cody, W.J., Meinardus, G., Varga, R.S. (1969): Chebyshev rational approximation to e^{-x} in $[0, \infty)$ and applications to heat-conduction problems. JAT *2*, 50–65

Collatz, L., Krabs, W. (1973): Approximationstheorie. Teubner, Stuttgart

Cromme, L. (1976): Eine Klasse von Verfahren zur Ermittlung bester nichtlinearer Tschebyscheff-Approximationen. Numer. Math. *25*, 447–459

Cromme, L. (1982a): Regular C^1-parametrizations for exponential sums and splines. JAT *35*, 30–44

Cromme, L. (1982b): A unified approach to differential characterizations of local best approximations for exponential sums and splines. JAT *36*, 294–303

Croom, F.H. (1978): Basic Concepts of Algebraic Topology. Springer, New York-Heidelberg-London

Cuyt, A., Wuytack, L. (1986): Nonlinear Numerical Methods: Theory and Practice. North-Holland, Amsterdam

Dennis, Jr., J.E., Gay, D.M., Welsch, R.E. (1981): An adaptive nonlinear least-squares algorithm. ACM Trans. Math. Software *7*, 348–368

Deuflhard, P., Apostolescu, V. (1980): A study of the Gauss-Newton algorithm for the solution of nonlinear least squares problems. In "Special Topics of Applied Mathematics" (J. Frehse, D. Pallaschke, and U. Trottenberg eds.), pp. 129–150. North-Holland, Amsterdam-New York-Oxford

Deuflhard, P., Heindl, G. (1979): Affine invariant convergence theorems for Newton's method and extensions to related methods. SIAM J. Numer. Anal. *16*, 1–10

Deutsch, F. (1965): Duality in Approximation Theory. Doctoral dissertation, Brown University

Deutsch, F. (1972): Theory of Approximation in Normed Linear Spaces. Lecture Notes, Penn. State University

Deutsch, F. (1980): Existence of best approximations. JAT *28*, 132–154

Deutsch, F., Maserick, P.H. (1967): Applications of the Hahn-Banach theorem in approximation theory. SIAM Review *9*, 516–530

DeVore, R. (1968): One-sided approximation of functions. JAT *1*, 11–25

Diener, I. (1986): On nonuniqueness in nonlinear L_2-approximation. JAT (to appear)

Doľženko, E.P. (1961): Some estimates concerning algebraic hypersurfaces and derivatives of rational functions. DAN SSSR *139*, 1287–1290 = Soviet Math. Dokl. *2*, 1072–1075

Donoghue, Jr., W.F. (1974). Monotone Matrix Functions and Analytic Continuation. Springer, Berlin-Heidelberg-New York

Dua, S.N., Loeb, H.L. (1973): Further remarks on the differential correction algorithm. SIAM J. Numer. Anal. *10*, 123–126

Dubovickii, A.Ja., Miljutin, A.A. (1965): Extremum problems in the presence of restrictions. Zh. Vyc. Mat. mat. Fiz. 5:3, 395–453 = USSR Comput. Math. math. Physics 5:3, 1–80

Dunham, C.B. (1968a): Necessity of alternation. Can. Math. Bull. *11*, 743–744

Dunham, C.B. (1968b): Continuity of the varisolvent Chebyshev operator. Bull. Amer. Math. Soc. *74*, 606–608

Dunham, C.B. (1969a): Characterizability and uniqueness in real Chebyshev approximation. JAT *2*, 374–383.

Dunham, C.B. (1969b): Chebyshev approximation by families with the betweeness property. Trans. Amer. Math. Soc. *136*, 151–157

Dunham, C.B. (1971): Degeneracy in mean rational approximation. JAT *4*, 225–229

Dunham, C.B. (1974): Failure of Loeb's method for rational L_1-approximation. Computing *13*, 235–237

Dunham, C.B. (1975): Chebyshev sets in $C[0, 1]$ which are not suns. Canad. Math. Bull. *18*, 35–38

Dyn, N. (1985): Generalized monosplines and optimal approximation. Constr. Approx. *1*, 137–154

Dyn, N. (1986/87): Uniqueness of least-norm generalized monosplines induced by log-concave weight-functions. (to appear)

Edelstein, M. (1970): A note on nearest points. Quart. J. Math. Oxford (2) *21*, 403–405

Edelstein, M., Thompson, A.C. (1972): Some results on nearest points and support properties of convex sets in c_o. Pac. J. Math. *40*, 553–560

Efimov, N.V., Stechkin, S.B. (1958a). Some properties of Chebyshev sets. [Russian]. DAN SSSR *118*, 17–19.

Efimov, N.V., Stechkin, S.B. (1958b): Chebyshev sets in Banach spaces. [Russian]. DAN SSSR *121*, 582–585

Efimov, N.V., Stechkin, S.B. (1959): Supporting properties of sets in Banach spaces and Chebyshev sets. [Russian]. DAN SSSR *127*, 254–257.

Efimov, N.V., Stechkin, S.B. (1961): Approximative compactness and Chebyshev sets. DAN SSSR *140*, 522–524 = Soviet Math. Dokl. *2*, 1226–1228

Ehlich, H., Zeller, K. (1966): Auswertung der Normen von Interpolations operatoren. Math. Ann. *164*, 105–112

Erdös, P. (1958): Problems and results on the theory of interpolation, I. Acta Math. Acad. Sci. Hungar. *9*, 381–388

Frobenius, G. (1881): Über Relationen zwischen den Näherungsbrüchen von Potenzreihen. J. für Math. *90*, 1–17

Ganelius, T. (1979): Rational approximation to x^α on $[0, 1]$. Analysis Math. *5*, 19–33

Ganelius, T. (1982): Degree of rational approximation. In "Lectures on Approximation and Value Distribution", (T. Ganelius, W.K. Hayman and D.J. Newman). Les Presses de l'Université de Montréal No. 79

Garkavi, A.L. (1961): Duality theorems for approximation by elements of convex sets [Russian]. Uspekhi Mat. Nauk 16:4, 141–145

Gauss, K.F. (1809): Theoria motus corporum coelestium. Opera 7, 225–245

Gerschgorin, S.A. (1931): Über die Abgrenzung der Eigenwerte einer Matrix. Isv. Akad. Nauk SSSR, Ser. Fiz. Mat. *6*, 749–754

Gončar, A.A. (1967): Estimates of the growth of rational functions and some of their applications. Mat. Sbornik *72*(114), 489–503 = Math. USSR Sb. *1*, 445–456

Gončar, A.A. (1978): On the speed of rational approximation of some analytic functions. Mat. Sbornik *105*(147), 147–163 = Math. USSR Sb. *34*, 131–145

Gragg, W. (1972): The Padé table and its relation to certain algorithms of numerical analysis. SIAM Review *14*, 1–62

Gutknecht, M. (1978): Non-strong uniqueness in real and complex Chebyshev approximation. JAT *23*, 204–213

Haar, A. (1918): Die Minkowskische Geometrie und die Annäherung an stetige Funktionen. Math. Ann. *78*, 294–311

Haverkamp, R. (1977): Beiträge zur Stetigkeit rationaler Interpolierender. Habilitationsschrift Münster

Hettich, R., Wetterling, W. (1973): Nonlinear Chebyshev approximation by H-polynomials. JAT 7, 198–211

Hettich, R., Zencke, P. (1982): Numerische Methoden der Approximation und semi-infiniten Optimierung. Teubner, Stuttgart

Hobby, C.R., Rice, J.R. (1965): A moment problem in L_1-approximation. Proc. Amer. Math. Soc. *16*, 665–670

Hobby, C.R., Rice, J.R. (1967): Approximation from a curve of functions. Arch. Rational Mech. Anal. *27*, 91–106

Jacobi, C.G.J. (1846): Über die Darstellung einer Reihe gegebener Werthe durch eine gebrochene Funktion. J. Reine Angew. Math. (Crelle J.) *30*, 127–156

James, R.C. (1947): Orthogonality and linear functionals in normed linear spaces. Trans. Amer. Math. Soc. *61*, 265–292

Jetter, K. (1978): L_1-Approximation verallgemeinerter konvexer Funktionen durch Splines mit freien Knoten. Math. Z. *164*, 53–66, for more details see: Approximation mit Splinefunktionen und ihre Anwendung auf Quadratur formeln. Habilitationsschrift, Hagen

Jetter, K., Lange, G. (1978): Die Eindeutigkeit L_2-optimaler Monosplines. Math. Z. *158*, 23–34

Johnson, G. (1986): A nonconvex set which has unique nearpoint property. JAT (to appear)

Johnson, R.S. (1960): On monosplines of least deviation. Trans. Amer. Math. Soc. *96*, 458–477.

Jones, R.C., Karlowitz, L.A. (1970): Equioscillation under nonuniqueness in the approximation of continuous functions. JAT *3*, 138–145

Jongen, H.Th., Jonker, P., Twilt, F. (1983): Nonlinear Optimization in $\mathbf{R}^n$. I. Morse Theory, Chebyshev Approximation. Peter Lang, New York-Bern-Frankfurt-Nancy

Kammler, D. (1973a): Existence of best approximations by sums of exponentials. JAT *9*, 78–90

Kammler, D. (1973b): Characterization of best approximations by sums of exponentials. JAT *9*, 173–191

Karlin, S. (1968): Total Positivity I. Stanford University Press, Stanford

Karlin, S. (1972): On a class of best nonlinear approximation problems. Bull. Amer. Math. Soc. *78*, 43–49

Karlin, S., Pinkus, A. (1976a): Gaussian quadrature formulae with multiple nodes. In "Studies in Spline Functions and Approximation Theory", pp. 113–142. Academic Press, New York-London

Karlin, S., Pinkus, A. (1976b): An extremal property of multiple Gaussian nodes. In "Studies in Spline Functions and Approximation Theory", pp. 143–162. Academic Press, New York-London

Karlin, S., Ziegler, Z. (1966): Tschebyscheffian spline functions. SIAM J. Numer. Anal. Series B, *3*, 514–543

Kaufmann, Jr., E.H., Taylor, G.D. (1978): Uniform approximation with rational functions having negative poles. JAT *23*, 364–378

Kilgore, T.A. (1978): A characterization of the Lagrange interpolating projection with minimal Tchebycheff norm. JAT *24*, 273–288

Kilgore, T.A., Cheney, E.W. (1976): A theorem on interpolation in Haar subspaces. aequationes math. *14*, 391–400

Kirchberger, P. (1903): Über Tschebyscheffsche Annäherungsmethoden. Math. Ann. *57*, 509–540

Klee, V.L. (1949): A characterization of convex sets. Amer. Math. Monthly *56*, 247–249

Klee, V.L. (1961): Convexity of Chebyshev sets. Math. Ann. *142*, 292–304

Kolmogorov, A.N. (1948): Remarks on Chebyshev polynomials deviating least from a given function. [Russian] Uspekhi Mat. Nauk 3:1, 216–221

Korneichuk, N.P. (1984): Splines in Approximation Theory. [Russian]. Nauka, Moscow

Krabs, W. (1967): Über differenzierbare asymptotisch konvexe Funktionenfamilien bei der nichtlinearen gleichmäßigen Approximation. Arch. Rational Mech. Anal. *27*, 275–288

Krein, M. (1951): The ideas of Tchebysheff and A.A. Markov in the theory of limiting values of integrals and their further developments. Uspekhi Fiz. Nauk VI.4, 3–120 = Transl. AMS Ser. *2*, 12, 1–122

Krein, M.G., Krasnosel'ski, M.A., Milman, D.P. (1948): On deficiency numbers of linear operators in Banach spaces and on some geometric problems [Russian]. Sb. Trudov Inst. Mat. Akad. Nauk SSSR *11*, 97–112

Kripke, B.R., Rivlin, T.J. (1965): Approximation in the metric of $L_1(X, \mu)$. Trans. Amer. Math. Soc. *119*, 101–122

Kritikos, M. (1938): Sur quelques propriétés des ensembles convexes. Bull. Math. Soc. Roumaine Sci. *40*, 87–92

Laguerre, E. (1898): Oeuvres 1. Gauthier-Villars, Paris

Lamprecht, G. (1967): Die Approximation von Funktionen in der L_q-Norm mit Hilfe rationaler Funktionen. Dissertation Münster

Lamprecht, G. (1970): Zur Mehrdeutigkeit bei der Approximation in der L_p-Norm mit Hilfe rationaler Funktionen. Computing *5*, 349–355

Laurent, P.J. (1972): Approximation et optimisation. Hermann, Paris

Lee, C.M., Roberts, F.D.K. (1973): Comparison of algorithms for rational l_∞-approximation. Math. Comp. *27*, 111–121

Levenberg, K.A. (1944): A method for the solution of certain nonlinear problems in least squares. Quart. Appl. Math. *2*, 164–168

Ling, W.H., Tornga, J.E. (1974): The constant error curve problem for varisolvent families. JAT *11*, 54–72

Loeb, H.L., Werner, H. (1974): Optimal numerical quadrature in H_p-spaces. Math. Z. *138*, 111–117

Löwner, K. (1934): Über monotone Matrixfunktionen. Math. Z. *38*, 177–216

Maehly, H.J. (1963): Methods for fitting rational approximations, Part. II. J. Assoc. Comp. Mach. *10*, 257–266

Maehly, H.J., Witzgall, Ch. (1960): Tschebyscheff-Approximation in kleinen Intervallen. II. Stetigkeitssätze für gebrochen rationale Approximationen. Numer. Math. *2*, 293–307

Magnus, A. (1986/87): Conjectured value of the solution of the 1/9-problem by the CFGT-method. (to appear)

Malozemov, V.N., Pevnyi, A.B. (1973): Alternation properties of solutions of nonlinear minimax problems. DAN SSSR *212*, 37 = Soviet Math. Dokl. *14*, 1303–1306

Markoff, A.A. (1898): On the asymptotic values of integrals in relation to interpolation [Russian]. Papers of the Academy of Science, St. Petersburg, Ser. 8, *6*

Markoff, V.A. (1916): Über Polynome, die in einem gegebenen Intervall möglichst wenig von Null abweichen. Math. Ann. *77*, 213–258. (posthum translation of an article in Russian, Akad. Nauk St. Petersburg 1892)

Marquardt, D.W. (1963): An algorithm for least-squares-estimation of nonlinear parameters. SIAM J. Appl. Math. *11*, 431–441

Meinardus, G. (1963): Invarianz bei linearen Approximationen. Arch. Rational Mech. Anal. *14*, 301–303.

Meinardus, G. (1964, 1967): Approximation von Funktionen und ihre numerische Behandlung (1964) = Approximation of Functions: Theory and Numerical Methods (1967). Springer, Heidelberg-New York

Meinardus, G. (1971): Über ein Problem von L. Collatz. Computing *8*, 250–254

Meinardus, G.: Approximation Theory. (in preparation)

Meinardus, G., Schwedt, D. (1964): Nichtlineare Approximation. Arch. Rational Mech. Anal. *17*, 297–326

Micchelli, C. (1972): The fundamental theorem of algebra for monosplines with multiplicities. In "Linear Operators and Approximation", (P.L. Butzer, J.P. Kahane and B.Sz. Nagy, eds.), pp. 419–430. Birkhäuser, Basel

Micchelli, C., Pinkus, A. (1977): Moment theory for weak Chebyshev systems with applications to monosplines, quadrature formulas and best one-sided L_1-approximation by spline functions with fixed knots. SIAM J. Math. Anal. *8*, 206–230

Milnor, J. (1963): Morse Theory. Annals of Math. Studies 51. Princeton University Press, Princeton

Motzkin, Th. (1935): Sur quelques propriétés caractéristiques des ensembles convexes. Rend. Accad. Naz. Lincei *21*, 562–567

Nemeth, G. (1977): Relative rational approximation of the function e^x. Mat. Zametki *21*, 581–586 = Math. Notes *21*, 325–328

Newman, D.J. (1964): Rational approximation to $|x|$. Michigan Math. J. *11*, 11–14

Newman, D.J. (1974): Rational approximation to e^{-x}. JAT *10*, 301–303

Newman, D.J. (1979): Approximation with Rational Functions. Regional Conference Series No 41.

AMS Providence, Rhode Island

Newman, D.J. (1982): Rational approximation versus fast computer methods. In "Lectures on Approximation and Value Distribution" (T. Ganelius, W.K. Hayman, and D.J. Newman). Les Presses de L'Université de Montréal No. 79

Newman, D.J., Shapiro, H.S. (1963): Some theorems on Čebyšev approximation. Duke Math. J. *30*, 673–682

Nikol'skii, S.M. (1974): Quadrature formulas [Russian]. Nauka, Moscow

Nürnberger, G. (1977): Schnitte für die metrische Projektion. JAT *20*, 196–219

Nürnberger, G., Braess, D. (1981/82): Nonuniqueness of best L_p-approximation by spline functions with free knots. Numer. Functional Anal. Optimiz. *4*, 199–209

Opitz, H.-U., Scherer, K. (1985): On the rational approximation of e^{-x} on $[0, \infty)$. Constr. Approx. *1*, 195–216

Ortega, J.M., Rheinboldt, W.C. (1970): Iterative Solution of Nonlinear Equations in Several Variables. Academic Press, New York-London

Osborne, M.R., Watson, G.A. (1969): An algorithm for minimax approximation in the non-linear case. Computer J. *12*, 64–69

Padé, H. (1982): Sur la représentation approchée d'une fonction par des fractions rationelles. Ann. Ec. Norm. Sup. *9*(suppl), 1–92

Palais, R.S. (1970): Critical point theory and the minimax principle. In "Global Analysis", Proc. Sympos. pure Math. XV, pp. 185–212. Amer. Math. Soc., Providence

Peisker, P. (1983): An alternant criterion for Haar cones. JAT *37*, 262–268

Peller, V.V. (1983): A description of the Hankel operators of class S_p for $p > 0$, an investigation of the rate of rational approximation, and other applications. Mat. Sbornik *122*(164) = Math. USSR Sb. *50*, 465–494

Perron, O. (1957): Die Lehre von den Kettenbrüchen. 3. Aufl. Teubner, Stuttgart

Petrushev, P.P., Popov, V.A. (1984): Relations between rational and spline approximation. Acta Math. Acad. Sci. Hung. *44*, 61–83

Petrushev, P.P., Popov, V.A. (1986/87): Rational Approximation of Real Functions. Encyclopedia of Mathematics, Cambridge University Press

Pinkus, A. (1976): A simple proof of the Hobby-Rice theorem. Proc. Amer. Math. Soc. *60*, 80–82

Pinkus, A. (1981): Bernstein's comparison theorem and a problem of Braess. aequationes math. *23*, 98–107

Popov, V.A. (1977): Uniform rational approximation of the class V_r and its application. Acta Math. Acad. Sci. Hungar. *29*, 119–129

Remes, E. (1934): Sur le calcul effectif des polynomes d'approximation de Tchebycheff. C.R. Acad. Sci. Paris *199*, 337–340

Rice, J.R. (1960): Chebyshev approximation by $ab^x + c$. J. Soc. Ind. Appl. Math. *8*, 691–702

Rice, J.R. (1961): Tchebycheff approximations by functions unisolvent of variable degree. Trans. Amer. Math. Soc. *99*, 298–302

Rice, J.R. (1962): Chebyshev approximation by exponentials. J. Soc. Ind. Appl. Math. *10*, 149–161

Rice, J.R. (1967): Characterization of Chebyshev approximations by splines. SIAM J. Numer. Anal. *4*, 557–565

Rice, J.R. (1969): The Approximations of Functions. II. Nonlinear and Multivariate Theory. Addison-Wesley, Reading, Mass

Riesz, F. (1918): Über lineare Funktionalgleichungen. Acta Math. *41*, 71–98

Rivlin, T.J., Shapiro, H.S. (1961): A unified approach to certain problems of approximation and minimization. J. Soc. Ind. Appl. Math. *9*, 670–699

Rubinstein, G.Sh. (1965): On an extremal problem in a normed linear space [Russian]. Sibirskii Mat. Zh. *6*, 711–714

Ruhe, A. (1979): Accelerated Gauss-Newton algorithms for nonlinear least squares problems. BIT *19*, 356–367

Rutishauser, R. (1963): Betrachtungen zur Quadratwurzeliteration. Monatshefte Math. *67*, 452–464

Ruttan, A. (1985): A characterization of best complex rational approximants in a fundamental case. Constr. Approx. *1*, 287–296

Saff, E.B. (1983): Incomplete and orthogonal polynomials. In "Approximation Theory IV" (C.K. Chui, L.L. Schumaker, and J.D. Ward, eds.), pp. 219–256. Academic Press, New York-London

Saff, E.B., Varga, R.S. (1977): Nonuniqueness of best approximating complex rational functions. Bull. Amer. Math. Soc. *83*, 375–377

Saff, E.B., Varga, R.S., Ni, W.-C. (1976): Geometric convergence of rational approximations to e^{-z} in infinite sectors. Numer. Math. *26*, 211–225

Schaback, R. (1985): Convergence analysis of the general Gauss-Newton algorithm. Numer. Math. *46*, 281–309

Schmidt, E. (1970a): Zur Kompaktheit bei Exponentialsummen. JAT *3*, 445–454

Schmidt, E. (1970b): Stetigkeitsaussagen bei der Tschebyscheff-Approximation mit Exponentialsummen. Math. Z. *113*, 159–170

Schmidt, E. (1979): Chebyshev approximation by sums of logarithmic functions. JAT *26*, 124–131

Schönhage, A. (1973): Zur rationalen Approximierbarkeit von e^{-x} über $[0, \infty)$. JAT *7*, 395–398

Schönhage, A. (1982): Rational approximation to e^{-x} and related L_2-problems. SIAM J. Numer. Anal. *19*, 1067–1080

Schumaker, L. (1968a): Uniform approximation by Tchebycheffian spline functions. J. Math. Mech. *18*, 369–378

Schumaker, L. (1968b): Uniform approximation by Chebyshev spline functions. II: Free knots. SIAM J. Numer. Anal. *5*, 647–656

Schumaker, L. (1969): On the smoothness of best spline approximations. JAT *2*, 410–418

Schwartz, J.T. (1969): Non-linear Functional Analysis. Gordon and Breach, New York-London-Paris

Shapiro, H.S. (1964): Some negative theorems of approximation theory. Michigan Math. J. *11*, 211–217

Singer, I. (1956): Caractérisation des éléments de meilleure approximation dans un espace de Banach quelconque. Acta Sci. Math. (Szeged) *17*, 181–189

Singer, I. (1970): Best Approximations in Normed Linear Spaces by Elements of Linear Subspaces. Springer, Berlin-Heidelberg-New York

Singer, I. (1974): The Theory of Best Approximation and Functional Analysis. Regional Conference Series, SIAM, Philadelphia

Stechkin, S.B. (1963): Approximative properties of sets in normed linear spaces. Rev. Roumaine Math. Pur. Appl. *8*, 5–18

Stieltjes, T.J. (1894): Recherche sur les fractions continues. Ann. Fac. Sci. Toulouse *8*, 1–122

Strauss, H. (1979): Optimale Quadraturformeln und Perfektsplines. JAT *27*, 203–226

Sz.-Nagy, B. (1942): Spektraldarstellung linearer Transformationen des Hilbertschen Raumes. Springer, Berlin

Taylor, G.D. (1972): An improved Newton iteration for calculating roots which is optimal. In "Numerische Methoden der Approximationstheorie", (L. Collatz and G. Meinardus, eds.), pp. 209–227 Birkhäuser, Basel-Stuttgart

Tihomirov, V.M. [Tikhomirov, V.M.] (1969): Best methods of approximation and interpolation of differentiable functions in the space $C[-1, +1]$. Mat. Sbornik *80* = Math. USSR Sb. *9*, 275–289

Tikhomirov, V.M. (1976): Some Questions of Approximation Theory [Russian]. Moscow State University, Moscow

Trefethen, L.N., Gutknecht, M.H. (1983a): Real vs. complex rational Chebyshev approximation on an interval. Trans. Amer. Math. Soc. *280*, 555–561

Trefethen, L.N., Gutknecht, M.H. (1983b): The Carathéodory-Fejér method for real rational approximation. SIAM J. Numer. Anal. *20*, 420–436

Trefethen, L.N., Gutknecht, M. (1985): On convergence and degeneracy in rational Padé and Chebyshev approximation. SIAM J. Math. Anal. *16*, 198–210

de la Vallée-Poussin, Ch.J. (1910): Sur les polynomes d'approximation et la représentation

approachée d'un angle. Académie Royale de Belgique, Bull. de la Classe des Sciences 12, 808–844
Varga, R.S. (1982): Topics in Polynomial and Rational Interpolation and Approximation. Les Presses de L'Université de Montréal No 81
Verfürth, R. (1982): On the number of local best approximations by exponential sums. JAT *34*, 306–323
Vjačeslavov, N.S. (1975): On the uniform approximation of $|x|$ by rational functions. DAN SSSR *220*, 512–515 = Soviet Math. Dokl. *16*, 100–104
Vlasov, L.P. (1961): Chebyshev sets in Banach spaces. DAN SSSR *141*, 19–20 = Soviet Math. Dokl. *2*, 1373–1374
Vlasov, L.P. (1967): Chebyshev sets and approximately convex sets. Mat. Zametki *2*, 191–200 = Math. Notes *2*, 600–605
Vlasov, L.P. (1973): Approximative properties of sets in normed linear spaces. Uspekhi Mat. Nauk 28:6, 1–66 = Russian Math. Surveys 28:6, 1–66
Walsh, J.L. (1931): The existence of rational functions of best approximation. Trans. Amer. Math. Soc. *33*, 668–689
Watson, G.A. (1980): Approximation Theory and Numerical Methods. John Wiley & Sons, Chichester-New York-Brisbane-Toronto
Werner, H. (1962a): Tschebyscheff-Approximation im Bereich der rationalen Funktionen bei Vorliegen einer guten Ausgangsnäherung. Arch. Rational Mech. Anal. *10*, 205–219
Werner, H. (1962b): Die konstruktive Ermittlung der Tschebyscheff-Approximierenden im Bereich der rationalen Funktionen. Arch. Rational Mech. Anal. *11*, 368–384
Werner, H. (1963): Rationale Tschebyscheff-Approximation, Eigenwerttheorie und Differenzenrechnung. Arch. Rational Mech. Anal. *13*, 330–347
Werner, H. (1964): On the rational Tschebyscheff operator. Math. Z. *86*, 317–326
Werner, H. (1966): Vorlesung über Approximationstheorie. Springer, Berlin-Heidelberg-New York
Werner, H. (1967): Die Bedeutung der Normalität bei rationaler Tschebyscheff-Approximation. Computing *2*, 34–52
Werner, H. (1969): Der Existenzsatz für das Tschebyscheffsche Approximationsproblem mit Exponentialsummen. In "Funktionalanalytische Methoden der numerischen Mathematik", (L. Collatz und H. Unger, eds.), ISNM *12*, pp. 133–143. Birkhäuser, Basel
Werner, H. (1972): Eine Faktorisierung der bei der rationalen Interpolation auftretenden Matrizen. Numer. Math. *18*, 423–431
Werner, H., Braess, D. (1969): Über den Zusammenhang von Interpolation und diskreter Tschebyscheff-Approximation mit rationalen Funktionen. Numer. Math. *13*, 112–128
Werner, H., Wuytack, L. (1983): On the continuity of the Padé operator. SIAM J. Numer. Anal. *20*, 1273–1280
Wolfe, J.M. (1974): On the unicity of nonlinear approximation in smooth spaces. JAT *12*, 165–181
Wolfe, J.M. (1983): Critical points in nonlinear L_1-approximation. JAT *37*, 147–154
Wulbert, D. (1971a): Uniqueness and differential characterization of approximation from manifolds of functions. Amer. J. Math. *93*, 350–366
Wulbert, D. (1971b): Nonlinear approximation with tangential characterization. Amer. J. Math. *93*, 718–730
Wynn, P. (1968): On the Padé table derived from a Stieltjes series. SIAM J. Numer. Anal. *5*, 805–834
Young, J.W. (1907): General theory of approximation by functions involving a given number of arbitrary parameters. Trans. Amer. Math. Soc. *8*, 331–344
Zhensykbaev, A.A. (1979): Characteristic properties of the best quadrature formulae. Sibirskii Mat. Zh. *20*, 49–68 = Sibirian Math. J. *20*, 34–49
Zhensykbaev, A.A. (1981): Monosplines of minimal norm and the best quadrature formulae. Uspekhi Mat. Nauk 36:4, 107–159 = Russian Math. Surveys 36:4, 121–181
Zolotarov, E.I. (1877): Application of elliptic functions to questions of functions deviating least and most from zero [Russian]. Zap. Imp. Akad. Nauk St. Petersburg 30 no. 5; reprinted in collected works II, pp. 1–59. Izdat. Akad. Nauk SSSR, Moscow 1932

Index

Main entries are given in *italics*